D0185746

Classic **Treks**

Classic Treks

The 30 Most Spectacular Walks in the World

General Editor Bill Birkett
Foreword by Chuck McGrady

David & Charles

FOREWORD

Over a hundred years ago, two men who barely knew each other climbed up into the Sierra Nevada and spent four days hiking the meadows and watching the wildlife; they saw elk and marmots, lupines and harebells. They didn't shave. They slept on beds of pine needles.

When they came down from the mountains, US President Theodore Roosevelt and Sierra Club founder John Muir were good friends. When they parted, they made a pact: they would work together to save America's special places.

We all have a special place – a place that speaks to us – whether it's the Grand Canyon or Vancouver Island, the Scottish Highlands, or the Alps, or what it was for John Muir – that "range of light" known as Yosemite. It's the place that puts your life into perspective, that helps you quiet the noise in your head, helps you hear your own voice again. It's the place that sums up for you the beauty and essence of our earth, that defines your nation...and your nature.

In John Muir's words, "Everyone needs beauty as well as bread, places to play in and pray in where Nature may heal and cheer and give strength to body and soul alike." His prescription for saving the earth was to get people out to explore. Muir believed that if one explored the "wild lands" that one would appreciate them and then work to protect them. I share that belief.

My appreciation of *Classic Treks* came from its description of the Grand Canyon's Kaibab Trail. When I first climbed the Canyon, I understood what people meant when they talked about coming back to a place you've never been before. It seemed like home, yet many of its visitors, like myself, came from so far away. German, Japanese and Spanish were heard as much as English, and I got the sense that I was part of a much larger community...a community of trekkers who shared values.

Classic Treks reminds me that I've got a lot more places to explore and many more cultures to experience. Some of the treks describe places I knew I needed to visit, but the book also describes places I really had no idea existed – places like the Toubkal Circuit and the Thorsborne Trail.

So use *Classic Treks* as your guide to the world's special places...go out and explore. My hope and expectation is that you will find yourself, and, like Teddy Roosevelt and John Muir, feel compelled to protect these "special places".

CHUCK MCGRADY
President, The Sierra Club

A DAVID & CHARLES BOOK

First published in the UK in 2000

ISBN 0-7153-1075-5

QUAR TREK

A catalogue record for this book is available from the British Library

Conceived, designed, and produced for David & Charles by
Quarto Publishing plc
The Old Brewery
6 Blundell Street
London
N7 9BH

Editor Kate Michell
Senior Art Editor Penny Cobb
Copy Editors Sue Gordon, Claire Waite, Peter Kirkham
Designer Pete Laws
Cartographers Julian Baker, Jonathan Young, Andy Mayers (JB Illustrations)
Locator Maps Andrew Green
Illustrator Cy Baker
Charts Dave Kemp
Indexer Dorothy Frame

Art Director Moira Clinch
Publisher Piers Spence

Printed in China

CONTENTS

Asia

Africa

Australasia

On a trek you may discover flora and fauna of the strangest kind, such as these giant lobelia and giant groundsel plants on the moorlands of East Africa.

INTRODUCTION

TREKKING, THE SIMPLE ACT OF WALKING FROM ONE PLACE TO ANOTHER. YET SUCH A BASIC EXERCISE CAN GIVE GREAT PLEASURES: THE CHANCE TO EXPLORE HIDDEN CORNERS OF THE WORLD, DISCOVER EXOTIC CULTURES AND OBSERVE NATURAL WONDERS; TO LIVE OFF ONLY WHAT YOU CAN CARRY ON YOUR BACK AND TO RISE TO A PERSONAL CHALLENGE. SO, THE CONCEPT OF TREKKING MAY BE AS STRAIGHTFORWARD AS FALLING OFF A LOG, YET THIS PASTIME HAS THE CAPACITY TO OFFER JOY, EXHILARATION AND IMMENSE SATISFACTION.

Never-ending vistas of snowclad peaks ahead, clear blue skies above and summer flowers blooming all around— just a few of the joys to be found on a summer trek.

The history of trekking is as old as mankind. Man's survival has often been dependent on his ability to migrate; to search for food, for a better climate or to find safety. Traveling improves our knowledge, enriches our lives, and satisfies our curiosity and quest for adventure. This primitive and basic instinct is engrained deep in our psychology.

As our knowledge has grown and technology has improved, travelling has taken increasingly sophisticated forms. Explorers have probed almost every nook and crann; mountaineers have scaled the highest mountains. In the sixteenth century the sailing ship took Europeans to the Americas; in the nineteenth century the wagon train led to America's wild west, and the steam railway opened up some of the most inhospitable parts of Europe and America; in the twentieth century the airplane has enabled us to visit, with relative ease, most parts of our earth. For the twenty-first century our technology will develop to such a degree that it may well become possible for our travels to extend beyond the confines of this planet.

Paradoxically, our extensive knowledge and technological development can, if used carelessly, act as a barrier between ourselves and the natural world, in effect distancing us from a true appreciation of its real worth and beauty. Trekking is a great way for us to break through this and return to our roots. Minimising the cumbersome entrapments of the material world, it keeps things simple. By using modern technology to best advantage and with the right equipment, you can stay reasonably comfortable for most of the time, and can get close to a natural wilderness without unduly diluting the experience.

With an eye to revealing something of the infinite variety and beauty of our remarkable planet, the treks in this book have been chosen from many corners of the earth. Most are in, or pass through, National Parks, world heritage sites, natural wildernesses and other specially protected areas. All the treks in this book cover places of exceptional scenic quality and cultural interest. Indeed, some pass through environments that are at the extreme limit of existence– where social and natural ecosystems

→ In autumn, a woodland trek may not only be quieter than during the summer months, but has the added attraction of the beautiful array of colours that the season creates.

Trekking can introduce you to cultures previously alien to you, just by passing through villages where the day-to-day life is far removed from your own. In Nepal, the vibrant Buddhist culture is inescapable, as flag-bedecked temples seem to feature in almost every village.

are hung in delicate balance. We must tread very carefully and look after all these places with the utmost care.

Classic Treks celebrates thirty of the greatest and most rewarding treks in the world. They range in difficulty from easy to demanding, and in duration from two days to two weeks or more, and roam over the five continents of the Americas, Europe, Asia, Africa and Australasia by mountain, river, sea and lake. From the awesome depths of the Grand Canyon to the huge snowclad heights of the Himalaya, by way of the serene tranquility of the English Lake District, through to the ancient civilisation of the Incas, and alongside the active volcanoes of New Zealand's North Island, the treks in this book represent an inspirational and informative journey through the most stunningly beautiful and exciting places on earth.

This indispensable guide, lavishly photographed, mapped, illustrated and backed by authoritative text, communicates evocatively the unique qualities of each route. Each contributor has personally tackled the selected routes described. Through words and pictures they are portraying something of their own experiences to you, and simultaneously evoking the spirit of the region: the place, flora, fauna, local character and customs.

Classic Treks also details all essential facts and figures, providing an invaluable planning resource to help you informatively choose and carefully assess any future trek. A wide range of aspirations and abilities are catered for. The joys of trekking know only the limitations of self. Many of the treks are suitable for people of average fitness and ability, both young and old.

Within each trek there is a degree of flexibility. Each trek is modular, often with easier and harder alternatives detailed, and the description has been structured in such a way that different sections can be chosen to suit your particular time scale. Although you can execute the treks in one continuous effort, in many cases it is possible to break them down into more manageable sections – the facts are detailed, but the choice is yours. You may wish to trek solo, with friends, or as part of an organised group led by a travel company providing varying degrees of back-up and comfort. Whatever your choice, choose carefully, be

prepared, and know your limitations. Of course, once the trek is complete and you are looking for the next great challenge, it is also very much a book to savour.

Among the expertise assembled here, no single person has, yet, completed all the treks detailed – with the aid and inspiration of this book you could be the first. I wish you luck. Perhaps more importantly, if you should get just a fraction of the wonder and enjoyment that I have experienced in its preparation, then my time will have been exceptionally well rewarded.

Our earth is a staggeringly beautiful planet. Let us enjoy it, understand its worth, and keep it in good shape. Happy trekking, go in peace, and may your God go with you.

Bill Birkett

BILL BIRKETT,
Dale View, Little Langdale, England 1999.

↑ *Tropical islands, where clear freshwater rivers flow over smooth rocks and through lush vegetation into warm seas, can provide treks that are as satisfying as those in the world's greatest mountain ranges.*

← *Meet the locals: some treks follow routes used daily by villagers. This Peruvian* campesino *in traditional dress carries her young child in a blanket sling on her back and hand-spins wool while she walks through the long grass of the Cordillera Raura.*

PRACTICAL ADVICE

To complete a trek successfully and safely, to enrich it with an awareness of all its many facets, you should do as much preparation as possible. Along with the specific guidebooks recommended, trekkers should also be familiar with the following: *The Backpacker's Handbook* by Chris Townsend (Ragged Mountain Press USA), *The Hillwalker's Manual* by Bill Birkett (Cicerone UK), *Medical Handbook for Mountaineers* by Peter Steele (Constable UK), *Mountain Sickness– Prevention, Recognition & Treatment* by Peter Hackett (American Alpine Club, USA/Cordee UK), *Medicine for Mountaineering* by James Wilkerson (The Mountaineers, Seattle/Cordee UK), The Lonely Planet Guides, and The Rough Guides to each specific area. The following breakdown to the important basics of trekking is useful information that should be taken on board.

Preparation & Execution: It cannot be over-emphasised that the greater amount of effort you put into properly preparing for a trek, the safer and more rewarding the actual trek will be. Although many treks are suitable for all hikers of average ability, everyone, no matter how experienced, should be physically and mentally prepared. Make sure you walk regularly before you go, jog if you want to, go to the gym, play sports– get yourself into reasonable physical shape. However, perhaps most important of all, is the considerable matter of mental preparation. If you want to succeed on a trek, and make that vital mental decision to do so, you will; if you want to enjoy the trek despite the unexpected, you will. Excellent equipment, the right survival techniques, the best guides, the kindest of circumstances, all the knowledge in the world– important as they all are– do not obviate the essential need for commitment.

Planning: Even if you put yourself into the hands of one of the excellent adventure travel companies it is important to plan your trek carefully. Study the factfiles presented here and ensure you have your own copy of the recommended maps and guidebooks. Go at the right time for the best climatic conditions, and allow plenty of spare time in the schedule– delays and frustrations are all part and parcel of trekking, so take it easy, and be flexible. Prepare a realistic budget– don't go short. If you are traveling independently, carefully cost your trip in detail, then double the amount.

Medical: Make sure you get the required vaccinations in time, and have a fully equipped first-aid kit.

Documentation: Check you have the required passport, visas, permits, travel insurance with adequate medical cover, and an International Health Certificate (with vaccinations kept up to date). An important tip is to keep a copy of the above items, along with credit card information, in a separate location in case the originals are lost or stolen.

Safety Techniques: Absolutely essential is a sound working knowledge of navigation techniques using map and compass. Before you go, you should have a practical knowledge of basic first aid. Rudimentary knowledge of survival techniques is also useful. Both may save your life. Trekkers in mountain areas will also be well served to

↑ *A good tent will keep you warm and dry, providing that its specifications suit the environment that you're trekking in.*

↑ *The importance of basic navigation and first-aid is not to be underestimated; stay safe and well by learning how to use these items properly.*

↓ *Forward planning will ensure that you have all the necessary documentation in duplicate, and increases your chances of a trouble-free trip.*

have a working knowledge of using an ice-axe, how to arrest a fall using the ice-axe braking techniques, and crampons.

Equipment: This must be selected for the specific trek, the weather, and the climatic conditions. You have to cut your cloth to suit, at the same time ensuring that you have the right equipment for the most likely extremes of conditions which might be encountered. Bearing in mind that you may have to carry everything you take, my advice is that you keep things simple and travel as lightly as possible. Another very important rule is to make sure you get the best equipment available– don't compromise on quality.

Footwear: This is the single most important item of equipment that you will be taking on the trek, and selection depends very much on the type of trek. Whatever you wear, make sure it is well-fitting and comfortable before you arrive at your destination. Buy from a reputable store, try the footwear on with the socks you will be using on the trek, walk around in it, take expert advice and take your time to choose. Once chosen, use before you go, even if it's just walking around your home or office. Suitable footwear may range from basic trainers, to Goretex-lined boots for mid-range trekking, to a suitably insulated mountain boot.

Clothing: This is obviously totally dependent on which treks you intend to do. For extreme conditions bear in mind the important principle of layering– you can always take off clothing unsuitable for the day. The full system is: Goretex jacket (with hood), overtrousers, and gloves and hat for outer skin protection; zipped fibrepile jacket/waistcoat for warmth, tracksuit and sweatshirt, followed by a base layer of thermal underwear. Don't forget a peaked or broad-rimmed sun hat. In snow, gaiters are invaluable as they keep the snow and damp out of your boots.

Rucksack: Get a leading brand, preferably of mountaineering design as they are the most practical. Make sure the pack fits you well– get if off the rack in the store and adjust to suit. If it isn't comfortable, irrespective of specification, don't buy it. A simple, though practical tip– because no rucksack is 100 per cent waterproof– is to use a large plastic sack within the rucksack shell to keep your spare clothing and sleeping bag dry.

Sleeping Bag: These come with many different specifications for climate and seasons. At the top of the range, suitable for mountaineering, comes the four-season down-filled bag. Most treks will not require such a high specification, and a zipped bag– allowing for variations in temperatures– with man-made insulation (which will in any case perform better than down in wet conditions), would often be a better choice. If you aim to camp or are sleeping rough don't forget the all-important ground insulation mat. If you don't have a tent or dry shelter, a Goretex bivi bag is invaluable.

Other Kit: A pair of lightweight, telescopic, trekking poles may prove invaluable and are today considered an essential piece of kit; they take a valuable percentage of loading off tired legs and maligned knees, and have a multitude of other uses. A plastic or aluminium lightweight water bottle. Sunscreen, sunglasses or snowglasses for altitude or trekking over snow (essential to prevent snow-blindness). Don't forget the head torch– apart from reading, when you really need a light, you also need both hands free.

Photography/Video: Although you may be on the trip of a lifetime and wish to record it for prosperity, you also want to enjoy it. To this end I recommend you choose something small and light. A huge selection of miniature 35mm cameras is available. Similarly with video, a wide range of lightweight high-quality gear is available. Pack your equipment so that you can access the camera or camcorder without having to stop and take off your rucksack or fiddle with your clothing.

Environmental Awareness and Responsibilities: The trekker should leave only footprints and take only photographs. Although trekking may be one of the purest and most liberating pastimes on earth, we all have a personal responsibility to others and to the environment. Respect other people's beliefs and customs. Learn about them and learn from them. Respect the environment and all living things, from the humblest blade of grass to the mightiest eagle. As far as possible take out what you take in. Bury human waste and toilet paper. Use sparingly of local resources. Being part of a large organised group in no way reduces your personal responsibilities. Never underestimate the awesome power of the natural world.

↑ *Comfortable footwear will bring a spring to your step, and trekking poles help ease tired legs. Trekking can be very energetic, so make sure you stay well-hydrated by having a water bottle to hand. A head torch is always useful.*

← *Rucksacks and sleeping bags come in all shapes and sizes– get ones that are suitable for you. A roll mat often proves invaluable when camping on hard or rocky ground.*

HOW TO USE THIS BOOK

Through maps, photographs, illustrations, charts and text, *Classic Treks* brings alive walks from all over the world. From the comfort of your own armchair you can follow in the footsteps of some of the world's most experienced trekkers, and be inspired to experience these treks for yourself. These two pages explain exactly how to follow each trek, what information you can expect to find within the main profiles of each trek and where to locate that information.

Photographs:
These have been carefully selected not only to capture the true spirit of the trek and the region and to highlight the scenic splendor, but also to provide informative visual information on key points of the trek.

Locator map:
Where in the world? Get a general indication of what part of the world the trek takes place in by using this locator map.

Degree of difficulty logos:
All the treks have been graded into levels of easy, moderate and difficult. An orange border around the relevant icon provides a basic guide as to how tough or simple the trek is.

Introductory text:
A colourful introduction leads to a descriptive analysis of the trek, which is broken down into manageable days or groups of days.

Itinerary text:
For each section the start and finish point is detailed along with the length of trail described. Where relevant, easier alternatives or shorter routes are suggested. In many instances, if the time you have available is limited, you can select only the portion you wish to complete.

AUSTRALASIA

the THORSBORNE TRAIL

QUEENSLAND, AUSTRALIA

Alan Castle

NOT FAR OFF THE COAST OF NORTH QUEENSLAND, ACROSS A SMALL CHANNEL, LIES HINCHINBROOK ISLAND, AN UNSPOILED WILDERNESS ISLAND FAMOUS FOR THE RICHNESS AND VARIETY OF ITS HABITATS. IT LIES WITHIN THE GREAT BARRIER REEF WORLD HERITAGE AREA, AND IS AUSTRALIA'S LARGEST ISLAND NATIONAL PARK.

The Thorsborne Trail is named after the late Arthur Thorsborne, who, along with his wife, Margaret, spent a lifetime dedicated to the conservation of Hinchinbrook Island and a study of its abundant and varied wildlife. The trek follows the island's east coast and offers a unique chance to savor a tropical island wilderness, stopping for refreshing swims in creeks and water holes, and enjoying the abundance of wildlife in the rain forest, mangrove swamps, and saltpans. Trekkers must be completely self-reliant and self-sufficient, but the rewards are immeasurable. There are a number of creek crossings to negotiate, including a couple of tidal creeks, and the terrain is sometimes rugged. Nevertheless, the route is well-waymarked, and, provided the weather conditions are not extreme, the trail should be within the capabilities of the average fit and well-prepared bushwalker.

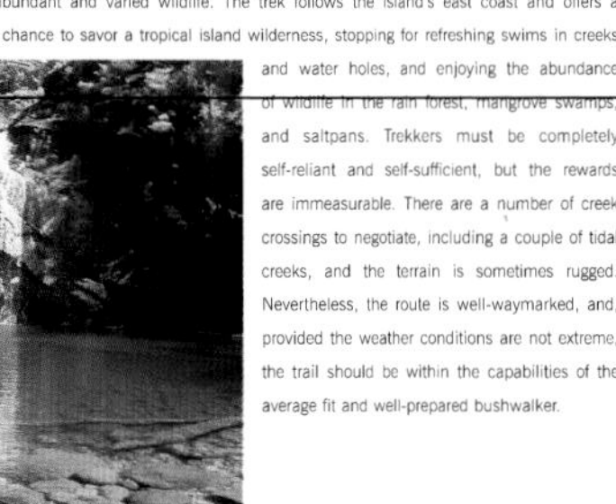

itinerary

•DAY 1 **4 miles (6km)**

Ramsey Bay to Little Ramsey Bay

The ferry carrying Thorsborne Trail trekkers usually leaves Cardwell at 9:00 A.M., crosses Hinchinbrook Channel, and enters an area of mangrove swamp, cruising along a channel until it arrives at a landing point close to Ramsey Bay. Upon disembarking, walk along the boardwalk to arrive on the pristine sand of Ramsey Bay.

Walk south along the beach to enter an area of tall open forest. About halfway along this section there is an opportunity to take an hour's detour and climb Nina Peak, on a rather steep, rugged path that leads off the main trail at its highest point between Ramsey and Nina Bays. The spectacular view from the summit of Nina Peak is worth this little bit of extra effort. Return to the main trail on the same path.

The trail reaches Nina Beach, where there is a permitted camping area. Some may wish to spend the night here, but otherwise it makes an ideal spot for a prolonged lunch, rest, and possible swim (but not during the box jellyfish season—see Factfile, page 185).

Rock-hopping around a rough headland makes up much of the next section, carefully following waymarks and cairns over Boulder Bay and on to Little Ramsey Bay, where you should arrive by mid-afternoon. This is the recommended site for your first night's camp, a gorgeous spot close to a freshwater lagoon. There's time to relax, cook and eat dinner, and perhaps have another cooling swim.

•DAY 2 **6½ miles (10km)**

Little Ramsey Bay to Zoe Bay

From the camping area, walk south along the beach to reach a tidal creek. This is usually fairly easy to cross, but on occasion can prove tricky. Continue to another small beach, from where you clamber over rocks to a larger sandy beach. From the other end of this beach, follow a gully up to the top of the ridge, from where a side path leads in to Banksia Bay. Descend on the main trail into Banksia Creek and continue to enter rain forest, open forest, and mangroves.

After crossing North Zoe Creek (beware of estuarine crocodiles), there follows a section through extensive palm swamps where it is essential to follow the waymarks carefully. Wet feet will be the order of the day as there are many creeks and swampy areas to cross. And beware of spider's webs stretching across adjacent tree trunks, as if you walk into one you may find its large, unhappy owner on your backpack or shoulder.

The trail eventually emerges onto the beach at Zoe Bay. The night's campground is at the southern end of this beach. Alternatively, follow South Zoe Creek through rain forest to another camping area, where there is good fresh water. After a rest, do take some time to walk upstream to Zoe Falls to enjoy the cascade and huge, clear pool into which the water tumbles.

↑ *Backpackers stride across Ramsey Bay, taking their first steps on the Thorsborne Trail. The pristine sands of the many unspoiled beaches on Hinchinbrook Island offer a sense of space, tranquility, and timelessness.*

← *Several of the creeks on this mountainous island have spectacular waterfalls, such as Zoe Falls. The falls are at their most dramatic during the wet season.*

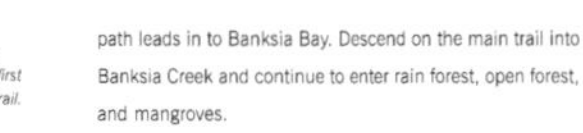

•DAY 3 **4¾**

Zoe Bay to M

The trail cros
Falls. Boulder
crossing gran
sure to look b
Creek is follo
take care and
(260m), the
A descent
side track to
provide an al
left Zoe Bay e
an extended
Diamantin
many creeks
necessary if
rains. Follow

Walk Profile:
To supplement the descriptive text, these charts assess the degree of difficulty of the trek on a daily basis in terms of ascent and descent. They have been drawn in sufficient detail to allow quick visual reference. They are a very useful and easy guide to help determine just how tough each day's trekking is going to be.

Map:
With the line of the route clearly marked, these have been drawn to provide an overall presentation of the complete trek. They show the key stages of the walk, start and finish points, major features and alternative routes. A comprehensive key clearly explains any symbols and markings used.

Illustrations:
Specially commissioned watercolours illustrate specific animals or plants that may be encountered on the trek.

AUSTRALASIA

THE THORSBORNE TRAIL

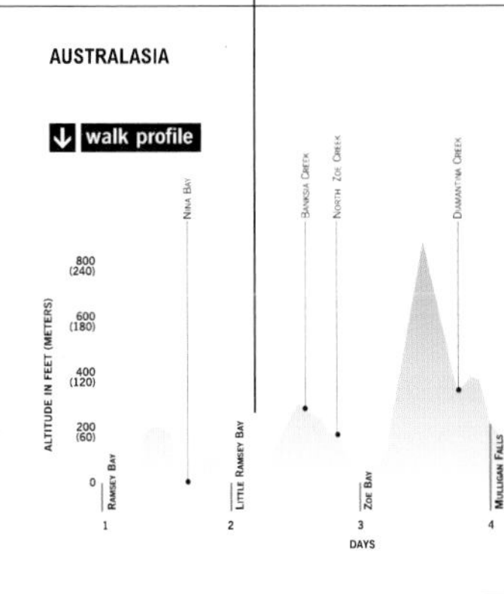

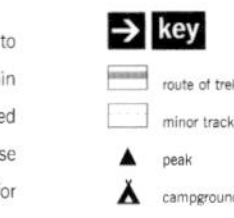

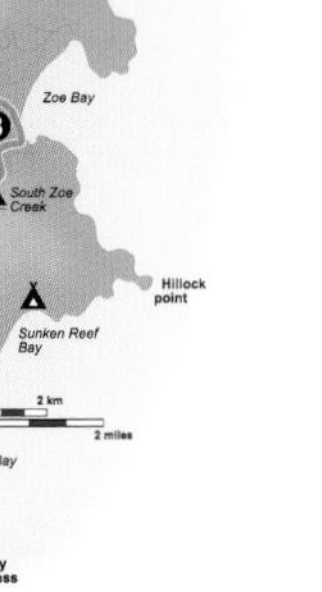

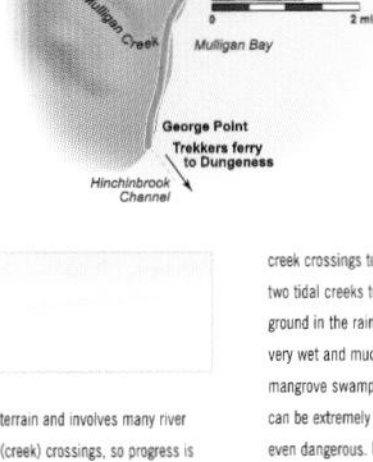

Factfile:
Carefully researched, this section is presented in a concise, easy reference format, and contains vital information for planning and completing the trek.
Overview: *An assessment of the route, including total distance and time required. The difficulty and altitude of a trek are two crucial factors in selecting the best trek for your abilities.*
Access: *This section informs you of the nearest international airports and transport details, particularly the best method to reach the start and return from the finish of the trek. The essential paperwork and formalities to be observed before accessing the country or trek are highlighted.*
Local Information: *Recommends specialist maps and guidebooks for the successful completion of each trek. Provides valuable advice on what sort of accommodation and range of supplies may (or may not!) be available en route, as well as supplying details on local currency and language. This section also advises on the best place to go for tourist information.*
Timing & Seasonality: *An optimum time to tackle the trek is suggested; and a general guide to weather conditions which may reasonably be expected is given.*
Health & Safety: *A number of very important considerations to your personal well-being are covered, and should be checked over carefully in the early planning stages of any intended trip.*
Highlights: *Points out many of the great things to look out for along the trek.*

Temperature and precipitation chart:
This chart provides monthly averages for maximum and minimum temperatures, and for rainfall and snowfall. The charts are organised to provide the information in Fahrenheit on the top level of each box, with the centigrade reading on the lower level; similarly, the precipitation figures are given in inches on the top and in millimetres below. These statistics have been gathered from the nearest available source to each trek; it must be noted that climatic conditions are affected by altitude.

North America

From the cold north to the balmy south, and from east to west, North America is a land of huge contrast and staggering proportion. Its wilderness and landscape, from the stunning depths of the Grand Canyon to the rugged heights of the Rocky Mountains, never fail to impress. The seven treks selected savour the very best of untouched North America.

The flow of water from the Nevada Falls has made smooth the steep side slopes of Liberty Cap.

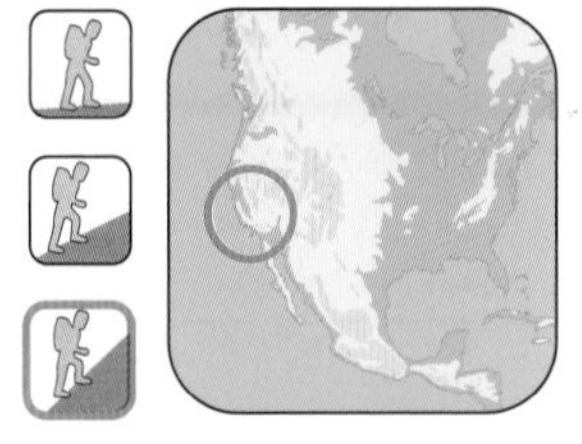

the JOHN MUIR TRAIL

CALIFORNIA, USA

Chris Townsend

THE JOHN MUIR TRAIL IS A WONDERFUL ROLLERCOASTER, RISING AND FALLING THROUGH THE INCREDIBLE VARIETY OF SCENERY OF THE INCOMPARABLE HIGH SIERRA – A GLORIOUS MOUNTAIN WORLD OF GLACIER-SCULPTED, GRANITE DOMES AND PEAKS, PRISTINE CONIFER FORESTS, TREELINE LAKES, HIGH WATERFALLS AND TUMBLING CREEKS.

The trail is named in honour of John Muir, the Scots-born environmentalist and mountaineer who explored these mountains from 1869 onward and whose campaign for their preservation resulted in the formation of the Sierra Club and the creation of Yosemite and Sequoia national parks. Muir called the Sierra Nevada the "Range of Light", an appropriate name for this beautiful world. The range is over 400 miles (640km) long and 60 to 80 miles (95–130km) wide. The basic form is that of a great, tilted block of granite, with the steep, scarp slope to the east, where the mountains drop abruptly 10,000 feet (3,000m) and more to the desert floor of Owens Valley. To the west, the range dips gently into wooded foothills.

Across the green expanse of Long Meadow the snowclad Cathedral Peak rises majestically.

itinerary

•DAYS 1–2 24 miles (38.5km)

Happy Isles to Tuolumne Meadows

The John Muir Trail starts in Yosemite Valley, where Muir himself lived for a time. It's a spectacular – if popular and crowded – place, with massive rock walls rising above the flat, valley floor and waterfalls tumbling from the heights. From Happy Isles at the east end of the valley, the trail climbs beside the Merced River to Nevada Falls, a curving sweep of water that rushes down granite slabs below the soaring cone of Liberty Cap. Leave the Merced River at Little Yosemite Valley, a popular campsite, to climb the valley of Sunrise Creek. The highlight of this section of the walk is the sight of the massive rock walls of Half Dome rising out of the forest. As the trail climbs through Long Meadow and up to 9,730-foot (2,965-m) Cathedral Pass, the graceful spires and rock turrets of Columbia Finger, Cathedral Peak, Echo Peaks, and Matthes Crest come into view. Once over the pass, the trail descends to Tuolumne Meadows, where there is a store, cafe, campsite, and visitor centre. Lembert Dome rises above the meadows that surround the Tuolumne River.

•DAYS 3–5 34 miles (54.5 km)

Tuolumne Meadows to Reds Meadow

For 8 miles (13km) beyond Tuolumne Meadows the trail runs through the meadows and forests of long, flat, curving Lyell Canyon before climbing to 11,056-foot (3,370-m) Donohue Pass. Next comes 10,200-foot (3,109-m) Island Pass, before the trail runs alongside beautiful Thousand Island and Garnet Lakes, with the rugged summits of Banner Peak and Mount Ritter rising to the west. South of these magnificent mountains, the jagged crest of the Minarets comes into view before the trail descends through forest to Reds Meadow and the Devil's Postpile National Monument. The latter is a massive cliff formed of vertical, regular rock columns. Reds Meadow boasts hot springs as well as a campsite, cafe and small store.

•DAYS 6–8 **32 miles (51km)**

Reds Meadow to Quail Meadows and Lake Edison

Climbing on to the slopes of Mammoth Mountain, the trail traverses high above Fish Creek, descends into the Cascade Valley, and then climbs to 10,900-foot (3,322-m) Silver Pass, where there are good views back to Banner Peak and Mount Ritter. A long descent leads to Quail Meadows, beside Mono Creek. Here the Mono Creek Trail leads for 2 miles (3km) to Lake Edison and, for those going that way, the ferry to Vermilion Resort.

•DAYS 9–11 **34 miles (54.5km)**

Lake Edison to Evolution Valley

Above Quail Meadows the trail crosses wooded Bear Ridge, then descends to Bear Creek before climbing back to 10,900-foot (3,322-m) Selden Pass. The scenery and the nature of the walk changes here. Behind lies open country with rolling, rounded mountains and long, wide, lake-filled valleys. Ahead the land is more rugged and austere, less green and more grey, crowded with steep mountains through which the trail winds a way over higher, narrower passes and along deeper, twisting valleys. Muir Trail Ranch lies 1 mile (1.5km) down a spur trail near the bottom of the descent from the pass to the South Fork San Joaquin River. If you go to the ranch, another spur trail (1½ miles/2.5km long) connects with the John Muir Trail higher up the valley. Eventually the route leaves the river for a long, slow ascent of Evolution Valley, which contains three beautiful meadows – Evolution, McClure, and Colby – which in early summer are filled with purple lupin and red paintbrush.

•DAYS 12–14 **37 miles (59.5km)**

Evolution Valley to South Fork Kings River

Leaving the forest for bare rocky slopes, the trail climbs past a series of lakes to 11,955-foot (3,644-m) Muir Pass, in the heart of very steep, mountainous country with only rock and snow in view. The pass is the way across the rugged Goddard Divide, which separates the San Joaquin River from the Middle Fork Kings River. The latter runs

The sharp peak of aptly named Columbia Finger is a distinctive landmark for the first few days of the trail.

through long Le Conte Canyon, in which lie magnificent Big Pete, Little Pete and Grouse meadows, with the great fractured block of The Citadel soaring overhead. The route then turns up beside Palisade Creek, above which rise the six 14,000-foot (4,200-m) plus, serrated summits of the Palisades. A tight set of steep switchbacks blasted out of the cliffs of the Palisade Creek Gorge, and known as the Golden Staircase, leads to the Palisade Lakes. A final, steep climb leads to Mather Pass (12,100 feet/3,688m), a great viewpoint for the Palisades to the north and the Upper Basin of the South Fork Kings River to the south, into which the trail descends.

↑ *The massive granite block known as The Citadel towers above Le Conte Canyon.*

•DAYS 15–17 **37 miles (59.5km)**

South Fork Kings River to Tyndall Creek

The trail now remains close to the Sierra crest, passing through perhaps the most spectacular and impressive scenery of the whole walk. From the South Fork Kings River the trail climbs past a string of shining lakes to 12,130-foot (3,697-m) Pinchot Pass and a superb view into the dark, forested Woods Creek valley. After descending to Woods Creek, the trail climbs again, at first through a magnificent forest of incense cedar, red fir and ponderosa pine, to reach Rae Lakes, reckoned by many to be the most beautiful →

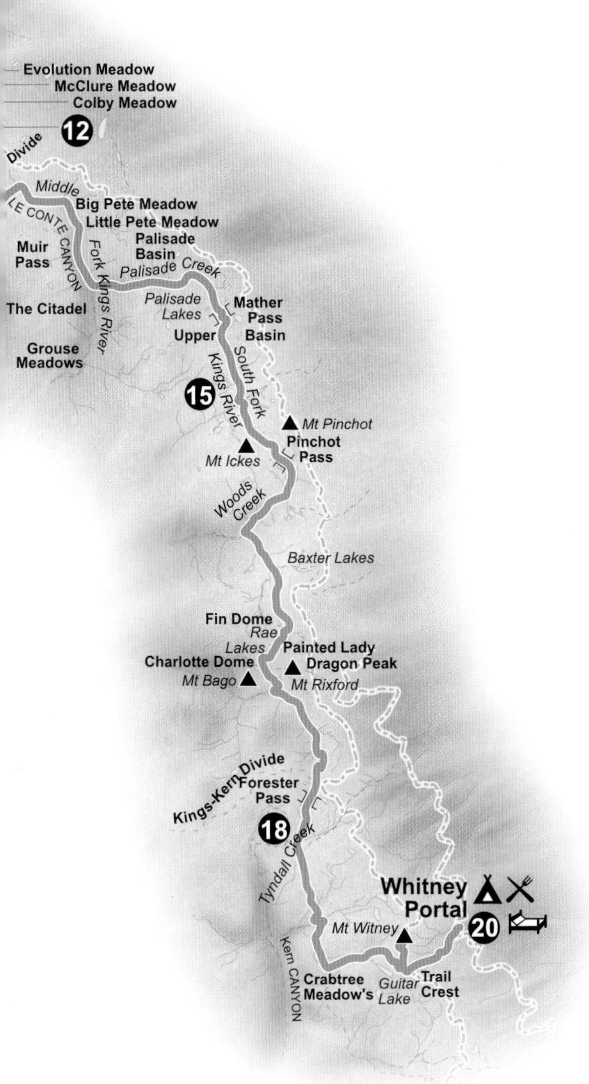

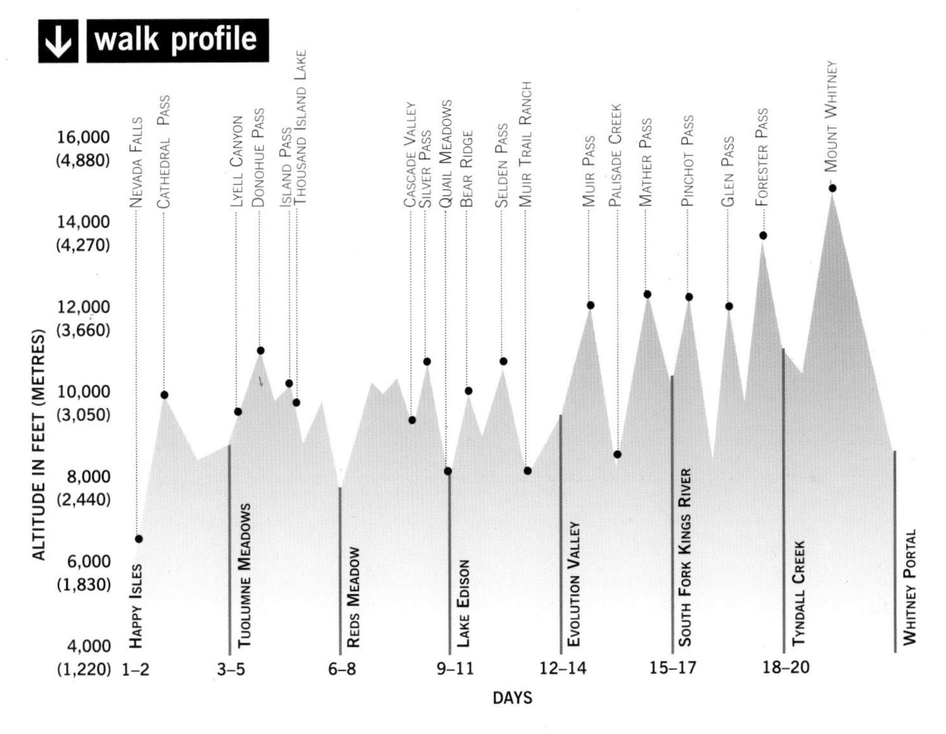

The start of the John Muir Trail is marked by the Half Dome and coniferous forests of the Yosemite Valley.

high-mountain lakes in the High Sierra. Above the lakes rises the fine tower of Fin Dome and the brilliant red-,gold-, and purple-banded cliffs of Painted Lady and Dragon Peak. Bare slopes lead to the notch of 11,978-foot (3,650-m) Glen Pass with, again, wonderful views all round. The south side of the pass is a steep cliff down which the path winds in a series of tight switchbacks. Farther down, the smooth curving slopes of Charlotte Dome can be seen rising out of the forest to the west. The highest pass on the trail, 13,180-foot (4,017-m) Forester Pass, is crossed next. This is the only route for walkers across the long, rugged Kings–Kern Divide. The views are spectacular, back north to the Palisades and south to the jagged Whitney peaks and the deep trench of the Kern Canyon. The south side of the pass is a huge cliff down which steep switchbacks have been built. Once on easier terrain, the trail leads down to Tyndall Creek.

•DAYS 18–20 25 miles (40km)

Tyndall Creek to Whitney Portal

Staying at treeline, the trail runs high above the Kern Canyon to Crabtree Meadows. Here the ascent of Mount Whitney starts, at first with a gentle walk to Guitar Lake. Steepening, the trail then cuts upwards across the south-west slopes of Whitney before ascending steep switchbacks to Trail Crest at 13,600 feet (4,145m). From here it's 2 miles (3km) to the summit, mostly on a narrow, winding, and spectacular trail that runs below pinnacles and above steep drops up the south ridge of the mountain. In places there are startling and dramatic views through notches in the ridge down 10,000 feet (3,000m) to the pale, sandy floor of Owens Valley. From the summit itself there is a sense of the vastness of the Sierra Nevada – a fitting climax to the John Muir Trail. All that remains is to wander back down to Trail Crest and descend east to Whitney Portal.

factfile

OVERVIEW

The John Muir Trail runs from Yosemite Valley south to Mount Whitney, the highest point in the 48 contiguous states, through the high country of the Sierra Nevada Mountains. En route it passes through three national parks, two wilderness areas and past a national monument. The length is 212 miles (341km). However, as the trail finishes on the summit of Mount Whitney, there's another 11 miles (17.5km) to walk to a trail head, making a total of approximately 223 miles (358km). Allow 2 to 3 weeks to hike the whole trail. The itinerary described here is an arbitrary one; camping is allowed almost everywhere (current restrictions are supplied with permits), so you can go as fast or slow as you like.

Start: Happy Isles, Yosemite Valley.

Finish: Whitney Portal – Lone Pine Road.

Difficulty & Altitude: The trail is well built and well maintained. However, in keeping with the wilderness spirit of the walk, there are few signposts at trail junctions and most creeks are not bridged. There is a great deal of ascent and descent too. The high point is 14,494 feet (4,418m) and much of the trail is above 10,000 feet (3,000m), so acclimatising to the altitude is necessary. Full wilderness backpacking gear is needed, as there are few facilities en route.

ACCESS

Airports: San Francisco and Los Angeles.

Transport: Trains run from San Francisco to Merced, from where there are buses to Yosemite Valley. The finish is 12 miles (19.5km) from Lone Pine, but lifts are easily picked up. From Lone Pine there are daily buses to Los Angeles.

Passport & Visas: Passports with visa or visa waiver form are required for all vistors to North America.

bighorn sheep

Permits & Restrictions: Wilderness permits are required. Numbers are limited, so book early. They can be reserved 24 weeks in advance of your start date. There is a fee. There are no access restrictions, but camping is not allowed in some heavily used areas, especially around lake shores and in easily damaged meadows.

LOCAL INFORMATION

Maps: *The John Muir Trail Map Book*, Tom Harrison, is a 13-page 1:63,360 topographic map set covering the whole trail. Forest Service Ansel Adams Wilderness, John Muir Wilderness, and Sequoia-Kings Canyon Wilderness 1:63,360 topographic maps cover the route from Tuolumne Meadows to Mount Whitney and Lone Pine. Trails Illustrated 1:100,000 topographic Yosemite National Park map covers the route from Yosemite Valley to Tuolumne Meadows.

Guidebooks: *Guide to the John Muir Trail*, Thomas Winnett (Wilderness Press); *Starr's Guide to the John Muir Trail & the High Sierra Region* (Sierra Club Books).

Background Reading: *My First Summer in the Sierra*, John Muir (various editions); *The Mountains of California*, John Muir (various editions); *A Sierra Club Naturalist's Guide to the Sierra Nevada*, Stephen Whitney (Sierra Club Books); *The High Sierra: Peaks, Passes and Trails*, R.J. Secor (The Mountaineers); flora and fauna identifier booklets published by the Nature Study Guild.

Accommodation & Supplies: Yosemite Valley and Lone Pine have everything from hotels to campsites. On the trail, however, you'll be camping. There are four places en route where supplies can be sent or bought. Tuolumne Meadows, 2 days from the start, has a store and a post office. 35 miles (56km) farther on is the Red's Meadow Resort. Supply parcels have to be dropped off here in person, but there is a cafe and a tiny store. Roughly half way along the trail lie Vermilion Valley Resort and the Muir Trail Ranch. The first is reached by a side trail to Lake Edison followed by a ferry. Supplies can't be bought at the resort but the owners will hold a parcel for a small fee. Muir Trail Ranch is a day or so farther on and lies a mile down the Florence Lake Trail. Again, there is no store but supply parcels will be held for a fee. Contact Vermilion Valley Resort, P.O. Box 258, Lakeshore, CA 93634, USA Muir Trail Ranch, P.O. Box 176, Lakeshore, CA93634, USA On the southern half of the trail, the only possibility for resupply is to cross Kearsage Pass, descend to Onion Valley and then get a lift to the town of Independence in Owens Valley, east of the mountains. It is easier to carry all your food.

great horned owl

Currency & Language: US dollars ($US). English.

Photography: No restrictions.

Area Information: Superintendent, P.O. Box 577, Yosemite National Park, CA 95389, USA; Superintendent, Sequoia and Kings Canyon National Parks, Three Rivers, CA93271, USA; Inyo National Forest, 873 N. Main Street, Bishop, CA 93514, USA. Websites: www.sequoia.national-park.com; www.gorp.com/pcta/jmt.htm; www.sierrawilderness.com/inyo.html.

TIMING & SEASONALITY

Best Months to Visit: Mid-July to September.

Climate: The Sierra Nevada has a very benign climate. The summer is usually hot and sunny but thunderstorms do occur and longer periods of rain are possible. Nights can be quite cold, especially at high altitudes. Heavy snow falls in the winter and often lingers well into the summer on the high passes. In heavy-snow years, snow may lie on the higher parts of the trail well into July.

HEALTH & SAFETY

Vaccinations: None required.

General Health Risks: No special risks. Health insurance should be taken out.

Special Considerations: Black bears abound along the John Muir Trail and are adept at stealing hikers' food. You should never cook or store food in your tent. The best way to protect food is to store it in a portable, bear-proof container. These weigh around 3lbs (1.3kg) and will hold 5 to 6 days' dried food. Food can be hung from branches, but this is no longer a secure method with Sierra bears. Some backcountry campsites have bear-proof storage lockers. Mosquitoes can be a problem, especially early in the summer.

Politics & Religion: No concerns.

Crime Risks: Low.

Food & Drink: Although much water in the backcountry is probably safe to drink, it may harbour giardia and other causes of stomach ailments. To protect against this, boil, filter or purify water with chemicals.

HIGHLIGHTS

Scenic: The Sierra Nevada is a beautiful and spectacular mountain range, and the trail is a scenic delight every mile of the way.

Wildlife & Flora: The size of the Sierra wilderness means that many animals and birds live here, including black bears, coyotes, wolverine, mule deer, bighorn sheep, marmots, pika, ground squirrels, red-tailed hawks, great horned owl and blue grouse. The great glory of the Sierra Nevada is, however, the coniferous forest – a wide variety of magnificent pines and firs fill the valleys and rise up the mountain walls. To appreciate the wildlife it's worth carrying a pair of mini-binoculars.

temperature and precipitation

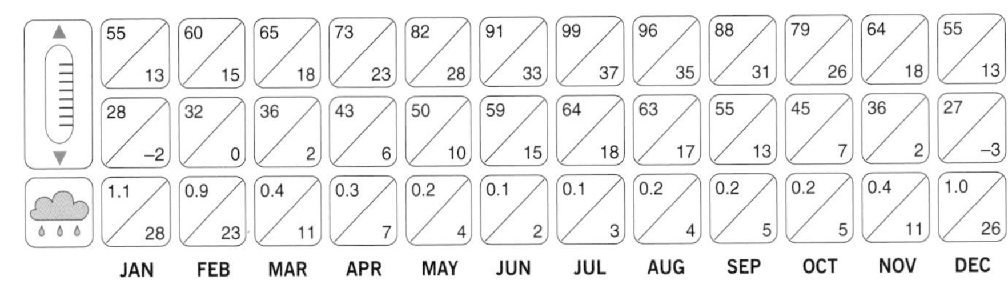

	JAN	FEB	MAR	APR	MAY	JUN	JUL	AUG	SEP	OCT	NOV	DEC
Max temp (°F / °C)	55 / 13	60 / 15	65 / 18	73 / 23	82 / 28	91 / 33	99 / 37	96 / 35	88 / 31	79 / 26	64 / 18	55 / 13
Min temp (°F / °C)	28 / –2	32 / 0	36 / 2	43 / 6	50 / 10	59 / 15	64 / 18	63 / 17	55 / 13	45 / 7	36 / 2	27 / –3
Precipitation (in / mm)	1.1 / 28	0.9 / 23	0.4 / 11	0.3 / 7	0.2 / 4	0.1 / 2	0.1 / 3	0.2 / 4	0.2 / 5	0.2 / 5	0.4 / 11	1.0 / 26

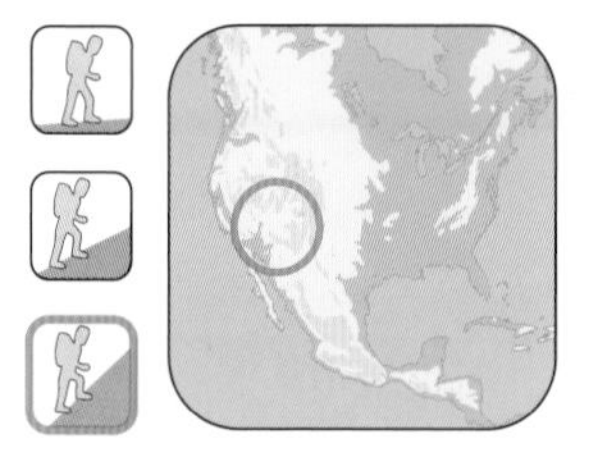

the KAIBAB TRAIL

GRAND CANYON, ARIZONA, USA

Chris Townsend

THE FIRST VIEW OF THE GRAND CANYON FROM THE SOUTH RIM IS ASTONISHING – MULTI-COLOURED ROCK TOWERS, BUTTRESSES AND TERRACES DROP INTO THE HIDDEN DEPTHS, THEN RISE AGAIN IN AN EVEN GREATER TANGLE TO THE DARK EDGE OF THE NORTH RIM, A LONG, LONG WAY AWAY; TO EITHER SIDE THE TWISTING CHASM STRETCHES OUT TO THE HAZY HORIZON.

The Grand Canyon is huge, some 280 miles (450km) long and varying from 4 to 16 miles (6.5 to 25.5km) in width. The South Rim ranges in height from 6,000 to 7,500 feet (1,830 to 2,290m), the North Rim from 7,500 to 8,500 feet (2,290 to 2,590m). The Colorado River, which cut the canyon, lies from 3,500 to 6,000 feet (1,050 to 1,830m) below. Inside this massive gorge there are side canyons that would be major landmarks elsewhere, and mountains that soar about 5,000 feet (1,500m) above the river, yet are not as high as the rims.

The South Kaibab Trail, built in 1924 and maintained by the National Parks Service, is one of the few trails into the canyon that follows ridges rather than side canyons, which means there are excellent views. You start in the cool forest of the South Rim, then descend through steep tiers of rock to the hot desert below, cross the Colorado River, then climb back to the flatlands above. In doing so, you go through all the life zones you would find on a journey from Mexico to Canada.

•DAY 1 7½ miles (12km)

Yaki Point to Bright Angel Campground

From the well-marked trail head at Yaki Point, the trail drops quickly through cliffs in a series of switchbacks for a couple of miles to reach Cedar Ridge. A long traverse down the east side of the ridge leads to more switchbacks through the steep Redwall Limestone formation. The bright colour of these cliffs is due to staining by iron oxide washed down from the rocks above. The trail is used by mule trains taking supplies and guests to Phantom Ranch. If you meet a mule train, stand to the side on higher ground to let it pass.

Below the Redwall, you reach the Tonto Platform, a gently sloping shelf about two-thirds of the way down the canyon. The Tonto Trail runs along the shelf, providing a link between rim to river trails. Cross the Tonto Platform, and

The awesome view across the Grand Canyon from the South Rim. Just right of the centre, the deep cleft of Bright Angel Canyon can be seen running towards the North Rim, following the route of this trail.

shortly afterwards start the last descent to the Colorado River at a dramatic point called the Tipoff. At the river, cross the Kaibab Suspension Bridge, one of only two bridges across the canyon. Follow the trail to Bright Angel Creek – the name was given by John Wesley Powell, who stopped here during the first descent of the canyon in 1869.

Bright Angel Campground is set among leafy cottonwood trees beside the sparkling waters of the creek, a welcome shady spot after the harsh brightness of the desert. Nearby, Phantom Ranch has a bar and dormitories.

•DAY 2 7½ miles (12km)

Bright Angel Campground to Cottonwood Campground

From Bright Angel Campground the North Kaibab Trail winds through a narrow gorge, 1,200 feet (370m) deep, called The Box, where the trail was built by blasting away the rock to avoid myriad creek fords. Beyond the gloomy confines of this cleft, you enter Bright Angel Canyon. The trail climbs gently beside the gurgling creek, and there are superb views of red and yellow rock buttes rising high above. Lush vegetation lines the creek, in stark contrast to the sparse desert plants not far away.

One mile (1.5km) before Cottonwood Campground, a cairned trail branches off to Ribbon Falls, a highly recommended diversion. The falls make a delicate tracery over a bright green, moss-covered cone of soft travertine rock below the cascade, built up by the lime-rich waters of Ribbon Creek. It is possible to climb the side of the falls and pass behind them on a wide terrace where the upper fall tumbles into a small pool on the lip of the cone. Cottonwood Campground is situated just above Bright Angel Creek.

•DAY 3 7 miles (11.5km)

Cottonwood Campground to North Kaibab Trail Head

For 2 miles (3km) the trail continues to climb slowly beside Bright Angel Creek. Looking back, you have an excellent view right down the valley and then back up to the South Rim. The canyon narrows as you approach the huge jutting ramparts of the Roaring Springs cliffs. To the right, Bright Angel Creek continues. To the left, Roaring Springs Creek →

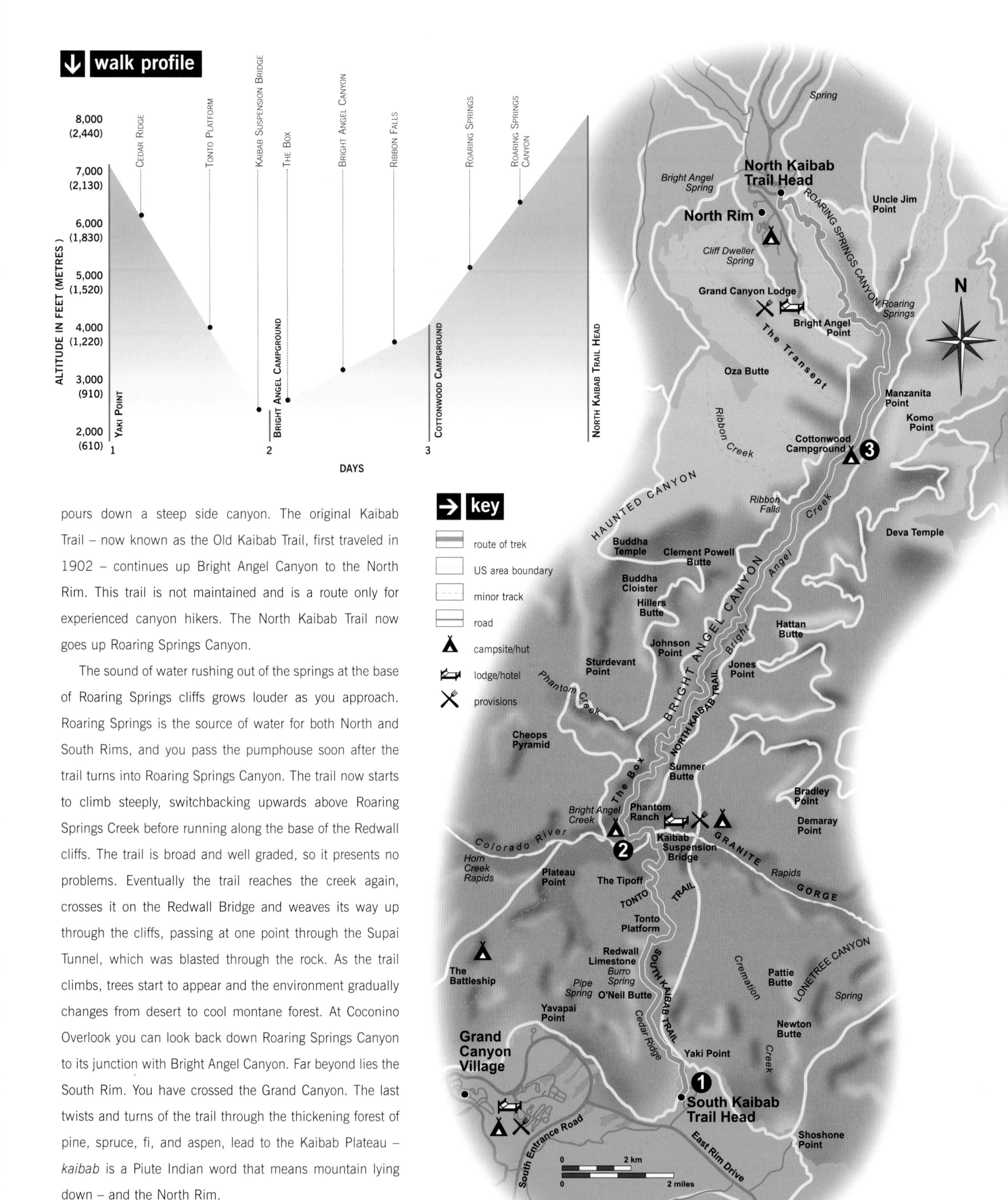

pours down a steep side canyon. The original Kaibab Trail – now known as the Old Kaibab Trail, first traveled in 1902 – continues up Bright Angel Canyon to the North Rim. This trail is not maintained and is a route only for experienced canyon hikers. The North Kaibab Trail now goes up Roaring Springs Canyon.

The sound of water rushing out of the springs at the base of Roaring Springs cliffs grows louder as you approach. Roaring Springs is the source of water for both North and South Rims, and you pass the pumphouse soon after the trail turns into Roaring Springs Canyon. The trail now starts to climb steeply, switchbacking upwards above Roaring Springs Creek before running along the base of the Redwall cliffs. The trail is broad and well graded, so it presents no problems. Eventually the trail reaches the creek again, crosses it on the Redwall Bridge and weaves its way up through the cliffs, passing at one point through the Supai Tunnel, which was blasted through the rock. As the trail climbs, trees start to appear and the environment gradually changes from desert to cool montane forest. At Coconino Overlook you can look back down Roaring Springs Canyon to its junction with Bright Angel Canyon. Far beyond lies the South Rim. You have crossed the Grand Canyon. The last twists and turns of the trail through the thickening forest of pine, spruce, fi, and aspen, lead to the Kaibab Plateau – *kaibab* is a Piute Indian word that means mountain lying down – and the North Rim.

Cascading water from Ribbon Falls comes as a welcome relief in the soaring temperatures of the Grand Canyon.

factfile

OVERVIEW

The Kaibab Trail runs from rim to rim across the incomparable Grand Canyon. The trail starts with a steep descent from the forests of the South Rim to the desert that flanks the Colorado River, and then moves more gradually upwards beside Bright Angel Creek to a final steep climb to the North Rim. The trail is 22 miles (33.5km) long and takes from 2 to 4 days.

Start: Yaki Point, 3 miles (5km) east of Grand Canyon Village.

Finish: North Kaibab trail head, North Rim, Grand Canyon.

Difficulty & Altitude: The biggest problems are heat and lack of water. There is no water between the start of the trail and the Colorado River. Although the trail is well maintained, it is very steep in places. It drops from 7,260 feet (2,210m) at the start to 2,480 feet (760m) at the Colorado River in just 6½ miles (10.5km), then climbs to the North Rim at 8,250 feet (2,510m).

ACCESS

Airports: Grand Canyon airport is 7 miles (11.5km) from Grand Canyon Village. There are flights from Flagstaff, Phoenix, and Las Vegas.

Transport: There are regular buses from Flagstaff to Grand Canyon Village. A shuttle bus runs every 15 minutes from Grand Canyon Village to Yaki Point. A shuttle bus also runs between the South Rim and the North Rim (closed during winter and spring).

Passport & Visas: Passports with visa or visa waiver form for all visitors to North America.

Permits & Access: Permits are required for all overnight camps within the Grand Canyon. There is a limited number of permits, so applications should be made several months in advance. For details contact: Backcountry Office, Grand Canyon National Park, P.O. Box 129, Grand Canyon, AZ 86023, USA. There is a permit fee.

LOCAL INFORMATION

Maps: United States Geological Survey topographic sheets 1:63,630 Bright Angel Point and 1:24,000 Phantom Ranch. Trails Illustrated Grand Canyon National Park 1:73,530 topographic map and the Earthwalk Press 1:48,000 topographic Hiking Map & Guide to Grand Canyon National Park have more information of use to hikers.

Kaibab squirrel

Guidebooks: *Official Guide to Hiking the Grand Canyon*, Scott Thybony (Grand Canyon Natural History Association); *Hiking the Grand Canyon*, John Annerino (Sierra Club Books).

Background Reading: *The Exploration of the Colorado River and its Canyons*, J.W. Powell (many editions) – story of the first journey through the Grand Canyon; *The Man Who Walked Through Time*, Colin Fletcher (Knopf) – story of the first end to end walk through the Grand Canyon; *A Field Guide to the Grand Canyon*, Stephen Whitney (Quill) – natural history.

Accommodation & Supplies: Grand Canyon Village has lodges, hotels, stores, and a campsite. At the bottom of the canyon Phantom Ranch has dormitory accommodation (book well in advance), a campsite, a bar, and meals can be provided if booked in advance. The North Rim has lodges, cabins and camping.

Currency & Language: U.S. dollars ($U.S). English.

Photography: No restrictions.

Area Information: Grand Canyon National Park, P.O. Box 128, Grand Canyon, AZ 86023, USA. Website: www.thecanyon.com/nps.

TIMING & SEASONALITY

Best Months to Visit: This walk can be done year round, but spring and fall are the best times. The summer is extremely hot in the bottom of the canyon, and in winter snow lies on the rims.

Climate: The inner Grand Canyon is a desert where summer temperatures often exceed 104°F (40°C). Storms do occur in the canyon, however, and rain is possible. The wooded rims are more temperate, with snow, heavy on the North Rim, in winter.

HEALTH AND SAFETY

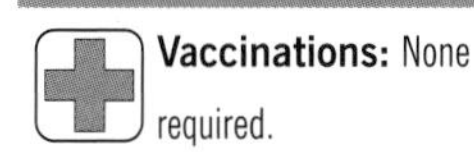

Vaccinations: None required.

General Health Risks: Heat exhaustion, dehydration, water intoxication, sunburn. To prevent these, drink, eat and rest frequently, and wear a sun hat and sunscreen. Water intoxication, caused by drinking lots of water without eating, is serious, so eat as well as drink. As US health care is expensive, good insurance is essential.

Politics & Religion: No concerns.

Crime Risks: Low.

Food & Drink: Carry plenty of food and drink.

HIGHLIGHTS

Scenic: The Grand Canyon is a unique gash in the surface of the Earth. The whole walk takes place in the most spectacular rock scenery, descending and ascending through brightly coloured rock formations.

Wildlife & Flora: On the rims, conifer forests are home to mule deer, coyotes and golden eagles. Black bears and Kaibab squirrels are also found on the North Rim. As you descend into the canyon, the trees give way to shrubs and then to desert plants such as yucca, prickly pear and other cacti. Animals include mule deer, mice, lizards, squirrels and rattlesnakes. Animals should never be fed, nor should food or litter be left where they can find it.

temperature and precipitation

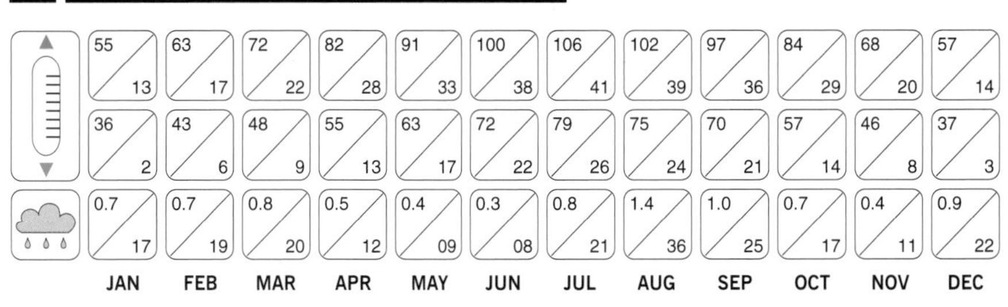

	JAN	FEB	MAR	APR	MAY	JUN	JUL	AUG	SEP	OCT	NOV	DEC
Max (°F / °C)	55 / 13	63 / 17	72 / 22	82 / 28	91 / 33	100 / 38	106 / 41	102 / 39	97 / 36	84 / 29	68 / 20	57 / 14
Min (°F / °C)	36 / 2	43 / 6	48 / 9	55 / 13	63 / 17	72 / 22	79 / 26	75 / 24	70 / 21	57 / 14	46 / 8	37 / 3
Precipitation	0.7 / 17	0.7 / 19	0.8 / 20	0.5 / 12	0.4 / 09	0.3 / 08	0.8 / 21	1.4 / 36	1.0 / 25	0.7 / 17	0.4 / 11	0.9 / 22

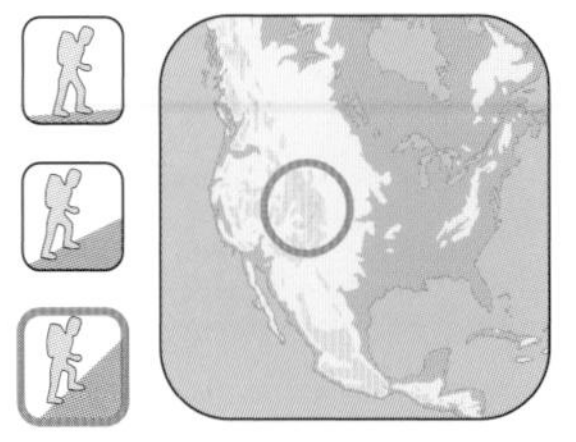

around the MAROON BELLS

COLORADO, USA

Chris Townsend

THE ELK RANGE IS FAIRLY SMALL AS MOUNTAIN RANGES GO, SOME 45 MILES (72KM) LONG AND 30 MILES (48KM) WIDE. THE HEART OF THE RANGE IS PROTECTED IN THE MAROON BELLS–SNOWMASS WILDERNESS, A WONDERFUL MIX OF RUGGED MOUNTAINS, BEAUTIFUL LAKES, FLOWER-FILLED MEADOWS AND FORESTED VALLEYS.

The area was first explored in the late nineteenth century by prospectors in search of silver. Huge amounts of high-grade ore were found, making the fortune of the town of Aspen. The boom ended in the 1900s, but renewal came after World War II, when Aspen became a major ski resort. Now it's busy year round, as are the mountains – at least near the trail heads. Venture into them, however, and the solitude and beauty of the wilderness still comes through. The peaks are high and steep – six rise above 14,000 feet (4,270m) – the valleys are narrow and deep. The most dramatic and well-known peaks are the Maroon Bells – North Maroon and Maroon Peak – named for their bands of dark red sedimentary rock. The region's compact size makes it suitable for circular walks that cover the highlights in a few days. This route crosses five passes at 12,000 feet (3,650m) or more, visits four beautiful treeline lakes, and encircles both the long Snowmass Mountain–Capitol Peak ridge and the Maroon Bells.

•DAY 1 5 miles (8km)

Maroon Lake to West Maroon Creek

The walk starts at the trail head at the end of the Maroon Creek Road. There is a campsite here and, unless you've been in the high country for a while, it's advisable to stay a night to start acclimatising to the altitude, already 9,600 feet (2,920m). The classic view of the Maroon Bells dominates the scene during the initial walk beside Maroon Lake and into the valley beyond. This is a popular day hike, and there will probably be many people around. Aspen and spruce line the trail along with multitudes of wildflowers.

Turning south, the trail reaches Crater Lake, a popular campsite. Rock walls line the route as it climbs into the West Maroon Creek valley. The first pass, West Maroon, is at 12,500 feet (3,810m) and there are no campsites between it and the second one, Frigid Air Pass, at 12,415 feet (3,785m), so a good place for a first wilderness camp is in the woods in this valley.

The easiest route up 14,156-foot (4,316-m) Maroon Peak runs from West Maroon Creek up to the south-east ridge. It's not a walk, however, so much as a tough scramble on steep, loose rock. Finding the route isn't easy either, and the traverse to 14,014-foot (4,272-m) North Maroon Peak is even harder. The Maroon Bells have the reputation of being two of the hardest of the 54 peaks over 14,000 feet (4,270m) in Colorado. There have been a large number of accidents on these peaks – so many that they are sometimes known as the "Deadly Bells" – which makes these climbs for the experienced only. They were first ascended in 1908.

↑ *The startling white peaks of Snowmass Mountain make a stunning background to the beautiful green of Geneva Lake.*

• DAY 2 12 miles (19.5km)

West Maroon Creek to Geneva Lake

Finally leaving West Maroon Creek, the trail leaves the last, sparse trees behind and climbs through flower meadows into a large rocky bowl. Marmots and pikas call shrilly from the rocks. A final steep climb leads to West Maroon Pass, a notch in a steep, castellated, red-rock ridge. Brilliantly coloured flower meadows line the trail on the high-level traverse to Frigid Air Pass, where there is a fine view of the Maroon Bells, and down to the green meadows of Fravert →

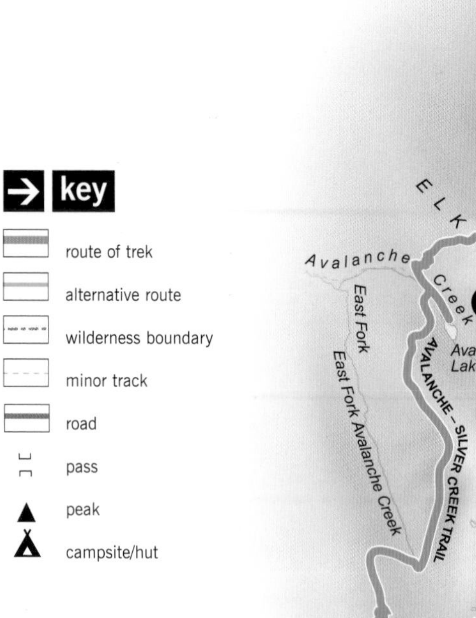

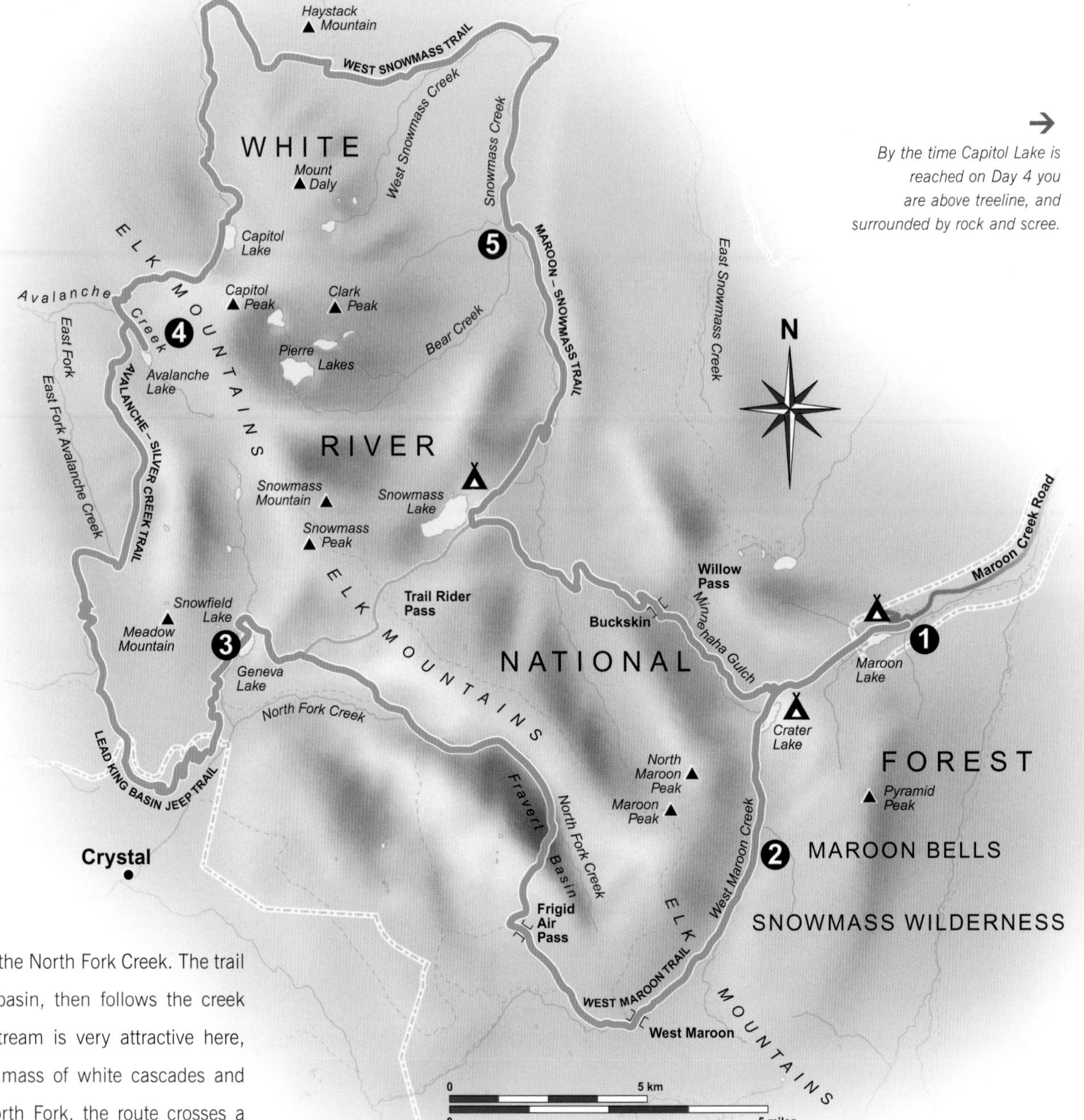

By the time Capitol Lake is reached on Day 4 you are above treeline, and surrounded by rock and scree.

Basin and the twisting line of the North Fork Creek. The trail switchbacks down into the basin, then follows the creek into the forest below. The stream is very attractive here, rushing down a ravine in a mass of white cascades and noisy rapids. Leaving the North Fork, the route crosses a ridge from which there is a good view south to Treasure Mountain in the Raggeds Wilderness. The trail descends from this ridge to Geneva Lake, a lovely deep green, treeline lake backed by pale screes and the cliffs of 14,092-foot (4,2965-m) Snowmass Mountain and 13,841-foot (4,219-m) Hagerman Peak. There are many flower meadows around the lake.

•DAY 3 12 miles (19.5km)

Geneva Lake to Avalanche Lake

From Geneva Lake, you could shorten the walk by heading east to Trail Rider Pass and Snowmass Lake. The longer route goes south down to the Geneva Lake North Fork Trail head, where you leave the wilderness briefly for the very rough Lead King Basin Jeep Trail. This leads to the Avalanche-Silver Creek trail head, where our route turns north and climbs to 12,440-foot (3,792-m) Silver Pass. As

walk profile

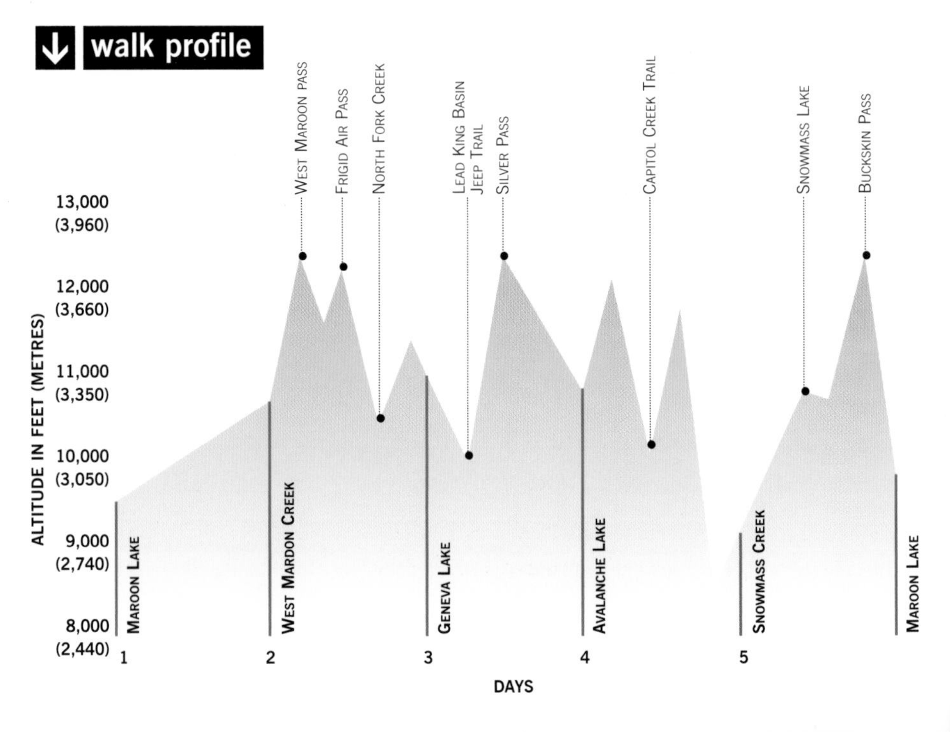

↑ *Day 5 takes you through the wide valley of Snowmass Creek, lined with conifers, aspen and cottonwoods.*

you reach the crest, there's a sudden and startling view of the impressive soaring rock cone of 14,130-foot (4,307-m) Capitol Peak. On the north side of the pass lies a great rugged and rocky bowl at the head of the East Fork Avalanche Creek. The trail, indistinct in some places but mostly marked by cairns, drops some 800 feet (240m) toward the East Fork, then makes a long, high-level traverse around the head of the bowl. After crossing a minor pass on the ridge east of the East Fork, the trail switchbacks steeply down to Avalanche Creek. Avalanche Lake lies a short distance upstream. Beyond the cool, clear, green-grey water, steep rock and scree slopes dotted with stands of dark green spruce rise into the sky.

•DAY 4 13 miles (21km)

Avalanche Lake to Snowmass Creek

A 12,040-foot (3,670-m) pass separates Avalanche Lake from Capitol Lake. This unnamed pass lies right at the foot of the cold rock monolith of Capitol Peak. Capitol Peak, along with Snowmass Mountain, are made from granite, a grey rock – although it glows a wonderful gold in early and late sunlight. Capitol Lake itself is situated just above treeline in a spectacular rocky setting, with the rock and scree slopes of 13,280-foot (4,048-m) Mount Daly rising to the north-east. The lake is the start point for the main route up Capitol Peak, which runs along the knife-edge ridge to the east. Again, this is not reckoned an easy ascent. The →

→ *Lush, green forest and dramatic peaks are inescapable scenic delights along the Maroon Bells trek.*

Capitol Creek Trail descends from the lake through forest and meadow to a junction with the West Snowmass Trail. This climbs to an unnamed 11,760-foot (3,585-m) pass, then descends to West Snowmass Creek. The trail isn't clear here and care is needed with route-finding. Eventually the trail leaves the wilderness briefly for cow pasture before fording Snowmass Creek to a junction with the Maroon–Snowmass Trail. Turning south, the route soon re-enters the wilderness and follows Snowmass Creek below huge, red-rock walls.

•DAY 5 **13 miles (21km)**

Snowmass Creek to Maroon Lake

The walk up the Snowmass Creek valley takes you through mixed forest – conifers, aspen, cottonwoods – and flower meadows, and past many beaver ponds, before leaving the creek for the climb to Snowmass Lake. This is another beautiful, treeline lake with Snowmass Mountain and Hagerman Peak rising above. It's also very popular, and not somewhere to camp if you like solitude. Part of the reason for this is that it's the base camp for the ascent of Snowmass Mountain. Compared with the Maroon Bells or Capitol Peak, Snowmass is a relatively easy ascent, although it still requires some scrambling.

Beyond Snowmass Lake the trail drops to a bridge over Snowmass Creek, then climbs steadily to the last pass of the walk, 12,500-foot (3,810-m) Buckskin. Look back and you have a splendid view of Snowmass Lake and Snowmass Mountain, with Capitol Peak in the distance. Ahead the view is down deep Minnehaha Gulch to Pyramid Peak, with North Maroon to the south. A steep descent into the gulch takes you through flower meadows into spruce forest, then through aspens to a view of Crater Lake. Soon you rejoin the West Maroon Trail for the walk back to Maroon Lake and the trail head.

factfile

OVERVIEW

A circular walk in the Maroon Bells–Snowmass Wilderness in the White River National Forest in central Colorado. The Maroon Bells are part of the rugged Elk Range, itself part of the Rocky Mountains. The distance is 55 miles (89km) and the walk takes 5 to 7 days. The itinerary described is an arbitrary one as there are plenty of opportunities for camping. Competent scramblers could climb some of the peaks along the way.

Start/Finish: Maroon Lake.

Difficulty & Altitude: The trails are mostly well maintained, but junctions are not signed and there are some unclear sections. They are steep and rocky in places, too, and rise well above treeline. The route crosses five passes at or over 12,000 feet (3,660m) and most of it is above 9,000 feet (2,740m), so the altitude is something that needs to be taken into account.

golden aspen

ACCESS

Airports: Denver is the nearest international airport. Flights are available from Denver to Aspen.

Transport: Shuttle bus from Aspen to West Maroon Portal.

Passport & Visas: Passports with visa or visa waiver form are required for all visitors to North America.

Permits & Restrictions: There are camping restrictions in certain areas to prevent damage and to allow over-used areas to recover.

LOCAL INFORMATION

Maps: Trails Illustrated 1:40,680 topographic map Maroon Bells/ Redstone/ Marble; USGS 1:24,000 topographic maps Maroon Bells, Snowmass Mountain, Capitol Peak, Highland Peak.

Guidebook: *Aspen Snowmass*

Trails: A Hiking Guide, Warren Ohlrich (Who Press).

Background Reading: *Colorado Fourteeners: Northern Peaks*, Louis Dawson (Blue Clover Press); *A Climbing Guide to Colorado's Fourteeners*, Walter R. Borneman and Lyndon J. Lampert (Pruett).

Accommodation & Supplies: A range of accommodation and supply stores is available in Aspen. Stock up on food and fuel in Aspen, as there are no facilities in the wilderness, and all supplies must be carried.

Currency & Language: US dollars ($US). English.

Photography: No restrictions.

Area Information: White River National Forest, Old Federal Building, Ninth St. and Grand Ave., P.O. Box 948, Glenwood Springs, CO 81601, USA. Websites: www.gorp.com/gorp/resources/US_Wilderness_Area/CO.MAROO.HTM.

coyote

TIMING & SEASONALITY

Best Months to Visit: Mid-July to mid-September. Snow may linger on the passes well into July and may start falling again in September.

Climate: The summer is usually warm and sunny with cool nights. Rain can fall, however, and thunderstorms are frequent after midday. Crossing high passes early in the day is a wise precaution.

HEALTH & SAFETY

Vaccinations: None required.

General Health Risks: No special risks. Health insurance should be taken out.

Special Considerations: This is a popular area and hikers should take great care to use minimum impact techniques. Some areas are closed to camping due to over-use. These are clearly indicated and should be avoided. Early in the summer mosquitoes can be a problem; insect repellent is worth carrying then.

Politics & Religion: No concerns.

Crime Risks: Low.

Food & Drink: Although much water in the backcountry is probably safe to drink, it may harbour giardia and other causes of stomach ailments. To protect against this, water can be boiled, filtered or purified with chemicals.

HIGHLIGHTS

Scenic: The whole walk is in spectacular scenery. Highlights are Geneva, Avalanche, Capitol, and Snowmass lakes and the twin peaks of the Maroon Bells themselves.

Wildlife & Flora: Black bears live in the mountains, but it is rare to see one and they don't usually bother hikers. Other animals include coyotes, mule deer, elk, beaver, marmots and porcupines. There are many birds too, including golden eagles, red-tailed hawks, blue grouse, ravens, Clark's nutcrackers, Stellar's jays, grey jays, dippers, various woodpeckers and many more. It's worth carrying a lightweight pair of binoculars for observing birds and animals. Masses of flowers decorate the meadows in summer, while the valleys are filled with trees, including stands of aspen that turn a glorious gold in autumn. To identify these take the little *Rocky Mountain Tree Finder* and *Rocky Mountain Flower Finder* booklets by Tom Watts (Wilderness Press).

beaver

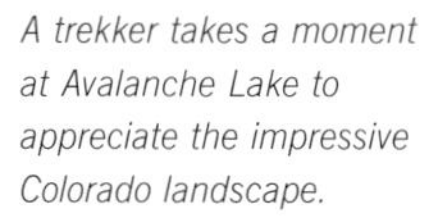

A trekker takes a moment at Avalanche Lake to appreciate the impressive Colorado landscape.

temperature and precipitation

	JAN	FEB	MAR	APR	MAY	JUN	JUL	AUG	SEP	OCT	NOV	DEC
Max temp (°F / °C)	32 / 0	36 / 2	41 / 5	50 / 10	61 / 16	72 / 22	79 / 26	77 / 25	70 / 21	60 / 15	43 / 6	34 / 1
Min temp (°F / °C)	0 / −18	3 / −16	12 / −11	21 / −6	28 / −2	34 / 1	41 / 5	39 / 4	32 / 0	23 / −5	14 / −10	5 / −15
Precipitation (in / mm)	1.2 / 32	1.0 / 26	1.4 / 35	1.1 / 28	1.5 / 39	1.3 / 34	1.7 / 44	1.8 / 45	1.3 / 34	1.4 / 36	1.2 / 31	1.2 / 32

Soon after the trail begins, you reach soaring rock walls in the narrow canyon of the Green River. Ahead lies Stroud Peak.

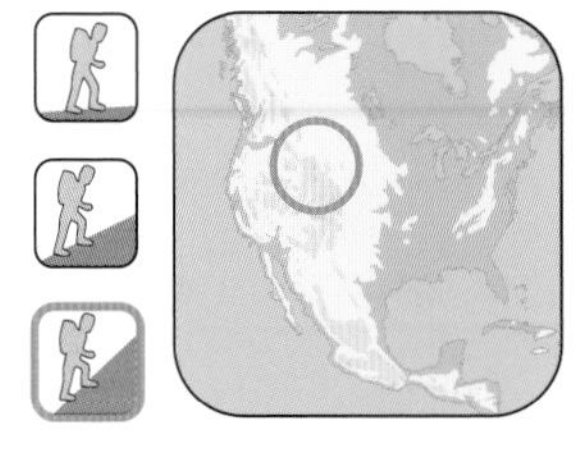

through the WIND RIVER MOUNTAINS

WYOMING, USA

Chris Townsend

THE WIND RIVER MOUNTAINS MAKE UP THE SOUTH-EASTERN END OF THE NORTHERN ROCKY MOUNTAINS, BEFORE THE GREAT DIVIDE BASIN SEPARATES THEM FROM THE ROCKY MOUNTAINS OF COLORADO. THE LONG LINE OF PEAKS IS A HUGE UPLIFTED GRANITE BLOCK THAT HAS BEEN CARVED INTO A COMPLEX MASS OF BROKEN AND TWISTED MOUNTAINS BY THE POWER OF GLACIER ICE AND EROSION.

As well as massive rock faces and soaring spires there are many glaciers – in fact seven of the ten biggest glaciers in the United States are found here. The Continental Divide runs down the crest of the range; the highest peak, 13,804-foot (4,207-m) Gannett Peak, is also the highest in Wyoming. Beautiful cirques, 1,300 or more lakes, hanging valleys, rushing streams and cool forests make this wonderful country for the wilderness-lover. Much of the walking is on or above treeline, with spacious views of the peaks. Rock is underfoot, or close under the thin soil, so ground vegetation is short and sparse, making for easy walking. Camps can be made almost anywhere, giving the walker a feeling of freedom not found where water supply, terrain or permit requirements restrict camping to certain spots. Side trips can be taken into the mountains, on to lower routes if storms threaten, as well as into the Bridger Wilderness which lies to the east.

itinerary

•DAY 1 13 miles (21km)

Green River Bridge to Trail Creek Park

Just beyond the trail head a bridge crosses the Green River to join the Highline Trail, running upstream along the east bank. Ahead rises the great block of Squaretop Mountain. Below this granite monolith the valley narrows to squeeze dramatically between the steep slopes on either side. The Green River is a sizeable stream that is indeed green in colour. It runs north here, just a few miles from its source, but soon turns south and, after 1,500 miles (2,400km), joins the Colorado River. Geographers regard the headwaters of the Green as the source of the Colorado.

The narrow, forested canyon opens out into the pleasant meadows of Beaver Park. A bridge takes you back across the Green River before the first climb of the walk to Trail Creek Park. This is a good place for the first night's camp as a decision has to be made here on which of two routes to take. The trail over Vista and Shannon Passes is the shorter and the more scenic, but also the higher and more exposed. In stormy weather the lower, longer, and more sheltered Highline Trail would be a better alternative.

•DAY 2 10 miles (16km)

Trail Creek Park to Fremont Crossing

Taking the high route the trail climbs to Vista Pass, from where there are superb views across the Green River canyon to the jagged rock summits of Ladd Peak, Mount Whitecap, Sulphur Peak, and Stroud Peak. At 10,120 feet (3,159m) Vista Pass is just on treeline and a great place for a camp. Beyond the pass the trail winds through a sparse landscape of naked rock as it climbs to Cube Rock Pass and then Shannon Pass. The route rejoins the Highline Trail just before cliff-rimmed Upper Jean Lakes and then descends to cross Fremont Creek. After two passes and some rugged walking, this is a good place to camp. From here a superb

key

- route of trek
- alternative route
- Continental Divide
- wilderness boundary
- minor track
- road
- pass
- peak
- campsite/hut
- lodge/hotel

Wonderful panoramas of peaks, lakes, and forests greet you at every turn when trekking in the heart of the northern Wind River Mountains.

side trip can be made to Island Lake and then into Titcomb Basin, a deep canyon whose steep rock walls are popular with rock climbers.

•DAY 3 11 miles (17.5km)

Fremont Crossing to Bald Mountain Basin

Excellent views of the jumble of spires, cliffs, and peaks around Titcomb Basin and along the Continental Divide make the trail to Island Lake and then Lester Pass a mountain-lover's delight. Across the pass, the scenery changes, becoming more plateau-like and spacious, the steeper mountains being farther away and less crowded together. At Cook Lakes you can leave the Highline Trail again for the higher Fremont Trail, which heads into the wide expanse of Bald Mountain Basin. There are plenty of potential campsites here, either in the shelter of the trees on the edge of the basin or beside one of the many lakes.

•DAY 4 14 miles (22.5km)

Bald Mountain Basin to Dream Creek

The unnamed pass at the head of the basin is another superlative viewpoint for the pinnacles and glaciers of Titcomb Basin. The trail stays in open country with expansive views as it climbs gently to wide Hat Pass then drops into forest to reach large North Fork Lake. Groves of conifers, meadows and masses of lakes cover the wide shelf along which the trail runs. A camp could be made almost anywhere – Dream Creek is a good choice.

Rising beyond Lower Green River Lake, near the start of the walk, is the impressive granite block of Squaretop Mountain (11,695 feet/3,508m).

walk profile

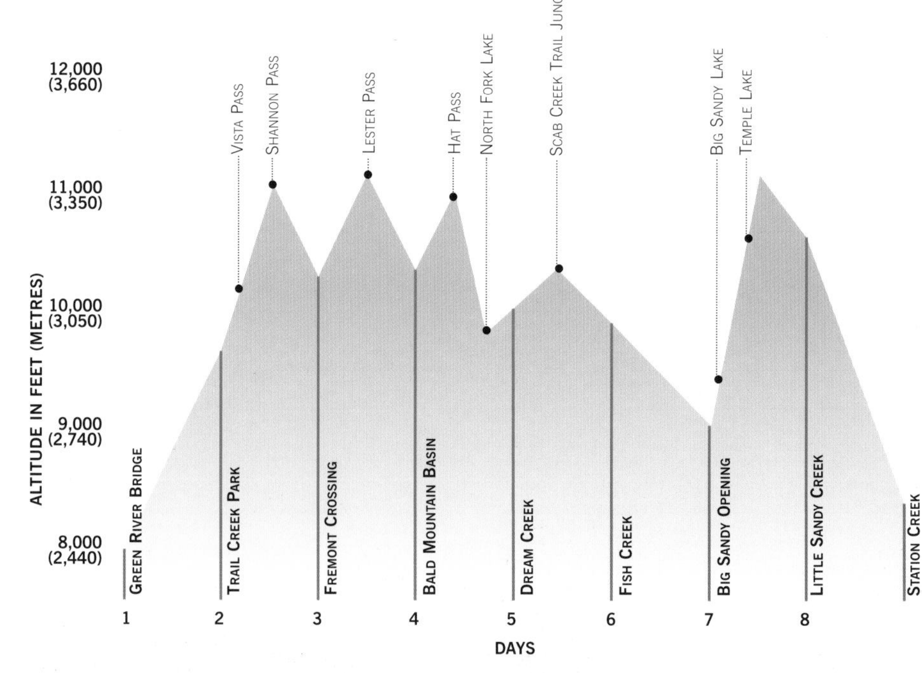

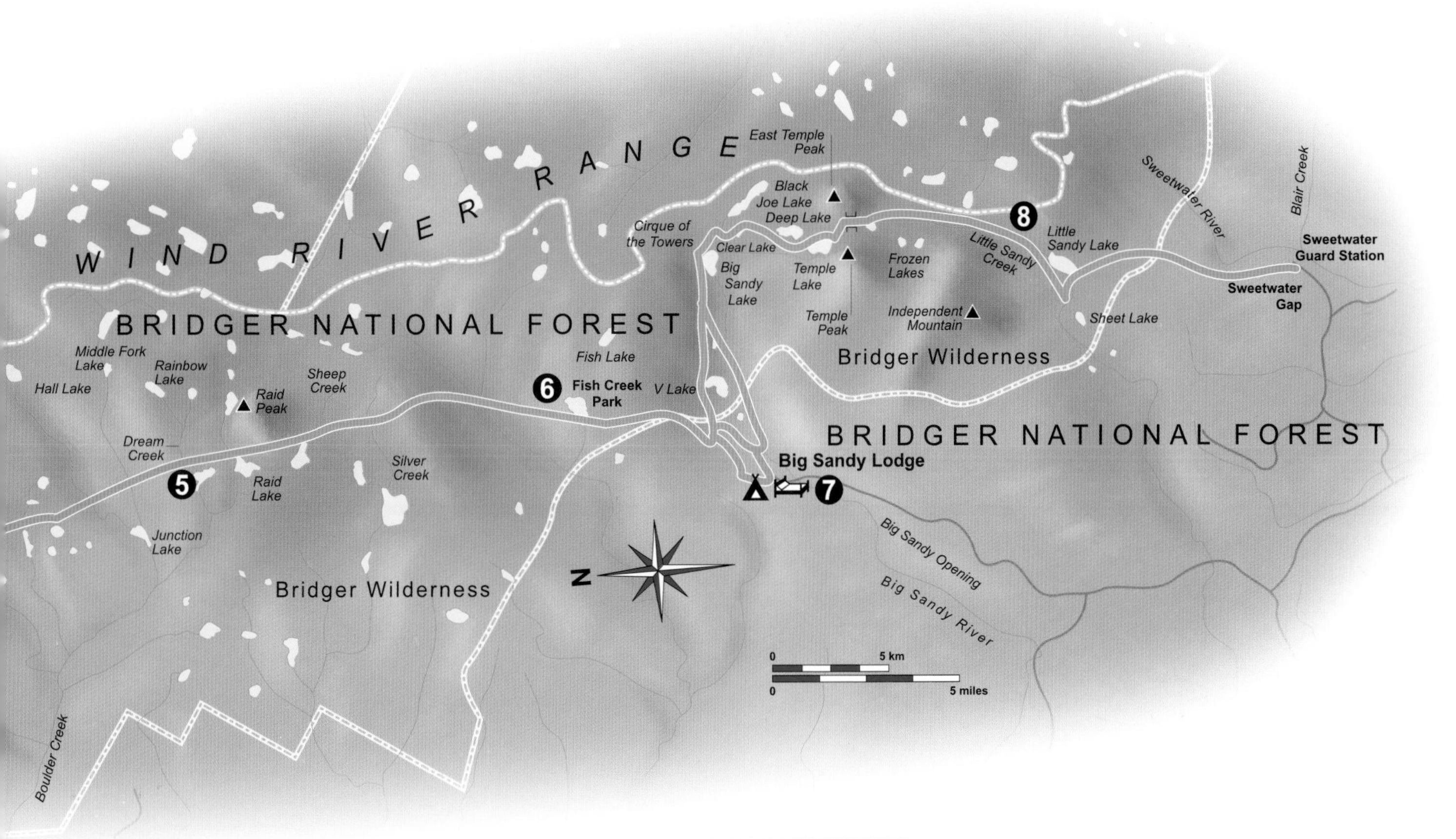

•DAY 5 13 miles (21km)

Dream Creek to Fish Creek

The rugged mountains of the Divide can be seen to the east, and across Sheep Creek, as the trail winds on through pleasant countryside, you have your first sighting of the spectacular Cirque of the Towers. Eventually, a slow descent begins to the forest and meadows around Fish Creek.

•DAY 6 6 miles (9.5km)

Fish Creek to Big Sandy Opening

A short day, mostly downhill and in forest, finishes at the trail head at Big Sandy Opening, where you could end the walk. There is a campsite and lodge here. Those who don't want to leave the wilderness can turn north-east instead of south-west at Meek Lake and follow the trail to Diamond Lake, at the far end of which you meet the trail from Big Sandy Opening.

•DAY 7 13 miles (21km)

Big Sandy Opening to Little Sandy Creek

After a forest walk to Big Sandy Lake, the trail leads into the very dramatic, glacier-sculpted scenery of the Temple Peaks →

Year after year experienced rock climbers return to attempt the difficult ascent of the jagged peaks of Cirque of the Towers.

region. Vast smooth slabs of granite curve down from the summits above Clear, Deep, and Temple lakes with sharp, steep, pointed East Temple and Temple peaks ahead. The 11,500-foot (3,502-m) pass between the two summits is the highest point on the walk. The trails are faint here, with cairns marking the route. On the rocky descent to Little Sandy Creek you get glimpses to the south of the sagebrush and desert land of the Great Divide Basin, beyond the end of the mountains.

•DAY 8 13 miles (21km)

Little Sandy Creek to Station Creek, Sweetwater Gap

The trek ends with a gentle forested descent. At Little Sandy Lake you can turn back for a last good look at the great rock peaks of the Wind River Mountains, your constant companions for the last week. This, on a gentle wooded pass, is the only point on the trek that you cross the Continental Divide. Keep an eye on the map during the descent as the trail is indistinct in places, and there are side trails and old jeep roads to cause confusion. Leave the Bridger Wilderness behind and then follow a jeep road for the last few miles to the trail head in Sweetwater Gap.

factfile

OVERVIEW

A traverse of the Wind River Mountains from end to end – from the Green River, at the northern end, southwards to Sweetwater Gap, on the edge of the Great Divide Basin – which separates the northern Rockies from the Colorado Rockies. The whole route is 93 miles (149km) long and takes from 1 to 2 weeks. A shortened version of the walk finishes after 72 miles (116km) at the Big Sandy Campground. The itinerary described is an arbitrary one as there are numerous excellent campsites along the route and camping is allowed anywhere in the wilderness.

Start: Green River.

Finish: Station Creek, Sweetwater Gap.

Difficulty & Altitude: This is a high mountain, wilderness-backpacking walk, all of it above 7,500 feet (2,290m) with a high point of 11,500 feet (3,450m). The walking is mostly easy but there are some steep ascents and descents and much rocky terrain. The entire route is on trails, but not all are signposted. Good route-finding skills and experience of wild camping are needed.

ACCESS

Airports: The nearest airports are at Jackson Hole and Rock Springs, to which there are flights from Denver and Salt Lake City.

←

On Day 7 the walk goes past Deep Lake and stark East Temple Peak, towards the highest pass on the route.

Transport: There is no public transport to the start or finish of the walk. US Highway 191 is the major road in the area, running west of the Wind River Range between Rock Springs and Jackson. Buses run along this road, stopping at Pinedale, the nearest large town to the mountains. The start of the trek lies at the end of a gravel forest road 50 miles (80km) from Pinedale. The finish is on a minor road 23 miles (37km) from Wyoming Highway 28 and 73 miles (117km) from Pinedale.

Passport & Visas: Passports with visa or visa waiver form for all visitors to North America.

Permits & Access: No permit required. Between July 1 and September 10 group size is limited to 10 and campfires are prohibited. You can camp anywhere as long as your site is out of sight of lakes or trails or at least 200 feet (60m) from them.

pika

LOCAL INFORMATION

Maps: Earthwalk Press 1:48,000 topographic maps Northern Wind River Range & Southern Wind River Range; USGS 1:24,000 topographic maps Green River Lakes, Squaretop Mountain, Gannett Peak, Bridger Lakes, Fremont Peak South, Horseshoe Lake, Mt. Bonneville (1:62,500), Big Sandy Opening, Temple Peak, Sweetwater Gap, Jensen Meadows, Sweetwater Needles.

Guidebooks: *Hiking Wyoming's Wind River Range*, Ron Adkison (Falcon Press); *Climbing & Hiking in the Wind River Mountains*, Joe Kelsey (Chockstone Press); *Guide to the Continental Divide Trail Volume 3: Wyoming*, James R. Wolf (Continental Divide Trail Society); *Wyoming Update*, James R. Wolf (CDTS).

Background Reading: *Wind River Trails*, Finis Mitchell (Wasatch); *Rocky Mountain Tree Finder*, Tom Watts (Nature Study Guild); *Rocky Mountain Flower Finder*, Tom Watts (Nature Study Guild).

Accommodation & Supplies: There are no supply points on the route. Accommodation and meals are provided at Big Sandy Lodge, 72 miles (116km) from the start.

Currency & Language: US dollars ($US). English.

Photography: No restrictions.

Area Information: Bridger-Teton National Forest, 340 North Cache, P.O. Box 1888, Jackson, WY 83001, USA. Website: www.gorp.com/gorp/resource/US_Wilderness_Area/WY.BRDGW.htm.

Clark's nutcracker

TIMING & SEASONALITY

Best Months to Visit: July through September.

Climate: Usually warm and sunny from June to September though rain is possible as are snow showers. Nighttime temperatures can drop to 25°F (-4°C). Snow can lie on the high passes until mid-July and streams can be high with snowmelt during June and July, making fords difficult and sometimes dangerous.

HEALTH AND SAFETY

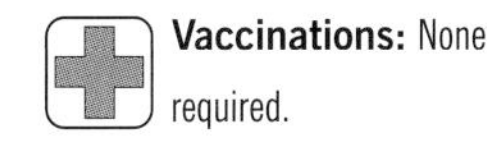

Vaccinations: None required.

General Health Risks: No special risks. Health insurance is essential.

Special Considerations: Mosquitoes, deerflies and horseflies are common, especially early in the summer, so insect repellent is essential. There are black bears, but these rarely cause problems and are unlikely to be seen. It is advisable, however, to hang food from a tree, cook away from your tent, and keep a clean camp. Rubbish should not be buried.

Politics & Religion: No concerns.

Crime Risks: Low.

Food & Drink: All food has to be carried. Although much of the water in the backcountry is probably safe to drink, it may harbour giardia and other causes of stomach ailments. As a precaution, water should be boiled, filtered or purified with iodine. Avoid drinking untreated water from streams below campsites or huts.

HIGHLIGHTS

Scenic: The impressive and distinctive granite scenery of the Wind River Range can be enjoyed throughout the walk. The initial approach along the Green River towards Squaretop Mountain stands out for its mix of water, forest and rock, while the stark cliffs and sharp peaks above Deep Lake are particularly spectacular. From various points on the route there are views of the dramatic granite spires and cliffs of Titcomb Basin and the Cirque of the Towers.

Wildlife & Flora: Moose, elk, bighorn sheep, mule deer, black bears, marmots, pika and beaver all live in these mountains. There are many birds, too, including grey jays and Clark's nutcrackers, which visit campsites in search of food. A small pair of binoculars is useful for wildlife watching. The lower sections of the walk are forested, mostly with spruce, fir and pine. Flowers abound in the meadows.

temperature and precipitation

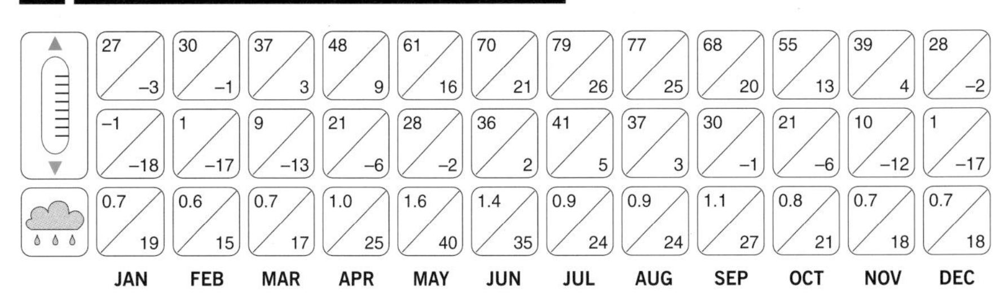

	JAN	FEB	MAR	APR	MAY	JUN	JUL	AUG	SEP	OCT	NOV	DEC
Max °F / °C	27 / -3	30 / -1	37 / 3	48 / 9	61 / 16	70 / 21	79 / 26	77 / 25	68 / 20	55 / 13	39 / 4	28 / -2
Min °F / °C	-1 / -18	1 / -17	9 / -13	21 / -6	28 / -2	36 / 2	41 / 5	37 / 3	30 / -1	21 / -6	10 / -12	1 / -17
Precipitation in / mm	0.7 / 19	0.6 / 15	0.7 / 17	1.0 / 25	1.6 / 40	1.4 / 35	0.9 / 24	0.9 / 24	1.1 / 27	0.8 / 21	0.7 / 18	0.7 / 18

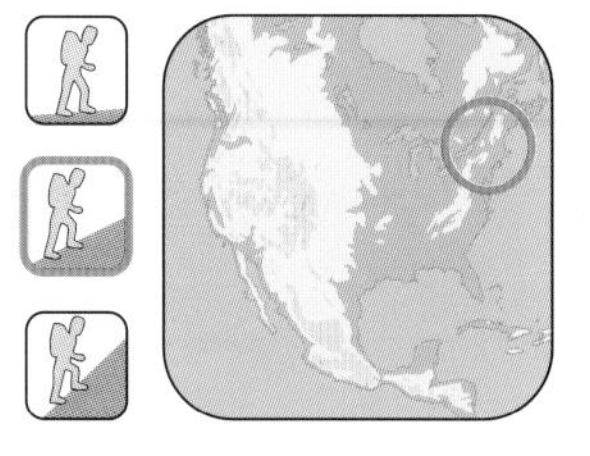

the LONG TRAIL

VERMONT, USA

Walt Unsworth

THE NEW ENGLAND STATE OF VERMONT IS LONG, NARROW, MOUNTAINOUS AND COVERED IN TREES. THEY SAY THE NAME IS A CORRUPTION OF THE FRENCH "VERT MONT" AND CERTAINLY THE PRINCIPAL RANGE IS CALLED THE GREEN MOUNTAINS. ONE OF AMERICA'S OLDEST TRAILS, THE LONG TRAIL FOLLOWS THE CREST OF THE GREEN MOUNTAINS THROUGHOUT ITS LENGTH.

From the Massachusetts line in the south to the Canadian border in the north, some 260 miles (420km). In addition to the Long Trail, begun in 1910, there are another 175 miles (280km) of subsidiary trails, many of which are short but well worth doing. Most of these subsidiary trails make a fine day out from the nearby towns, as do sections of the Long Trail itself. Completion of the whole trail merits an End to End Certificate from the Green Mountain Club (G.M.C.).

→ *By the time you reach this point of the trail – the final rocky ascent of Mount Mansfield – you may have mixed feelings of tiredness and jubilation, considering the great distance you will have travelled over the last 10 days; but there are still 3, albeit easier, days to go.*

itinerary

• DAY 1 **13½ miles (21.5km)**

Williamstown Railroad Station to Congdon Camp

From just north of the former railroad station, the Pine Cobble Trail (blue waymarks) leads steeply to the summit of Pine Cobble and continues for a couple of miles (3km) to a rocky knoll, where it joins the Appalachian Trail. The Long Trail and Appalachian Trail follow the same route for the next 97 miles (155km), until they part near Rutland (Division 5). Continue past Lake Hancock to the camp.

• DAY 2 **21½ miles (34.5km)**

Congdon Camp to Story Spring Shelter

From the camp the path crosses Highway Vt. 9, and then the City Stream by the William A. MacArthur Memorial Bridge, named in honour of one of the trail pioneers. It then climbs to pass through the curious Split Rock to Maple Hill where, despite the wooded crest, there are wide views. The town of Bennington is down to the west and the Hoosac Range to the east. The path crosses a swamp before climbing to Little Pond Lookout along a very narrow ridge, then to the densely wooded summit of Glastenbury Mountain, where there is a reconditioned fire tower affording a panoramic view which some consider unequalled anywhere else on the trail. It takes in the states of Massachusetts and New York – miles of pure wilderness. Moving north from the summit, begin a long descent to Story Spring Shelter.

• DAY 3 **19 miles (30.5km)**

Story Spring Shelter to Bromley Camp

The path climbs to Stratton Mountain (3,936 feet/1,200m) where there is a manned fire tower offering superb panoramas. The mountain has played a significant part in the history of American hiking for it was on its summit that Benton MacKaye is said to have conceived the idea of the Appalachian Trail and it is also where James Taylor thought of the Long Trail. To the west of the mountain lies Stratton Pond, a very popular place for day-trippers

The ascent of Camel's Hump is one of the highlights of the trail. It is a fairly difficult hike, but a very popular one, and care must be taken not to damage the rare Arctic vegetation that grows there.

and campers. To the west is the Lye Brook Wilderness, an area of small ponds interspersed with pleasant trails. (A permit, free, is required for Lye Brook Wilderness and can be got from the G.M.C. office and various local Forest Service offices.)

The trail climbs up to Prospect Rock, giving an aerial view down on to Manchester – a pretty New England village. A trail descends direct to the village but the Long Trail turns north to cross Vt. 11 and continues to the camp.

•DAY 4 23½ miles (38km)

Bromley Camp to Greenwall Shelter

Climb through the Bromley ski area to Bromley Mountain (3,268 feet/996m) where there is an observation tower. Beyond lies a steep descent to Mad Tom Notch. After a climb from the Notch it is an easy hike from Styles Peak to Peru Peak (3,429 feet/1,045m) before a descent to Griffith Lake and the ascent of Baker Peak. The climb to the summit of this fine viewpoint is on open rock, easy but more exposed than anything met so far on the trail. In case of bad weather there is a bypass route avoiding the summit. Beyond, the trail continues north, crossing a couple of forest roads until it comes to the popular Little Rock Pond, a favourite picnic area easily accessible from Wallingford.

The trail misses the summit of White Rocks, but a side trail leads to White Rocks Cliff and gives a bird's-eye view of the valley below. The detour takes about 20 minutes. The main trail continues to the lean-to shelter.

•DAY 5 23 miles (37km)

Greenwall Shelter to Long Trail Inn

The Long Trail follows a low ridge to the Clarendon Gorge, which is crossed by a suspension bridge to reach a major road. Just beyond this is one of America's historic highways, the Crown Point Military Road, built in pioneering days when there were wars between the French and Indians.

The trail now approaches one of the most famous ski areas in the USA, Killington. The peak itself, at 4,235 feet (1,291m), is the second highest in the range after Mount Mansfield. The summit is bare rock and, although littered with ski lifts and other man-made structures, offers a wide panorama taking in a large part of central Vermont, the Adirondack Mountains and the White Mountains.

The route moves swiftly down over Pico Peak (3,957 feet/1,206m) to meet Route 4 at Sherburne Pass. At this point there is the Long Trail Inn, one of the inns for which Vermont is famous, and anyone doing the whole route could do worse than take a break here – some of the best, and toughest, parts of the trail lie ahead.

Also from the inn the little Deer Leap Trail can be done – only a couple of hours' walking but an interesting circular route that includes some scrambling. It is here, too, that the Appalachian Trail veers off to the east.

•DAY 6 **18¾ miles (30km)**

Long Trail Inn to Sunrise Shelter

The Long Trail continues north. The route heads through woods, and passes the Chittenden Reservoir. It continues over a series of ridges – Mount Carmel, Bloodroot and Farr Peak – to descend to the shelter.

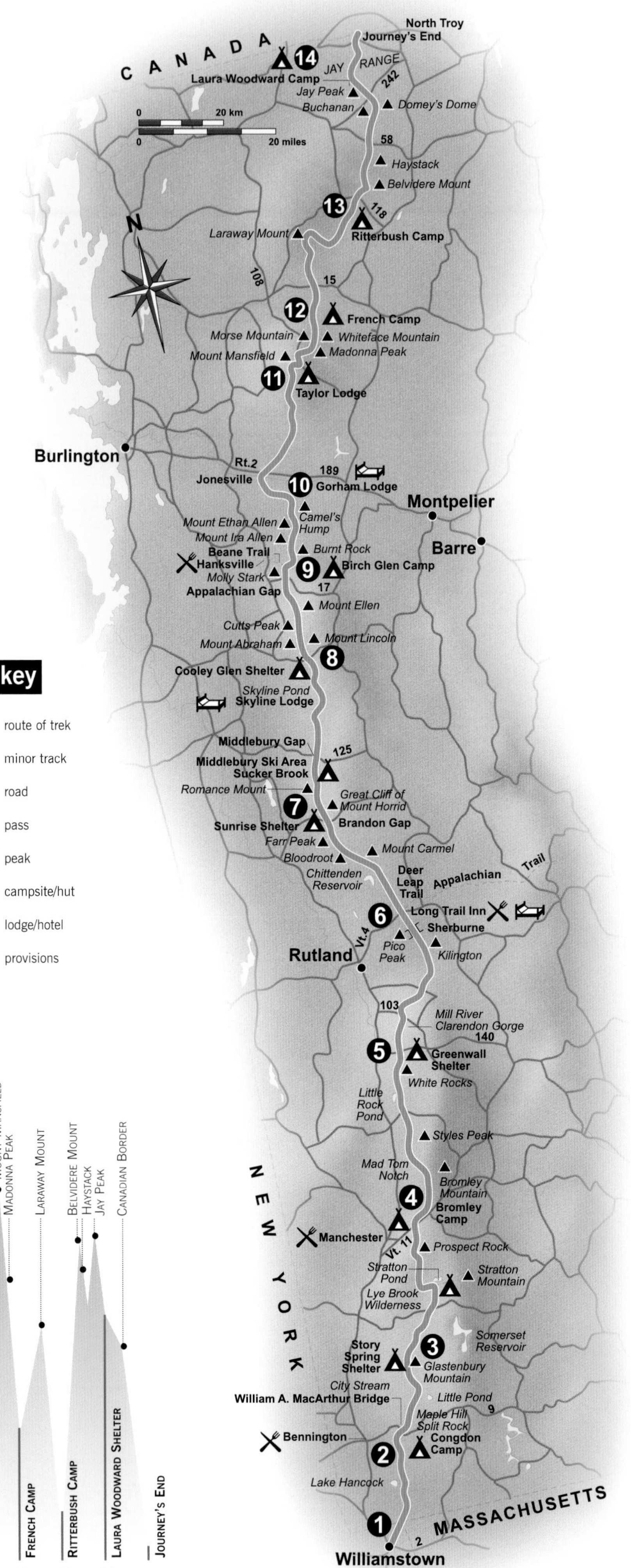

key

- route of trek
- minor track
- road
- pass
- peak
- campsite/hut
- lodge/hotel
- provisions

walk profile

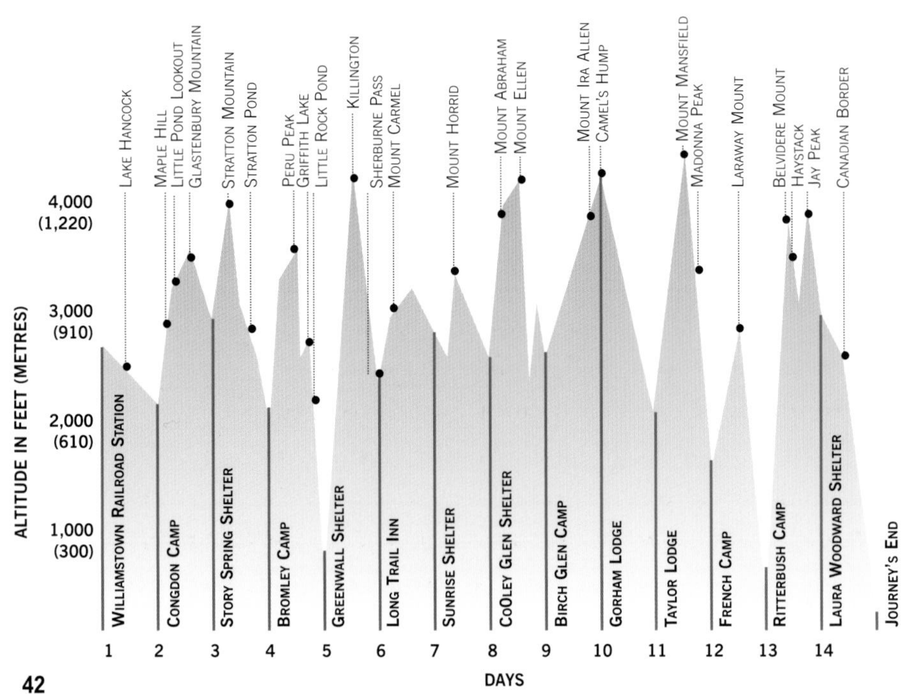

↑ *Sunlight filters through the forest trail that climbs to Mount Horrid. In autumn, these paths are covered in a blanket of flame-coloured leaves.*

•DAY 7 **23 miles (37km)**

Sunrise Shelter to Cooley Glen Shelter

The path meets the road at Vt. 73, at the Brandon Gap, from where it climbs up the Great Cliff of Mount Horrid (3,216 feet/980m) with beaver dams below and grey rocks poking through the dense forest. The going becomes hard – with lots of ups and downs – and boggy, sometimes protected by duckboards. After Romance Mount, the trail descends steeply to the little hut at Sucker Brook and then climbs a long slow ridge to meet the Middlebury Ski Area and Middlebury Gap. This whole region is popular with skiers in winter.

Beyond the Gap the trail hovers at about 3,500 feet (1,050m) as it crosses various summits, many named after American presidents. It reaches the pretty Skyline Pond, where there is a good lodge; Skyline Lodge is owned by a private club but open to all hikers. The path continues to the Cooley Glen Shelter.

•DAY 8 **18½ miles (30km)**

Cooley Glen Shelter to Birch Glen Camp

The peaks you will cross today are the highest between Killington and the Camel's Hump, and include the several summits of Mount Lincoln: Mount Abraham, at 4,006 feet (1,221m), Cutts Peak at 4,020 feet (1,225m) and Mount Ellen at 4,083 feet (1,244m). A ski lift reaches this last peak, but better views are had from the open rock summit of Mount Abraham. The path drops to the Appalachian Gap and then crosses the Molly Stark summits to Birch Glen Camp. From here the Beane Trail leads to Hanksville (1 mile/1.5km) where fresh supplies can be picked up.

•DAY 9 **10½ miles (17km)**

Birch Glen Camp to Gorham Lodge

This day provides one of the most fascinating sections of the Long Trail because it includes what is arguably the finest peak of the Green Mountains, Camel's Hump (4,083 feet/1,244m). The section is more rugged than most, especially from Burnt Rock (3,160 feet/963m) over Mount Ira Allen (3,460 feet/1,055m), Mount Ethan Allen (3,588 feet/1,094m), and the Camel's Hump itself. This section is all rocky and at one point involves climbing a ladder. The trail goes steeply up and down, dodging in and out of woods, until suddenly confronted by the imposing rock face of Camel's Hump. It seems impregnable, but a way leads off left and then, by cracks and easy slabs, up to the summit.

The summit supports rare Arctic tundra vegetation, and one should keep to the marked path. Like Mount Mansfield, the Camel's Hump can be tackled as a day hike from the valley; as over 10,000 people a year are said to do this, there is some concern over the environment. The lodge is just beyond the peak. A short day of walking allows for exploration of this stunning area.

•DAY 10 **21½ miles (34.5km)**

Gorham Lodge to Taylor Lodge

The path descends to Jonesville, where it crosses the railroad and Rt. 2 before continuing to Bolton Lodge. This next section is also one of the best parts of the Long Trail, tackling, as it does, Mount Mansfield, which, at 4,393 feet →

(1,339m), is the highest summit in the Green Mountains. The trail climbs from Bolton Lodge to Taylor Lodge, which can be regarded as the start of Mount Mansfield proper.

The mountain has two summits, The Nose (4,062 feet/1,239m) and The Chin (4,393 feet/1,340m), just over a mile apart and a road reaches the saddle between them, culminating at a radio station and cafe. The view from The Chin is very wide. Care should be taken to keep to the trails and avoid damaging the sensitive Arctic tundra.

•DAY 11 **20 miles (32km)**

Taylor Lodge to French Camp

The trail continues north over the summits of Mount Mansfield, crosses the Vt. 108, then climbs three attractive peaks – Madonna Peak, Morse Mountain, and Whiteface Mountain – before descending to French Camp.

•DAY 12 **21½ miles (34.5km)**

French Camp to Ritterbush Camp

The trail continues to cross various peaks. The best of these is Laraway Mount (2,795 feet/852m), where the trail turns sharply east then north-east to follow the ridge line.

•DAY 13 **22½ miles (36km)**

Ritterbush Camp to Laura Woodward Shelter

Once across Vt. 118, the peaks gather themselves up again to finish with a flourish: Belvidere Mount and Haystack are over 3,200 feet (975m); Buchanan Mount and Domey's Dome are both over 2,900 feet (884m); and last but not least comes Jay Peak at 3,861 feet (1,177m). Climb steeply to its open top, which gives fine views back along the Green Mountains to Camel's Hump and over to the Adirondacks and White Mountains. Descend to Laura Woodward Shelter.

•DAY 14 **8½ miles (13.5km)**

Laura Woodward Shelter to Journey's End

The trail follows the ridge over various summits of the Jay Range and over Burnt Mountain before descending to the Canadian border and the end of the Long Trail.

factfile

OVERVIEW

A traverse of the Green Mountains of Vermont, from south to north. Much of the route is deep in the forests, but in places it breaks free of the trees, climbs a rocky vantage point, and offers superb views. In the autumn the forest becomes a carpet of red and gold as the leaves take on glorious tints. The Long Trail is split into 12 divisions, each of which is looked after by a section of the Green Mountain Club (G.M.C.) who have responsibility for the path, along with the US Forest Service. In total, the trail is 265 miles (427km) long, and it would take some 14 days to complete as one long hike; many people approach it in bits, and there is easy access at several points.

chipmunk

Start: Williamstown Railroad Station, 1 mile (1.5km) north-east of Williams College.
Finish: North Troy, 5¼ miles (8.5km) from the end of the trail.
Difficulty & Altitude: The path is generally unreconstructed, but kept fairly clear and marked with white flashes (side trails carry blue flashes). The walking can be of the rollercoaster sort, with plenty of ups and downs. Most of the mountain peaks are over 3,000 feet (910m), with the highest, Mount Mansfield and Mount Killington, rising to 4,393 feet (1,340m) and 4,235 feet (1,291m) respectively. The guidebook sums up the route thus: "The Long Trail is steep, boggy, and rugged".

The summit of Killington offers a wide prospect over central Vermont, the Adirondacks and the White Mountains.

ACCESS

Airports: Nearest international airport is Boston, with linking service to Rutland.
Transport: Bus along US 7 to the start; return by bus along Vt. 100 to Rutland. Contact Vermont Transit Company, 135 St. Paul Street, Burlington VT 05401.
Passport & Visas: Passport with visa or visa waiver form for all visitors to North America.
Permits & Access: About half the route is on public land and the rest is private – there are ongoing agreements with the Green Mountain Club over access. No permits are required. No camping permitted outside the G.M.C. shelters, except by special permission of the landowner.

LOCAL INFORMATION

Maps: The 1:190,000 maps in the Green Mountain Club guidebook are adequate.
Guidebook: *Guide Book of the Long Trail* (Green Mountain Club).
Accommodation & Supplies: There is accommodation at the end of each section described here, as well as plenty en route; all clearly marked and listed in the guidebook. There are some 70 Green Mountain Club shelters on the trail, open to all. These are of three types: open lean-to, enclosed hut and lodges. Some of the lodges are serviced and a small fee is payable for this. Food must be carried, but it is easy to leave the trail at various points to pick up supplies en route.

Tree-shrouded peaks are frequently seen all along the trail; Mount Ethan Allen is only one of the many mountains encountered.

The guidebook shows where this is possible.

Currency & Language: US dollars ($US). English.

Photography: No restrictions.

Area Information: Green Mountain Club, 4711 Waterbury–Stowe Road, Waterbury Center, VT 05677, USA. Websites: http://www.greenmountainclub.org

TIMING & SEASONALITY

Best Months to Visit: This is a summer/autumn hike. It would be too cold and snowy in winter, too muddy in spring.

Climate: Vermont's weather is traditionally unpredictable, and temperature and precipitation vary greatly at different elevations. In general, however, the summers are cool and pleasant, with the occasional hot humid day. Poor conditions, including rain, fog and low temperatures, can occur at any time.

HEALTH & SAFETY

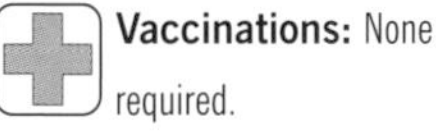

Vaccinations: None required.

General Health Risks: Health insurance is essential. Pack out all trash. Light fires only in authorised fireplaces. Use the toilet facilities provided at shelters; failing this, keep away from streams.

Special Considerarions: This walk traverses deep woods and, despite waymarking, it is easy to go off route. This can be quite scary. If you get "lost", try to backtrack to your last known point of reference. Stay alert and take compass checks from time to time as a precaution. Lightning is also a serious risk for the trekker, and summits and ridges should be avoided during storms.

Politics & Religion: No concerns.

Crime Risks: Low.

Food & Drink: Boil, filter, or purify all drinking water. A serious problem is porcupines, who are persistent food stealers. Keep food, boots, etc. out of harm's way.

HIGHLIGHTS

Scenic: The Long Trail crosses four designated wildernesses, and all the glories of nature that go with them. In the autumn, magnificent tree foliage turns the forest into a leafy carpet of fiery reds and yellows.

porcupine

Wildlife & Flora: Apart from porcupines (see above), deer, racoons, chipmunks and mice are common, and bears have been reported but are rarely spotted. On the ground, keep your eye out for the colourful trillium, bloodroot and jack-in-the-pulpit plants . On the summits of Mount Mansfield three fragile ecosystems support rare Arctic–alpine tundra and vegetation.

Some rocky sections of the Long Trail are overcome by the use of ladders, such as this one on Mount Mansfield.

temperature and precipitation

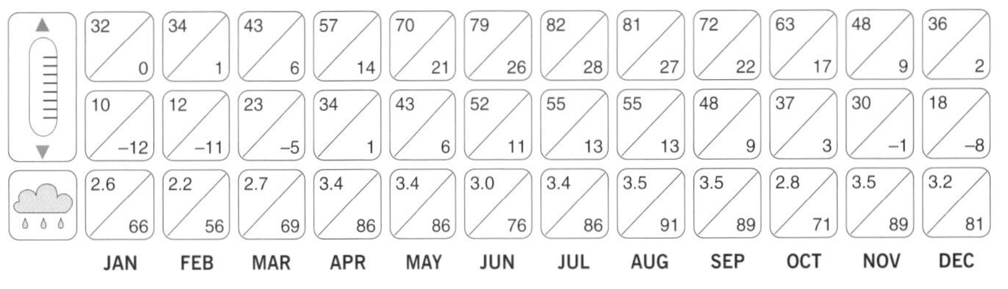

	JAN	FEB	MAR	APR	MAY	JUN	JUL	AUG	SEP	OCT	NOV	DEC
Max temp (°F / °C)	32 / 0	34 / 1	43 / 6	57 / 14	70 / 21	79 / 26	82 / 28	81 / 27	72 / 22	63 / 17	48 / 9	36 / 2
Min temp (°F / °C)	10 / −12	12 / −11	23 / −5	34 / 1	43 / 6	52 / 11	55 / 13	55 / 13	48 / 9	37 / 3	30 / −1	18 / −8
Precipitation (in / mm)	2.6 / 66	2.2 / 56	2.7 / 69	3.4 / 86	3.4 / 86	3.0 / 76	3.4 / 86	3.5 / 91	3.5 / 89	2.8 / 71	3.5 / 89	3.2 / 81

→

Views of the towering pyramid of Mount Assiniboine are one of the highlights of the early part of this trail.

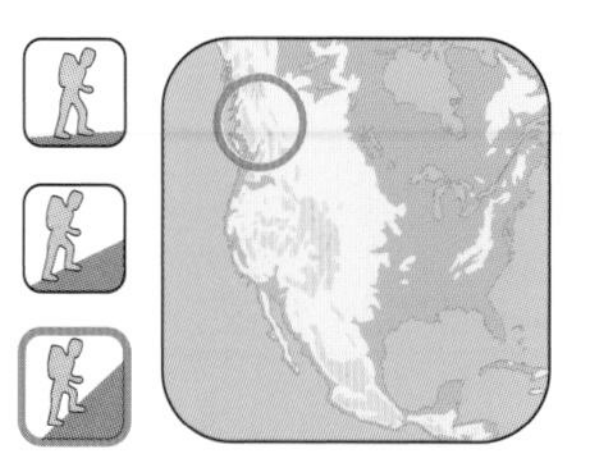

the GREAT DIVIDE

BRITISH COLUMBIA & ALBERTA, CANADA

Chris Townsend

THE GREAT DIVIDE, THE WATERSHED OF NORTH AMERICA, RUNS ALONG THE CREST OF THE ROCKIES AND FORMS THE BORDER BETWEEN ALBERTA AND BRITISH COLUMBIA. THE WALK CROSSES THE DIVIDE THREE TIMES BEFORE HEADING WEST TO THE VERMILION RANGE. ON THE WAY IT PASSES THROUGH MOUNT ASSINIBOINE PROVINCIAL PARK AND BANFF, KOOTENAY AND YOHO NATIONAL PARKS.

Although the Great Divide Trail was first proposed in 1967, it has not yet been formally made into a trail, and may never be so. In the main, this trek follows the suggested southern section of the trail as first described by Jim Thorsell in *The Canadian Rockies Trail Guide*. Linking together several trails through the vast wilderness of the Canadian Rockies, the route climbs from deep forested valleys to treeline lakes and high alpine passes. Much of the walking is on or above treeline, in a beautiful and dramatic alpine world. It is a walk for the wilderness backpacker; the lover of wild, unspoiled places who wants to be away from signs of civilisation and to live for a while in the natural world. Although the terrain is rugged, with many steep ascents and descents, all the walking is on good trails and bridges.

←

Helmet Falls plunges 1,200 feet (370m) into a huge amphitheatre. You can hear the roar of the water long before you see the falls themselves.

↓ itinerary

•DAY 1 **9 miles (14.5km)**

Watridge Creek to Bryant Creek Campground

Beginning rather inauspiciously with a walk through an old logging area, the trail soon enters dense forest and the Bryant Creek valley, which leads to the Bryant Creek Warden Cabin and Campground. This is a short first day, which will probably be welcome with a heavy load, especially as there is much climbing to come.

•DAY 2 **9 miles (14.5km)**

Bryant Creek Campground to Lake Magog

The glories of the route are revealed as the trail contours high above dark blue, tree-shrouded Marvel Lake and then climbs steeply to Wonder Pass, where the Great Divide is crossed. During the ascent, the views open up to the peaks and glaciers rising beyond the head of the lake. From the pass, the vista is extensive and spectacular, with mountains all around. To the north you can see all the way up the Og Valley and beyond to the peaks around Sunshine Meadows.

The descent past Gog Lake to Lake Magog is dominated by the soaring rock and snow cone of the 11,870-foot (3,618-m) Mount Assiniboine. Known as the "Matterhorn of the Rockies", it is one of the most impressive peaks in the whole of the Canadian Rockies, ringed by its eponymous provincial park. No roads run into the park, although there are helicopter flights to and from the lodge.

•DAY 3 **14 miles (22.5km)**

Lake Magog to Howard Douglas Lake

Beyond Lake Magog, the trail winds gently through groves of subalpine trees and flowery meadows to Og Lake. Here, the terrain changes dramatically as the lake lies at the end of the strange Valley of the Rocks, where the forest is dotted with apparently randomly placed, oddly shaped boulders. This is, in fact, a huge rockslide, containing roughly one billion cubic yards (m^3) of limestone. The trail twists and turns through the boulders, before the terrain becomes less →

rough in the Golden Valley. A steep climb leads to 7,740-foot (2,322-m) Citadel Pass – another crossing of the Great Divide – and enters Banff National Park. Ahead are the Sunshine Meadows: a rich, rolling subalpine landscape of flowers and forest groves through which the Great Divide runs. Not far from the pass lies Howard Douglas Lake.

•DAY 4 **11 miles (17.5km)**

Howard Douglas Lake to Egypt Lake

From the lake, the trail winds through the meadows with superb views back to the now distant, but still dominant, Mount Assiniboine. The peaceful beauty of Sunshine Meadows is interrupted by the Sunshine Village Ski Area, giving easy access to the area via a gondola. Here the nature of the walk changes, as the open landscape is left for the closed-in terrain of a series of high passes and deep valleys. From Wawa Ridge, directly above the resort, there is a last, tremendous, expansive view of Sunshine Meadows with Mount Assiniboine behind them.

Now the trail descends through forest to unmemorable Simpson Pass, named after Sir George Simpson, governor of the Hudson's Bay Company, who crossed it in 1841 while searching for a new route across the Rockies for fur traders. More interesting is Healy Pass, with good views all around. To the west lies the cluster of summits known as Pharaoh Peaks, and below is Egypt Lake Campground, a popular site with a shelter cabin in it. Lying in the centre of an area of attractive lakes – Scarab, Mummy, Pharaoh, Sphinx, and more – it is a good place for a rest day.

•DAY 5 **13 miles (21km)**

Egypt Lake to Hawk Creek

Above Egypt Lake lies Whistling Valley, a long, rocky pass separating the Pharaoh Peaks from the Great Divide. The boulder slopes in the pass are home to marmots, whose shrill warning whistles give the valley its name. From the crest, there are views north to the peaks of the Ball Range and down to Haiduk Lake, where you are heading. Below the lake is Ball Pass Junction Campground, much quieter than Egypt Lake, and very scenic. About 1,000 feet (300m) above lies the narrow gap of Ball Pass, where the route

← *The Rockwall Trail runs for almost 20 miles (32km) along the base of a line of dramatic limestone cliffs. Numa Creek provides water music as background to the alpine scenery.*

crosses the Great Divide for the last time. From here, the view takes in bulky Mount Ball, capped with its distinctive, curving white glacier. A pounding descent follows – 2,900 feet (870m) in 6 miles (9.5km) – down a mostly dry and rocky trail to Hawk Creek and the Banff–Radium Highway. A campsite just before the highway makes a good place to stay, as a steep climb comes next.

•DAY 6 7 miles (11.5km)

Hawk Creek to Floe Lake

A 2,350-foot (700-m) climb through forest brings you to Floe Lake, where you may want to stay overnight in the fairly formal campsite in order to fully appreciate its beauty. Nestling below the 3,000-foot (910-m) high cliffs of the Rockwall, an escarpment that stretches unbroken for nearly 25 miles (40km), the pale blue lake water is dotted in summer with small ice floes which break off the ribbon of glacier running along the base of the cliffs above. Dawn in particular is breathtaking, as the rays of the rising sun first light distant snowy peaks to the south, then gradually slide along the cliffs, turning the grey rock a warm gold and the dull glacier snow shining white.

•DAY 7 11 miles (17.5km)

Floe Lake to Tumbling Creek

From Floe Lake, the Rockwall Trail runs north for nearly 20 miles (32km) along the base of the limestone cliffs. It is a →

→ *The Sunshine Meadows are filled with wildflowers in summer. The rolling tundra is ringed with conifers. All around, the snowy peaks provide a dramatic contrast.*

The Rockwall Trail starts from Floe Lake, providing memorable views of snowy mountains reflected in the calm waters of the lake.

marvellous rollercoaster of a trail, past waterfalls, flower-filled cirques and hanging glaciers. Over Numa Pass, under the pyramid of Foster Peak, down to Numa Creek, back up to Tumbling Pass, and down to Tumbling Creek goes the trail. By the last creek there is a campsite on the edge of a meadow, from where the evening light can be watched as it highlights the curving arêtes and blocks of snow high on the Tumbling Glacier.

•DAY 8 8 miles (13km)

Tumbling Creek to Helmet Falls

After climbing to Wolverine Plateau, the trail stays high over Rockwall Pass and Limestone Summit, then drops down to Helmet Creek. During the descent, a growing roar can be heard, the sound of the 1,200-foot (370-m) Helmet Falls tumbling down a vast amphitheatre which marks the northern end of the Rockwall.

•DAY 9 8 miles (13km)

Helmet Falls to McArthur Creek

The wide, flower-filled meadows of Goodsir Pass lie 1,500 feet (460m) above Helmet Creek. The pass marks the boundary between Kootenay and the last national park on this walk, Yoho. Ahead tower the twin peaks of Mount Goodsir, while to the south the Rockwall stretches into the distance. Leaving the views behind, the trail drops through dense forest to the Ottertail River and McArthur Creek. From the banks of the river, the north-east face of Mount Goodsir can be seen rising 6,200 feet (1,860m) from Goodsir Creek.

•DAY 10 16 miles (25.5km)

McArthur Creek to Cataract Brook

Once across McArthur Pass, the last pass on the walk, it is a short walk to Lake O'Hara in the heart of wonderful alpine scenery and an extensive trail system. Beyond the buildings and campsite a trail continues down the Cataract Brook valley to the Trans-Canada Highway. This gentle descent through forest is a relaxing way to finish the spectacular Great Divide Trail.

factfile

OVERVIEW

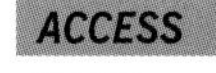

The Great Divide Trail is still only a concept, but this route links already existing trails to stay close to the watershed of Canada. It runs through three national parks: Banff, Kootenay, and Yoho, and one provincial park, Mount Assiniboine. The total distance is 106 miles (168km), and the time required is 1 to 2 weeks. There are plenty of campsites along the way.

Start: Watridge Creek.

Finish: Cataract Brook, Lake O'Hara Access Road, Trans-Canada Highway.

Difficulty & Altitude: The trails are well marked and well maintained. Many ascents and descents: the highest point at 7,725 feet (2,318m) is Numa Pass. There are few facilities en route and full backpacking gear is needed.

ACCESS

Airports: Calgary, Alberta.

Transport: No public transportation to the start. Buses run from Calgary to Banff. Take a cab from there. At the finish, buses run to Lake Louise or Banff.

Passport & Visas: Passports and visas for all visitors to North America.

Permits & Access restrictions: Wilderness permits are required for the national parks. Numbers are limited, so book in advance. Pay on site for use of Naiset Cabins and Lake Magog Campsite.

grizzly bear

LOCAL INFORMATION

Maps: 1:50,000 Canadian topographic maps: Spray Lakes Reservoir, Mount Assiniboine, Banff, Mount Goodsir, Lake Louise. The National Park; 1:200,000 Banff-Yoho-Kootenay map is good for an overview.

Guidebook: *The Canadian Rockies Trail Guide* by Brian Patton & Bart Robinson (Summerthought).

Background Reading: *The Handbook of the Canadian Rockies* by Ben Gadd (Corax) – superb natural history and general guide; *High Summer: Backpacking the Canadian Rockies* by Chris Townsend (Oxford Illustrated Press/Cloudcap) – the first hike along the length of the Rockies.

Accommodation & Supplies: This is a wilderness backpacking trip: apart from basic cabins and the lodges, there is no accommodation en route. To book at the lodges, contact Mt. Assiniboine Lodge, Box 1527, Canmore, Alberta, TOL OMO, Canada, or Lake O'Hara Lodge, Box 1677, Banff, Alberta, TOL OCO, Canada. There are no shops on or near the trail and all supplies need to be carried. Meals are available at Sunshine Village ski resort, 35 miles (56km) into the walk, and at Lake O'Hara Lodge, 8 miles (13km) from the end. For most of the route camping is only allowed at campsites.

bald eagle

Currency & Language: Canadian dollars ($Can.); English.

Photography: No restrictions.

Area Information: Mount Assiniboine Provincial Park, Wasa Lake Park, P.O. Box 118, BC, VOB2KO, Canada. Kootenay National Park, P.O. Box 220, Radium Hot Springs, B.C., VOA 1MO, Canada. Banff National Park, Box 900, Banff. Alberta, TOL OCO, Canada. Yoho National Park, Box 99, Field, B.C., VOA 1GO, Canada. Websites: www.worldweb.com/parkscanadaBanff/Guide/; www.worldweb.com/parkscanada-kootenay/index.html.

TIMING & SEASONALITY

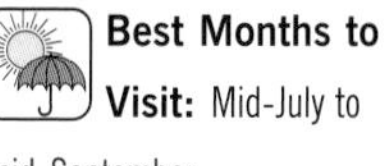

Best Months to Visit: Mid-July to mid-September.

Climate: The Canadian Rockies is a northern mountain range with a continental climate. Summers can be hot, but it can rain and the mountains are only free of snow for a few months. Snow may linger on above-treeline trails into July and the first snow can fall again in early September, though the hiking season may last until late October. The weather can change quickly. Warm and waterproof clothing should be carried.

HEALTH AND SAFETY

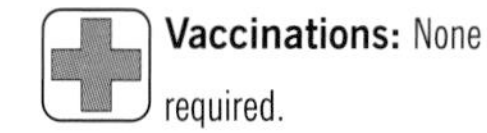

Vaccinations: None required.

General Health Risks: No special risks. Medical insurance should be taken out.

Special Considerations: The Canadian Rockies is home to both grizzly and black bears. While you are very unlikely to even see one of these magnificent animals precautions should be taken. Stay alert when hiking and never cook or store food in your tent. At night, food supplies should be hung up. Backcountry campsites have bear poles for doing this.

Politics & Religion: No concerns.

Crime Risks: Low.

Food & Drink: Although much water in the backcountry is probably safe to drink, it may harbour giardia and other causes of stomach ailments. Boil, filter or purifiy with iodine.

HIGHLIGHTS

Scenic: The whole route passes through magnificent mountain scenery. However, the soaring pyramid of Mount Assiniboine, the unbroken cliff of the Rockwall, Floe Lake and Helmet Falls stand out.

Wildlife & Flora: This hike runs through a vast northern wilderness. The lower areas are covered with conifer forest. Above treeline lies alpine tundra and permanent snowfields. In summer the meadows are packed with wildflowers. Animal life includes grizzly bear, black bear, grey wolf, wolverine, lynx, moose, elk, beaver, and porcupine. Over 270 species of birds include bald and golden eagles and seven species of owl.

At the beginning of Day 8 you climb to the Wolverine Plateau, and spend the day high in the mountains before dropping down to Helmet Creek and the northern end of the Rockwall.

temperature and precipitation

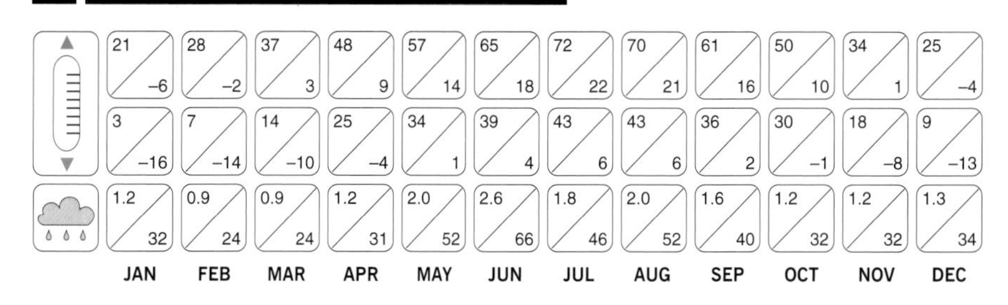

	JAN	FEB	MAR	APR	MAY	JUN	JUL	AUG	SEP	OCT	NOV	DEC
High	21 / –6	28 / –2	37 / 3	48 / 9	57 / 14	65 / 18	72 / 22	70 / 21	61 / 16	50 / 10	34 / 1	25 / –4
Low	3 / –16	7 / –14	14 / –10	25 / –4	34 / 1	39 / 4	43 / 6	43 / 6	36 / 2	30 / –1	18 / –8	9 / –13
Precipitation	1.2 / 32	0.9 / 24	0.9 / 24	1.2 / 31	2.0 / 52	2.6 / 66	1.8 / 46	2.0 / 52	1.6 / 40	1.2 / 32	1.2 / 32	1.3 / 34

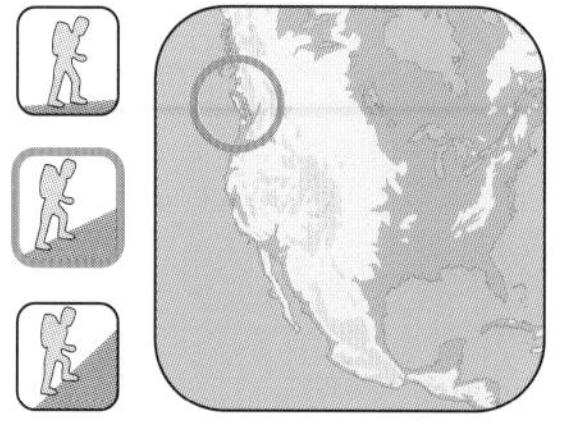

the WEST COAST TRAIL

VANCOUVER ISLAND, BRITISH COLUMBIA, CANADA

Philip Stone

ONCE FEARED AS THE "GRAVEYARD OF THE PACIFIC", THE SECTION OF VANCOUVER ISLAND'S COAST THAT THE WEST COAST TRAIL NOW FOLLOWS SAW THE DEMISE OF MORE THAN SIXTY SHIPS, WITH THE LOSS OF HUNDREDS OF LIVES. WITHIN THE PACIFIC RIM NATIONAL PARK, IT NOW SERVES A DUAL ROLE: PROVIDING RECREATION FOR HIKERS WHILE PROTECTING THE TEMPERATE RAIN FOREST ECOSYSTEM.

The trail began its modern role in 1890 as a telegraph line to the coastal community of Bamfield from the bustling colonial capital of Victoria. After the tragic loss of the SS *Valencia* in 1906, when 117 people were pinned by a relentless storm against a rocky headland, improvements began to the route to assist future rescues. It served shipwrecked sailors for half a century until the advent of modern navigational aids diminished its importance, and in 1954 the federal government ceased maintaining the route. The vigorous rain forest began quickly to reclaim the route until a renewed interest in the trail began in the 1960s. A series of upgrades culminated in 1993, when the West Coast Trail Unit was officially created. Now, crashing surf, ornate sandstone sea caves, lush temperate rain forest, some of the world's tallest trees, tumbling waterfalls and numerous sandy beaches await the hiker along this classic 53-mile (85-km) trek.

The wild Pacific coast of Vancouver Island, part of which the West Coast Trail follows, takes on a tropical feel in high summer.

itinerary

•DAY 1 7½ miles (12km)

Pachena Bay to Michigan Creek

From the Parks Canada headquarters on the sandy beach of Pachena Bay, the West Coast Trail immediately heads inland through the forest to avoid the headlands and cliffs between Pachena Bay and Michigan Creek. Following the old supply road to the Pachena Point lighthouse, the trail is wide and, for the most part, dry. Several spur trails lead down to coves among the headlands where camping is possible, but most hikers prefer to keep on track for a first night at "Michigan".

•DAY 2 8 miles (13km)

Michigan Creek to Tsusiat Falls

Between Michigan Creek and Klanawa River the usual route is along the beach. The trail continues to parallel the shoreline and in foul weather would be the preferable option. A number of alternative campsites are passed, notably at Darling River and Orange Juice Creek. The exquisite waterfall at Darling River is worth a quick jaunt upstream from the beach. Just before Tsocowis Creek, hikers must leave the beach and take the trail as far as Trestle Creek to avoid the impassable Valencia Bluff. A lookout at the site of the *Valencia* tragedy gives a chilling view of the ocean swell crashing against the rocks. Grey whales may be spotted feeding in the rich kelp beds below the cliffs. From Trestle Creek, the usual route returns to the beach, passing a staffed Parks Canada cabin tucked in the forest but clearly marked by numerous fishing floats strewn in a tree. Klanawa River must be crossed by the cable car. There is excellent camping at Klanawa, but most people prefer to push on to the spectacular camp at Tsusiat Falls. From the cable car, stay on the trail to avoid an impassable headland. Ladders climb from and then descend back to the beach at Tsusiat Falls, arguably the most scenic point on the trail. Here, a wide, freshwater fall cascades into a deep pool in the middle of a sandy beach.

• DAY 3 **10½ miles (17km)**

Tsusiat Falls to Cribs Creek

At high tide, the small rock promontory and archway at Tsusiat Point, Hole-in-the-Wall, may not be passable, so you should opt for the trail as far as Nitinat Narrows. However, at low tide the archway is well worth a visit and the beach route is preferable as far as the boundary of the Tsuquadra reserve at Tsuquadra Point. When passing through the reserve lands, hikers must not leave the trail. There is access onto the beach east of the reserve, but an impassable headland a short distance farther on makes travel on this beach pointless, except to explore. The trail itself climbs up round this headland and then proceeds inland to meet the water again at Nitinat Narrows. To cross Nitinat Narrows, a fast-moving waterflow from the tidal Nitinat Lake, a ferry must be taken. It is also possible to leave the trail by the same ferry to the Ditidaht village at the far end of Nitinat Lake. From the eastern shore of Nitinat Narrows, the trail passes through two Ditidaht First Nations reserves, Whyac and Clo-oose. Again, hikers must not leave the trail, and must pass through without disturbing anything. There is no camping permitted on reserve lands. Having passed Clo-oose, the trail reaches Cheewhat River. The water in the Cheewhat is unpalatable – better water may be found from a small nearby stream. A bridge crosses the Cheewhat and hikers may quickly return to the beach. There are numerous campsites along the beach, but

walk profile

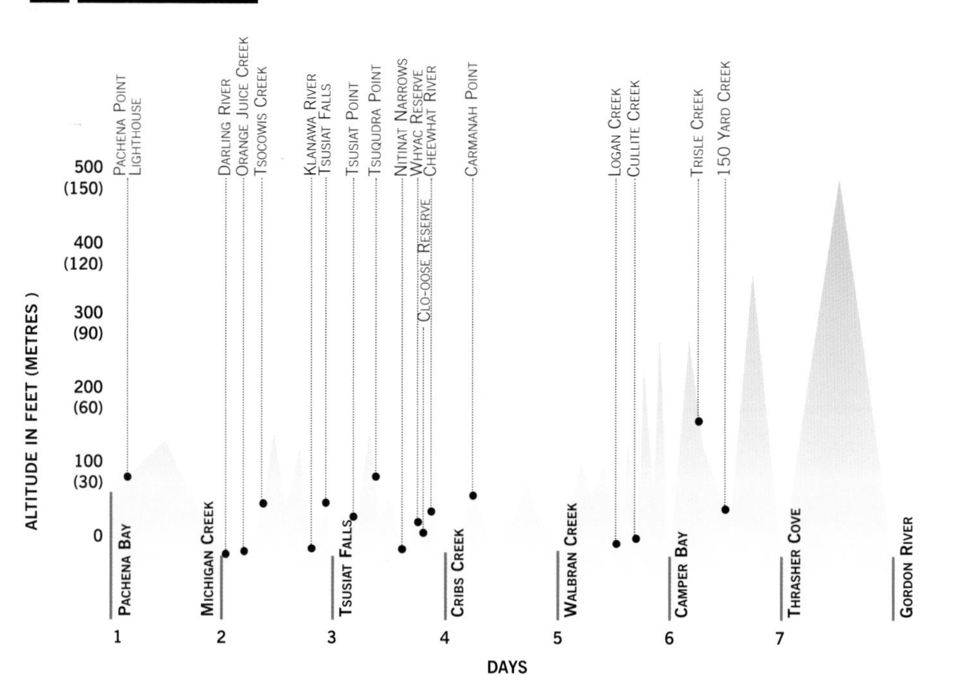

Hikers, young and old, carefully pick their way along the beach as they try to avoid getting their feet wet – an often impossible task on this trail.

←
Cable cars, boardwalks, suspension bridges and ladders all assist the hiker along the West Coast Trail.

water continues to be scarce. A long flight of stairs at the eastern end of the beach takes the trail up to avoid Dare Point. Combined with a surge channel to the east of it, Dare Point at certain tides presents an obstacle worth avoiding. The trail returns to the beach at Cribs Creek, where there is a campsite.

•DAY 4 **13 miles (21km)**

Cribs Creek to Walbran Creek

From Cribs Creek, you can hike the beach for a short distance to the east end of the Bay of Cribs. Here, rejoin the trail in the forest to bypass Carmanah Point, where a lighthouse erected in 1891 stands guard over this part of the coast. The trail returns to the beach, and from here to Vancouver Point the beach is the usual route of travel. A cable car crosses Carmanah Creek, but during the summer months wading is usually possible. The Carmanah and adjacent Walbran valleys are provincial parks set aside to preserve the gigantic trees that reside there. Carmanah is home to the largest and oldest sitka spruce trees in the world, which reach an astonishing 310 feet (95m). A bypass trail avoids Vancouver Point, which can be an obstacle at certain tides, and then either the trail or the beach can be followed to Walbran Creek. A long cable-car crossing traverses Walbran Creek to an excellent, if busy, campsite.

•DAY 5 **5½ miles (9km)**

Walbran Creek to Camper Bay

From Walbran Creek, ladders lead up into the forest and mark the beginning of the more difficult and less scenic part of the hike. You can follow the beach route from Walbran to Logan Creek, but at Adrenaline Creek, a notorious surge channel – which has seen its share of mishaps – blocks the route, and should be tackled only by the confident. At Logan Creek, a bridge crosses the creek and a campsite can be found below on the beach. The trail continues to be the best option all the way to Camper Bay, unless you are blessed with an extremely low tide and have a penchant for wading. Highlights along this part of the trail include the bog around Cullite Creek and the lush surrounding forest. As the name suggests, there is good camping at Camper Bay.

→

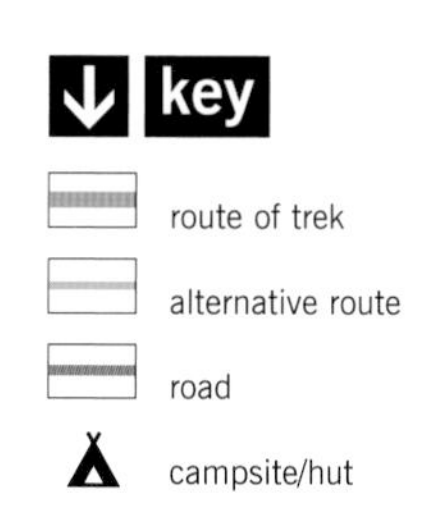

Cheewhat River
Cribs Creek
Dare Point
4
Bay of Cribs
Carmanah Point
Carmanah Creek
Walbran Creek
5
Vancouver Point
Adrenaline Creek
Logan Creek
Cullite Creek
Camper Creek
6
Camper Bay
Tristle Creek
150 Yard Creek
7
Thrasher Cove
Port San Juan
Port Renfrew
Gordon River
Parks Canada Information Centre
0 4 km
0 4 miles

The freshwater Tsusiat Falls thunder straight onto the beach, and make a wonderful camping spot. These are the most impressive falls found along the trail.

•DAY 6 5½ miles (9km)

Camper Bay to Thrasher Cove

Camper Creek can be difficult to wade across, so it is advisable to use the cable car. Ladders lead up into the forest, and the trail makes a wide arc round Trisle Creek. From Camper Bay to Thrasher Cove, it is possible to follow the beach in places, but the combination of a surge channel near the *William Tell* wreck site and promontories impassable at mid- to high tides make the trail a better option. Hikers choosing to use the shoreline here should pay particular attention to the tides. At 150 Yard Creek, the trail heads inland and climbs gradually, eventually reaching a high point of around 600 feet (175m) above sea level, the highest point on the trail. The trail is muddy and is slow-going in places. Above Thrasher Cove the trail winds down to the cove, where a great campsite with ample water awaits.

•DAY 7 3 miles (5km)

Thrasher Cove to Gordon River

Retrace the steps up from Thrasher Cove to the high point, and then east along the trail. The route descends to Gordon River gradually, and will be muddy and slick in many places. A ferry awaits at Gordon River to take hikers across to the Parks Canada Information Center, and the end of the West Coast Trail.

factfile

OVERVIEW

The West Coast Trail is a coastal hike along a 53-mile (85-km) section of the rugged outer coast of Vancouver Island. Wholly protected within the 64,100-acre (25,640-ha) West Coast Trail Unit of the Pacific Rim National Park, the trek is an accessible wilderness experience, and a rite of passage for local hikers. Following native gathering and trading routes, an old telegraph line, a marine rescue trail, and numerous beaches, the route is usually completed in 7 days.

Start: Pachena Bay, Bamfield.

Finish: Gordon River, Port Renfrew.

Difficulty & Altitude: There is relatively little elevation gain and loss, the highest point on the route being 600 feet (175m). Mud, long ladders, fine sand and long sections of boardwalk present the principal difficulties. Foul weather and its situation can quickly combine to make the West Coast Trail a remote, inhospitable place. Hikers must be prepared to travel in severe weather, and to be able to maintain shelter and nourishment for several days at a time in poor conditions. The trail is wide, for the most part well-maintained, and comprised of an astonishing number of bridges, cable cars, suspension bridges, and miles and miles of boardwalk. In many sections the beach is the preferred route, although the trail runs parallel through the forest. An awareness of tides and the ability to read tide tables is essential.

ACCESS

Airports: International airports in Vancouver and Seattle, USA. Connections to Victoria airport on Vancouver Island's southern end. Regional airports at Port Alberni and Nanaimo.

Transport: From Vancouver and the British Columbia mainland, or from Seattle in the USA, access onto Vancouver Island is only by air or ferry. Ferries leave Vancouver from Tsawwassen and Horseshoe Bay. Routes arrive in Swartz Bay, Victoria (the provincial capital). and at two terminals in Nanaimo, located about a third of the way "up island" from Victoria. Regular bus services operate between Victoria and Port Renfrew and Victoria and Port Alberni. Several shuttle services operate from Victoria to either Port Renfrew or Bamfield. From Port Alberni, the *MV Lady Rose* provides a regular service to Bamfield. Travel between Port Renfrew, Nitinat Village, and Bamfield is by Pacheenaht First Nation Bus Service, taking around 3 hours; or there is a water-taxi service between Bamfield, Nitinat, and Port Renfrew. Shuttle services also make the short run between the town centres in Port Renfrew

and Bamfield to the trail heads.

Passport & Visas: Passport required with visa for all non-North Americans.

Permits & Restrictions: A permit is required from Parks Canada to hike the trail between May 1 and October 1. A fee and a mandatory orientation session are required. The number of hikers permitted to start the trail is restricted to 60 a day – 26 from either end and 8 from Nitinat. Groups sizes are limited to 10 or less, and Parks Canada recommends four to six as ideal. Hikers should minimise impact on the trail and the surrounding environment by camping only on the beach and not in the adjacent forest; lighting fires only if essential; and packing out all rubbish. No pets are permitted on the trail. Do not disturb any plants or wildlife on the trail.

LOCAL INFORMATION

Maps: 1:50,000 topographical map available at Parks Canada Headquarters.

Guidebooks: *Blisters and Bliss A Trekker's Guide to the West Coast Trail*, David Foster and Wayne Aitken, illustrated by Nelson Dewey (B & B Publishing); *The West Coast Trail and Other Great Hikes*, Tim Leadem (Greystone Books); *Canadian Tide and Current Tables Volume 6* – an annual publication from Fisheries and Oceans Canada. Parks Canada also sells tide tables.

Accommodation & Supplies: All nights are spent camping. There are no supply points along the trail.

Currency & Language: Canadian dollars ($Can.); US dollars ($US) are accepted. English and French are the two official languages of Canada, and Trail information is provided in both languages. English predominates across British Columbia.

Photography: No restrictions are posted, although the trail does pass through a number of indigenous First Nations' reserves and culturally sensitive sites.

Area Information: Pacific Rim National Park Headquarters, Pachena Bay (tel.: 001 250 728 3234); Gordon River (tel.: 001 250 647 5434); Nitinat Lake (tel.: 001 250 745 8124). Websites: www.westcoasttrail.com; www.bcferries.bc.ca; www.ladyrosemarine.com; www.parkscanada.pch.gc.ca.

Hikers must venture into the lush understorey of Vancouver Island's temperate forest when the shoreline becomes impassable.

TIMING & SEASONALITY

Best Months to Visit: May to September inclusive. The later summer months are generally better, as the trail should have dried out. There are no parks services from October to May, and thus there is no ferry service at Nitinat Narrows or from Port Renfrew to Gordon River.

Climate: Vancouver Island is under the influence of a temperate marine climate. Winters (December to March) are characterised by heavy rainfall (snow at higher elevations), strong south-east winds, and temperatures between 15 and 55°F (-10 and 12°C) at sea level. Spring (April to June) is typically warm, with sporadic rainstorms. Summer (July to September) can be dry and hot or rainy and cool. Autumn (October and November) can be dry for several weeks, but then the rains begin. The shoreline will be cooler, wetter, and windier than just a short distance inland. Storms can blow in off the Pacific Ocean at any time of year, and high rainfall is to be expected.

HEALTH & SAFETY

Vaccinations: None required.

General Health Risks: None.

Special Considerations: Consumption of shellfish during a period of "red tide" may result in paralytic shellfish poisoning. Mosquitoes, black flies and "no-seeums" (tiny midges) are abundant during the summer months. Blackheaded and yellow jacket hornets can be a problem in the very late summer. Ticks are a problem in the early season (March to May) and carry Lyme's disease; they should be removed very slowly and carefully, and kept for analysis in case symptoms develop. Larger animals may also be encountered, such as cougars, black bears and wolves. If you encounter cougars or bears, maintain a strong demeanour and do not turn or run from the animal. Instead, adopt a stance as large and imposing as possible, and calmly retreat from the animal's presence. Hanging food out of reach at night will help to avoid nocturnal bear encounters. Under no circumstances should food be stored inside a tent or a rucksack. Mice will gnaw through anything for a snack.

sea lion

Politics & Religion: No concerns.

Crime Risks: Low.

Food & Drink: Water is more than plentiful, but some rivers flow from brackish cedar swamps and water is discoloured and unpalatable. Usually the smaller streams will have better water. At busy campsites, boil or treat water.

HIGHLIGHTS

Scenic: The surging waters of the Pacific Ocean, towering evergreen forests lining the shoreline and tumbling waterfalls provide a wonderful landscape for this 7-day hike.

Wildlife & Flora: Watch out for bald eagles, bears, gray whales, sea lions, harbour seals, dolphins, porpoise and orca (killer) whales. Cloaking the trail is some of the oldest and largest coniferous forest in the world. The Western Red Cedar is one giant of the area. A true delight are the lush understorey plants: ferns of every hue, salmon berry, devil's club (avoid), mosses, Oregon grape, Solomon's seal and skunk cabbage.

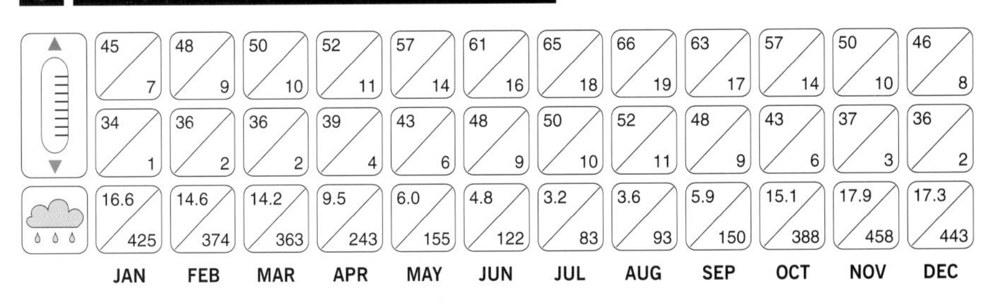

South America

The Andes, the longest chain of mountains in the world, stretch along the entire west coast of South America. Their profound influence flavours each of these three treks: the world-famous, historically intriguing Inca mountain stronghold of Machu Picchu, the highlands of the Cordillera Huayhuash – source of the mighty River Amazon – and the Patagonian Towers of Paine.

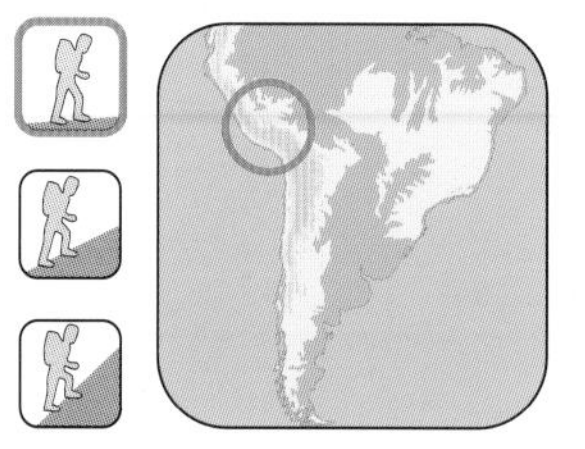

the INCA TRAIL

PERU

John Cleare

THIS IS A UNIQUE TREK. IT HAS EVERYTHING: THE REMAINS OF A LOST CIVILISATION SET AMONG A RICH DIVERSITY OF STUNNING LANDSCAPES. EVERY TREKKER WILL MARVEL AT THE INCA ENGINEERING AND THE AESTHETIC SITING OF THEIR RUINS.

Many will consider this journey almost a mystical experience. Machu Picchu, the culmination of the journey, remains a fascinating and awe-inspiring place, despite having lost much of its magic to international tourism. Much remains to be discovered, and archaeologists are still at work.

The most popular of several starts to the Inca Trail is described, but longer and more ambitious routes are possible that cross the spine of the Cordillera Vilcabamba, passing below the great snow peaks – and possibly attempting to climb one – before joining the Inca Trail.

The environment traversed by the Inca Trail is exceedingly fragile, and when the Trail was little known some 20 years ago, serious problems occurred with inexperienced trekkers unused to the disciplines of low-impact camping. Its popularity today has prompted the authorities to attempt some sort of benevolent regulation, and rules may change from year to year. One plan under consideration is to scrutinise and list each trekker's gear on entry and levy fines on leaving the park for every can or bottle not carried out.

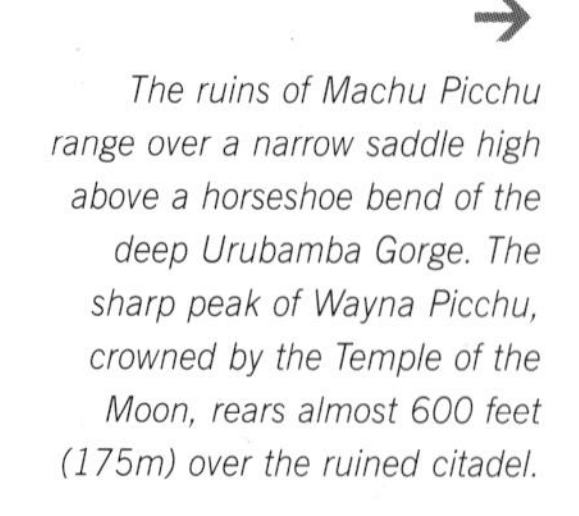

→ *The ruins of Machu Picchu range over a narrow saddle high above a horseshoe bend of the deep Urubamba Gorge. The sharp peak of Wayna Picchu, crowned by the Temple of the Moon, rears almost 600 feet (175m) over the ruined citadel.*

itinerary

•DAY 1 **1 mile (1.5 km)**

Cuzco to Llactapata via Km.88

From San Pedro Station near the Covered Market, catch a "local" Machu Picchu–Santa Ana train. Make sure that the guard knows that you are bound for the Km.88 halt. The journey lasts between 3 and 4 hours, the train at first climbing to the Anta Plateau at 12,000 feet (3,660m) before making a tortuous descent into the Urubamba Valley, the Sacred Valley of the Incas. Chilca is the final station before the train plunges into the deep gorges carved by the Urubamba on its journey toward the lowland jungles and the distant Amazon. The train stops at Km.88 only momentarily.

An I.N.C. (Instituto Nacional de Cultura) official (if he is genuine, he is likely to be wearing a uniform hard-hat) will check you in and collect your entry fee. A swinging, wire suspension bridge crosses the tumultuous Urubamba. The trail now leads back above the south bank through eucalyptus woods to Llactapata, where the campsite lies above the confluence of the tributary Cusichaca River, among impressive Inca terraces.

VARIATION: 8 miles (13km). Chilca to Llactapata. Chilca, the road head at the western end of the Sacred Valley, can be reached by road or rail from Cuzco via any of the valley's numerous Inca ruins, including Ollantaytambo. The trail, fairly level until it crosses a final deep ravine, descends through the hot and arid valley, following the south bank of the Urubamba River, at first through eucalyptus groves and then along stony slopes scattered with bushes and cacti.

•DAY 2 **10 miles (16km)**

Llactapata to Llullucha Pampa

A well-used trail follows the true right bank of the Cusichaca River, at first very steeply, then climbing gradually. Now, narrowing between steep, rumpled hillsides, the *quebrada* (valley) holds areas of cultivation and occasional cottages, →

and there are superb views back toward the graceful snow peak of Veronica (19,042 feet/5,804m). After some 5 miles (8km), you reach the small village of Wayllabamba at 9,000 feet (2,740m), where a friendly schoolmaster runs a primitive school and where it may be possible to buy bottled soft drinks. This is the last serious habitation on the trail.

Now the route strikes steeply up, into the narrow tributary Llulluchayoc glen, starting the long, continuous climb towards the first pass. There are mossy woods, clearings where pigs, cattle or *burros* (donkeys) graze, several river crossings, and even some shade before the trees fall away and an area of flat moorland or *puno* provides a good campsite below craggy hillsides. Llullucha is an edible moss.

•DAY 3 13 miles (21km)

Llullucha Pampa to Phuyupatamarca

The trail climbs up the left side of the bare *quebrada* to the col visible at its head, bleak First Pass (13,800 feet/4,200m) or Abra de Warmiwanusca (Pass of the Dead Woman). A zigzag trail, rugged and eroded, drops steeply into the Quebrada Pacamayo, the Sunrise Valley, full of waterfalls.

Across the river, the trail ascends again through scrub and bare hillsides to Runkuragay, the Egg Hut, a ruined oval-shaped Inca watchtower, the first such building on the trail not discovered by the Spanish conquistadors. The original Inca pathway is now much in evidence. Carefully cobbled, paved or stepped, it continues up hillsides of yellow grass, past two tarns, and over a false summit to Second Pass, Abra de Runkuragay, at 13,120 feet (4,000m). Salcantay, at 20,546 feet (6,262m), and the snowy giants of the Cordillera Vilcabamba dominate the western horizon.

The most strenuous sections of the Inca Trail are now completed, and a descent of almost 1,000 feet (300m) passes a green tarn and leads to the spectacular ruin of Sayajmarca. This magical place remained undiscovered until Hiram Bingham located it in 1915; its appropriate name, given in 1940, means Inaccessible Place. The building, hung with wildflowers and moss, contains many small rooms and four ceremonial baths.

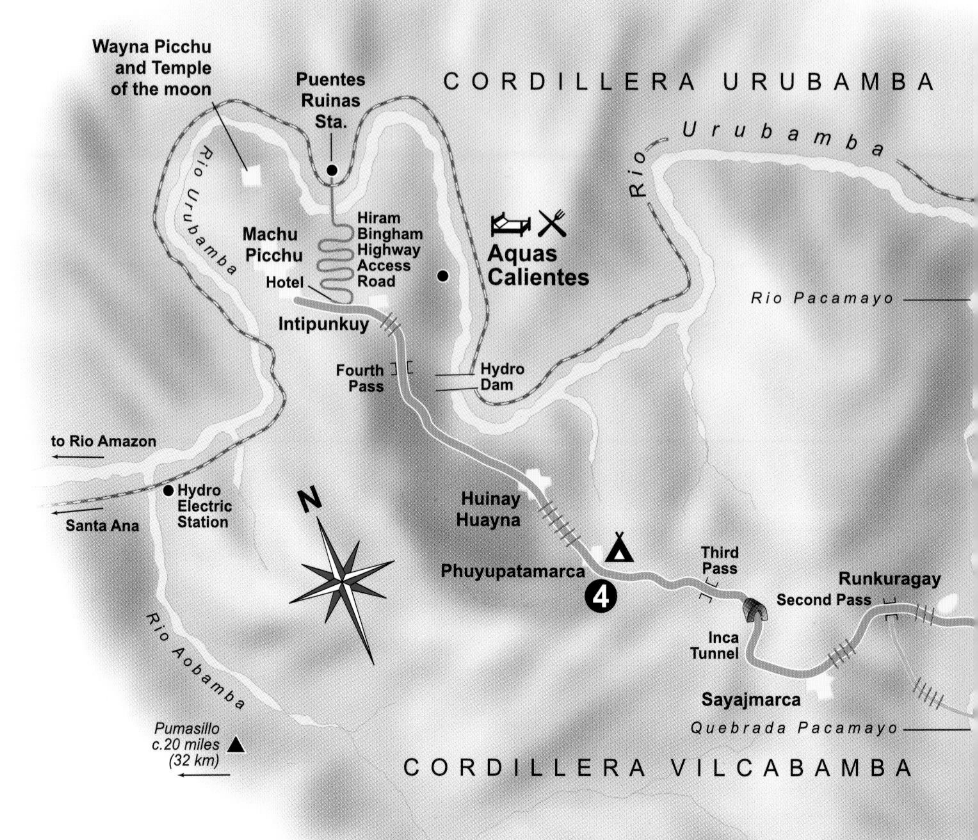

The first full day of walking leads from Llactapata up the Quebrada Cusichaca, from where there are superb views back towards the graceful Nevado Veronica (19,042 feet/5,804m), the highest peak in the Cordillera Urubamba.

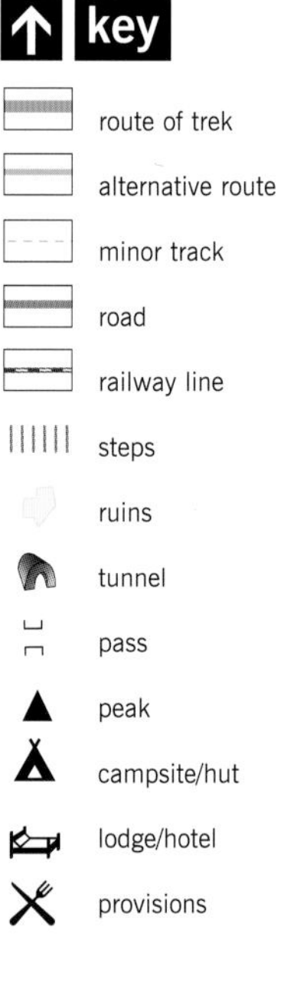

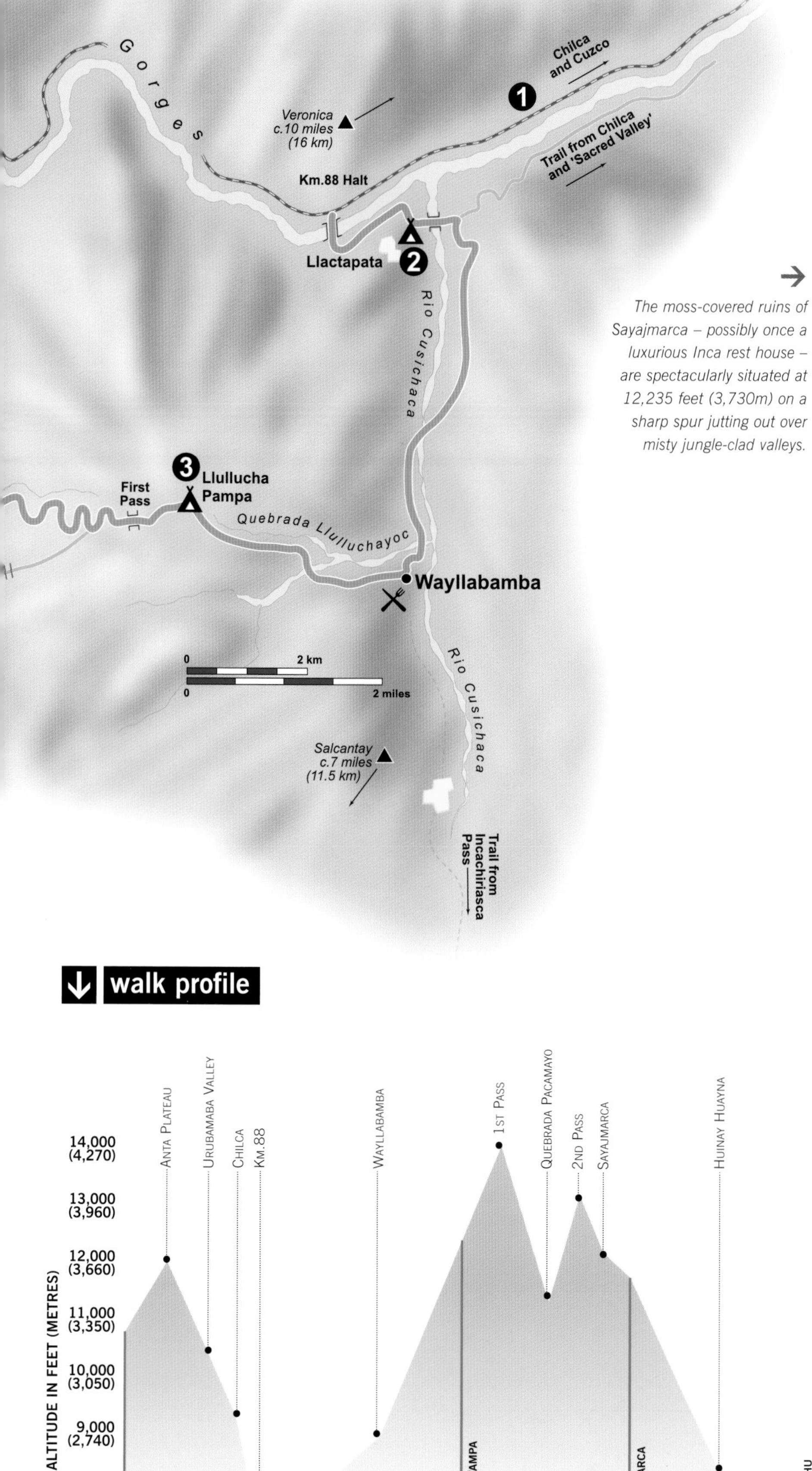

The moss-covered ruins of Sayajmarca – possibly once a luxurious Inca rest house – are spectacularly situated at 12,235 feet (3,730m) on a sharp spur jutting out over misty jungle-clad valleys.

The path now curves away round the hillside, along the upper limit of the cloud forest to reach a sheer rock buttress through which the Incas constructed a short tunnel. Third Pass is a mere grassy saddle a little way farther, and soon the ridge-top trail reaches Phuyupatamarca.

VARIATION: 12 miles (19.5km). Llullucha Pampa to Phuyupatamarca. An exciting, more direct, but more strenuous route branches left some 800 feet (240m) below First Pass. A narrow, little-used path contours steep hillsides, ascending and descending steep crags on superbly engineered Inca stairways, round the head of the Quebrada Pacamayo, to rejoin the regular route on the crest of Second Pass.

walk profile

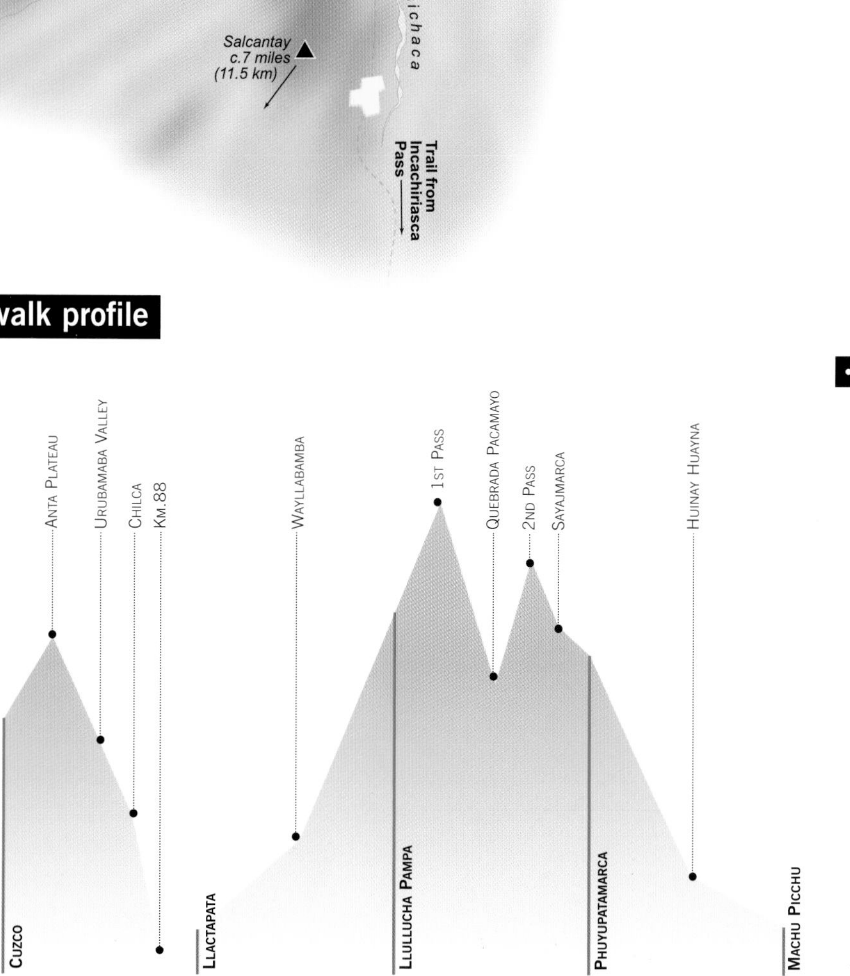

•DAY 4 **6 miles (9.5km)**

Phuyupatamarca to Machu Picchu

The ruins of Phuyupatamarca (The Place on the Edge), another of Hiram Bingham's discoveries, lie just below the campsite. There are spectacular views from the hilltop above the camp to the Vilcabamba where icy Pumasillo, the Puma's Claw, holds pride of place. The route, which was only discovered in the 1980s, drops precipitously from Phuyupatamarca down more than 1,000 Inca steps and through a tunnel to Huinay Huayna, the Place of Eternal Youth, a cluster of intriguing buildings terraced into almost vertical forest and seeming to overhang the Urubamba Gorge far below. A modern path traverses the hillside to a recently developed, ugly and inappropriate trekker's hostel, where an old route is rejoined. It continues through forest,

This is one of the two tarns that nestle below the crest of the Second Pass. First Pass and its surrounding hillsides are reflected in its waters.

crosses the hardly noticed Fourth Pass, to approach the Inca gatehouse of Intipunkuy, the Gateway of the Sun, up a final flight of steps. At last the trekker looks down on Machu Picchu itself, a mere 30-minute descent.

At the gatehouse, your trail pass must be stamped and regulations may require you to leave your rucksack. The first tourist train arrives at Puente Ruinas Station mid-morning, from where the shuttle buses grind up the tortuous Hiram Bingham Highway. An early start from Phuyupatamarca is recommended to be able to explore Machu Picchu before the tourists arrive and the spell is broken. Wayna Picchu, the tooth-like peak above the ruins, is the only place of escape. A strenuous climb of 1,100 feet (340m) leads to the summit and the Temple of the Moon.

The last bus down leaves the ruins mid-afternoon, and final trains depart for Cuzco soon afterwards. Trekkers may prefer to stay inexpensively in the small township of Aguas Calientes, a mile (1.5km) upstream from the station, and savour the hot springs and one more day at Machu Picchu.

A worthwhile and more exciting alternative route, but considerably higher and more strenuous, involves a further 3 or 4 days' walking. This trail starts at Mollepata, a road head just off the Cuzco–Abanca highway. After climbing the Soray Valley to the Incachiriasca Pass (16,150 feet/4,920m) across the main divide of the Cordillera Vilcanota, it descends the Quebrada Pampacahuna to Wayllabamba, where it joins the regular Inca Trail.

First sight of Sayajmarca: a Quechua guide points out the distant ruins across the head of Quebrada Yanachoca from the alternative high-level route from First Pass on Day 3.

factfile

OVERVIEW

A high-altitude hike through the "brow of the jungle" foothills of the Cordillera Vilcabamba to the ancient Inca citadel of Machu Picchu. The usual route of some 30 miles (48km) will normally take 4 days to complete, Cuzco to Cuzco, and most of it lies within the Machu Picchu National Archaeological Park. There are many variations, but the regular route is the most famous trek in South America and deservedly popular.

Usual Start: Km.88, a request halt on the Cuzco–Santa Ana railroad.

Finish: Machu Picchu.

Difficulty & Altitude: Technically this is a straightforward trek along well-paved trails and ancient Inca staircases, but it does involve several long and steep ascents and descents. The entire route lies above 7,500 feet (2,290m) and, with two major passes of 13,800 feet (4,200m) and 13,120 feet (4,000m), altitude is a serious consideration. Adequate acclimatization is essential. The terrain is typically cloud forest and scrub at lower elevations, and rocky hillsides cloaked in long grass higher up. The region is subtropical and there is no shade at the higher elevations. After First Pass the route is committing and there are no easy escapes.

ACCESS

Airports: Cuzco, Peru's second city, is served by regular jet flights from Lima International Airport.

Transport: The Cuzco–Santa Ana railroad runs down the Urubamba Gorge past Puentes Ruinas, the Machu Picchu station. Only local trains stop at Km.88, and previous booking is advisable. Trains do not run on Sundays. Alternative starting points are

accessible by road (buses, taxis, hire-vans) from Cuzco.

Passport & Visas: All visitors require passports; only Western Europeans and North Americans don't need visas. A Tourist Card, issued on arrival, must be shown on demand and upon departure.

Permits & Access: The area is not only a national park but also a UNESCO World Heritage Site, with no development (e.g. formalized campsites, toilet blocks, etc.) permitted. A nominal fee is charged at Km.88 to enter the park and visit Machu Picchu. Regulations governing camping may change from time to time (details from Cuzco Tourist Office). Opening hours at Machu Picchu itself are 7:00 A.M to 5:00 P.M.

LOCAL INFORMATION

Maps: Official IGN maps are non-existent. At the time of writing, the best map is an A3 sheet produced originally for the Peruvian adventure travel firm Explorandes, bearing on one side a pictorial map with every landmark noted, on the other a practical route description. Other privately published maps are occasionally available.

Guidebooks: *The Inca Trail: Cuzco & Machu Picchu*, Richard Danbury (Trailblazer); *Backpacking and Trekking in Peru and Bolivia*, Hilary Bradt (Bradt Publications/Hunter); *Apus & Incas*, Charles Brod (Charles Brod).

Background Reading: *Lost City of the Incas*, Hiram Bingham (Atheneum); *Peru – A History*, H. Mellersh (Plata); *Peru – A Travel Survival Kit*, Rob Rachowiecki (Lonely Planet).

Accommodation & Supplies: On the trail itself, food and accommodation are non-existent and water supplies infrequent, except at designated campsites. All tents, food, and fuel must be carried; porters are often employed. Supplies are readily available in Cuzco, where you can also savour local specialties such as *cuy*, roast guinea pig, and *lomo salato*, diced meat and potatoes in a tomato and onion sauce.

Currency & Language: Peruvian Inti of 100 centimos; US dollars ($U.S) are also useful. Spanish is the national language, but Quechua – the language of the Incas – is still the first language of the Indian mountain peasant (*campesino*). English is rarely understood.

Photography: No restrictions.

Area Information: The Government Tourist Information Office in Cuzco.

TIMING & SEASONALITY

Best Months to Visit: October to April is the rainy season. The prime season for trekking is June until mid-September, but other times are quite feasible. Wildflowers are at their best in May.

Climate: This is mountain country; rain and even snow are possible at any time. In clear weather, temperatures can rise to over 80°F (27°C) and night frost may be expected. At this latitude, nights in June last some 11 hours.

HEALTH AND SAFETY

Vaccinations: Typhoid and tetanus are a matter of routine; polio and hepatitis are wise precautions.

General Health Risks: Diarrhoea, a possible hazard, is more likely to be contracted in Cuzco than on the trail. Don't eat unpeeled fruit or unwashed salads. In Peru, colds – resulting from rapid temperature changes – are common, and at altitude can degenerate into painful coughs or even bronchitis unless properly treated.

Special Considerations : Hypothermia is a risk for those who trek without carrying appropriate spare warm and weatherproof clothing. Proper precautions should be taken against sunburn, sunstroke and heat exhaustion. Altitude is likely to affect all who go high, and the consequent initial headache can be treated with aspirin. Acute Mountain Sickness (A.M.S.) is the penalty for those who go too high too quickly. Cuzco stands at 11,000 feet (3,350m), and several days should be spent acclimatising and taking gentle exercise before embarking on the Inca Trail. Read up on altitude sickness before leaving home. Both medical and baggage insurance are wise precautions. There are no formalised rescue facilities.

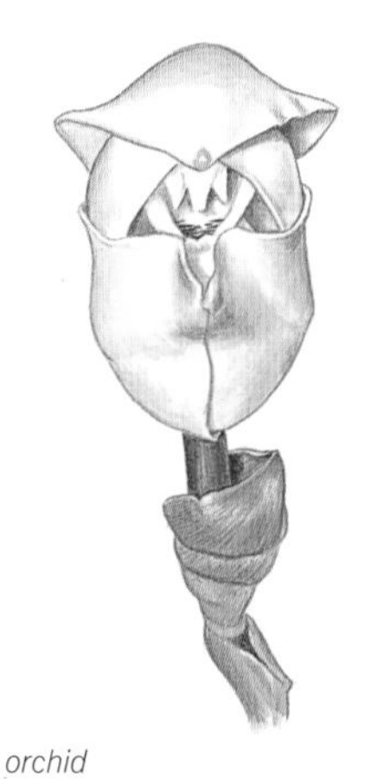

orchid

guinea pig

Politics & Religion: In the past, Peru suffered from terrorist activity and guerrilla fighting, but in recent years the dangers have receded to practically nothing. However, independent travelers should seek local advice before setting off into the hills. Although Peru is essentially a Roman Catholic country, Peruvian Indians tend to leaven Christianity with their ancient beliefs.

Crime Risks: The threat of theft must be taken seriously, but armed robbery is rare. Incidents are more likely in Lima, Cuzco or at popular tourist venues than on the trail itself.

Food & Drink: Pasta, pizza, and rice dishes are safe and inexpensive local fare, and the numerous (for historical reasons) Chinese restaurants or *Chifas* always serve good food. Local beer is good or try the national drink, a white grape brandy called Pisco. The many soft drinks include the ubiquitous Incacola. Do not drink tap water.

HIGHLIGHTS

Scenic: The colourful dress of the local people, the charming colonial architecture of old Cuzco, the yellows, browns, and ochres of the landscape, and always the backdrop of great snow peaks along the horizon, create abiding memories.

Wildlife & Flora: Large mammals are rarely seen on the trail, but various deer and even spectacled bears are a possibility. Exceptions are the odd llama or alpaca bought up to Machu Picchu for the tourists. However, birdlife is abundant and likely sightings include hummingbirds and condor. Ninety different orchid varieties have been identified in the National Park, and the subtropical scrub contains flowering bushes such as fuchsia and berberis. Many trekkers will remember hillsides of long, yellow grass and stands of blue and white lupins.

temperature and precipitation

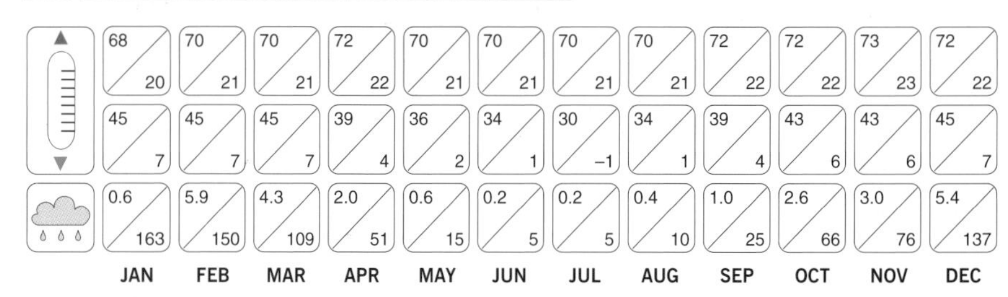

	JAN	FEB	MAR	APR	MAY	JUN	JUL	AUG	SEP	OCT	NOV	DEC
Max. temp. (°F / °C)	68 / 20	70 / 21	70 / 21	72 / 22	70 / 21	70 / 21	70 / 21	70 / 21	72 / 22	72 / 22	73 / 23	72 / 22
Min. temp. (°F / °C)	45 / 7	45 / 7	45 / 7	39 / 4	36 / 2	34 / 1	30 / −1	34 / 1	39 / 4	43 / 6	43 / 6	45 / 7
Precipitation (in / mm)	0.6 / 163	5.9 / 150	4.3 / 109	2.0 / 51	0.6 / 15	0.2 / 5	0.2 / 5	0.4 / 10	1.0 / 25	2.6 / 66	3.0 / 76	5.4 / 137

→

The west face of Jirishanca – Icy Beak of the Hummingbird – rears nearly 7,000 feet (2,130m) over the head of Quebrada Jahuacocha.

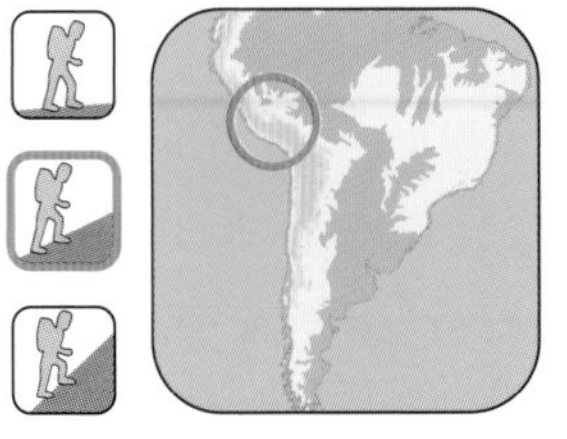

the CORDILLERA HUAYHUASH

and the Source of the Amazon

PERU

John Cleare

THE CORDILLERA HUAYHUASH IS A COMPACT GROUP OF FORMIDABLE ICE-HUNG PEAKS ASTRIDE THE CONTINENTAL DIVIDE. ADJOINING IT IS THE CORDILLERA RAURA, A SMALLER AND MORE GRACEFUL GROUP OF ICE-MOUNTAINS, IN WHICH LIES THE SOURCE OF THE MIGHTY RIVER AMAZON.

The highest summit of the Cordillera Huayhuash, Yerupajá, is, at 21,759 feet (6,632m), the second highest in Peru, while Jirishanca (20,099 feet/6,126m), whose Quechua name means Icy Beak of the Hummingbird, is as imposing as they come. Hardly explored by visitors until the 1950s, the range lends itself to a circumnavigational trek that has become a popular goal for hardy trekkers. This is a long and committing route, and the necessary logistics require more than casual organisation. Unknown, however, except to a few *cognoscenti* is the exciting and slightly shorter variation described here. Having sampled the best of the Cordillera Huayhuash, the route traverses into the Cordillera Raura to finish at the source of the Amazon (as determined by the Royal Geographical Society in 1950), the tiny Laguna Niñococha, into which the glaciers flow.

←

Trekkers ascend out of the bare Rio Huayhuash valley, past a poor campesino's potato patch, en route to the Cordillera Raura.

itinerary

DAY 1 13 miles (21km)

Chiquian to Llamac

From the dusty and unprepossessing little town of Chiquian, a long, stony trail descends into the deep valley of the Rio Pativilca through green, terraced fields and past clumps of cacti. The snows of the Huayhuash rise along the horizon, giving promise of what lies ahead before disappearing as the valley narrows. A rough road was being built along the rocky valley bottom and it may well be possible to hire transport for part of the way. After some 6 miles (9.5km) the Llamac trail forks left up the similar valley of the tributary Rio Achin and left again up the Rio Llamac Valley. Llamac is an adobe village clustered round an earthy plaza and a primitive earthquake-cracked church.

DAY 2 9 miles (14.5km)

Llamac to Laguna Jahuacocha

It is possible to save a day by continuing up the valley, but hardy trekkers will prefer to take the path that climbs southwards. A long, steep ascent leads past goatherds' cottages before the angle eases and arid, stone-scattered pampas dotted with cacti and flowers lead to the wide saddle of Punta Llamac at 14,100 feet (4,298m). The view opens out with great ice peaks rising above dry, tawny hills.

A descending traverse crosses rocky hillsides dotted with gnarled, red-flowered queuña trees into the upper reaches of the Rio Achin Valley. Waterfalls enliven the river, there is a settlement surrounded by irrigated fields and green meadows, with craggy hillsides rising on either side. At the valley head, the ice-hung walls of Rondoy and Jirishanca frown →

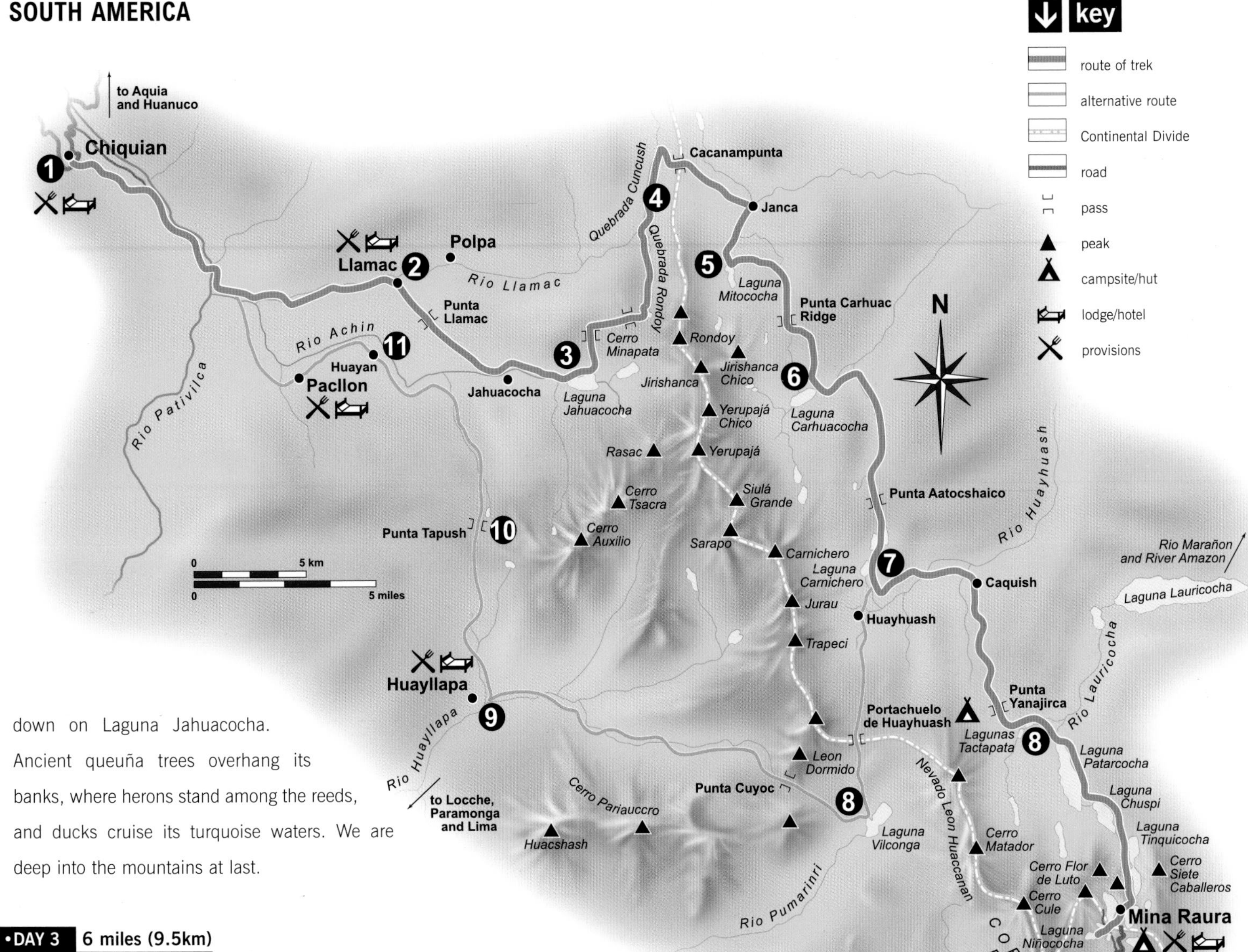

down on Laguna Jahuacocha. Ancient queuña trees overhang its banks, where herons stand among the reeds, and ducks cruise its turquoise waters. We are deep into the mountains at last.

•DAY 3 **6 miles (9.5km)**

Laguna Jahuacocha to Quebrada Cuncush

A rough path enters the cwm above the lake head and ascends the bare hillside above, providing fine views of the tangled Yerupajá icefall. After 2 hours a faint trail breaks off eastwards from a wide saddle, traversing scree across the northern flanks of Cerro Minapata to a second col. Now, rocky slopes, awkward at first, drop steeply into a shadowed basin where a tiny green tarn is overhung by the ice of Rondoy's awesome West Face. This is the head of the lovely Quebrada Rondoy, whose flowery meadows are followed down to its confluence with the Rio Llamac. Beyond the village, crags and grey limestone pinnacles wall the empty valley of the Quebrada Cuncush .

•DAY 4 **10 miles (16km)**

Quebrada Cuncush to Laguna Mitococha

A well-constructed track, probably of pre-Inca origin, zig-zags easily upward for some 2 hours to Cacanampunta (15,420 feet/4,700m), the pass over the main chain of the Andes – the Continental Divide. The pass itself – a knife-

walk profile

ALTITUDE IN FEET (METRES): 16,000 (4,890); 15,000 (4,570); 14,000 (4,270); 13,000 (3,960); 12,000 (3,660); 11,000 (3,350); 10,000 (3,050); 9,000 (2,740)

Chiquian, Rio Pativilca, Llamac, Punta Llamac, Laguna Jahuacocha, Cerro Minapata, Quebrada Cuncush, Cacanampunta, Laguna Mitococha, Puta Carhuac, Laguna Carhuacocha, Punta Aatocshaico, Laguna Carnichero, Caquish, Laguna Tactapata, Mina Raura, Laguna Ninococha

DAYS: 1 2 3 4 5 6 7 8

↑

A string of burros *(donkeys) carrying expedition supplies crosses Cacanampunta, one of only two passes over the main chain of the Cordillera Huayhuash. The narrow limestone crest is the Continental Divide – on the right of which the waters flow eastwards to the Amazon, on the left westwards to the Pacific.*

edge of grey rock and wiry grass – is set between two Dolomite-like towers, an imposing place overlooking the serried foothills of the Amazon flank of the Huayhuash. An easy descent into a wide, green strath leads to the settlement of Janca. From here the river is followed upstream over boggy ground to Laguna Mitococha at the mouth of an impressive cirque of great peaks – Jirishanca Chico, Jirishanca, Rondoy, and Ninashanca.

•DAY 5 **11 miles (17.5km)**

Laguna Mitococha to Laguna Carhuacocha

This is an easy day. It starts by rounding the hillside east of the lake to ascend a wide valley of rumpled, grassy slopes, dotted with sheep and cattle, to the saddle of Punta Carhuac at 15,000 feet (4,570m). The sharp summits of the main Huayhuash peaks rise ahead. Immediately above the pass is the Cerro Pucapuca ridge, with its limestone pavement and little pools. This pass is worth climbing for the view before descending to camp on the lake shore.

•DAY 6 **7 miles (11.5km)**

Laguna Carhuacocha to Laguna Carnichero

When dawn catches the Amazon faces of Siulá, Yerupajá, Yerupajá Chico, and Jirishanca, the prospect across the lake is one of the finest in the Andes. Although the river below the lake is deep and awkward to cross, the trail beyond, leading southwards up a gentle, green strath along the eastern foot of the range, is well used and easy to follow. Above several small tarns a gloomy defile crosses the summit of a low pass – Punta Aatocshaico at 15,100 feet (4,600m) – before the valley opens out and drops past two more lakes and a few cottages to the marshy shores of Laguna Carnichero, the Butcher's Lake (14,600 feet/4,450m). The name actually refers to the savage peak above.

•DAY 7 **8 miles (13km)**

Laguna Carnichero to Laguna Tactapata

Now the route divides. The regular circuit returning to Chiquian continues southwards, but the route to the Amazon source strikes off initially eastwards, crosses the mini-gorge of the Rio Huayhuash on a natural bridge, and emerges further down the valley among the potato fields of Caquish township (13,200 feet/4,020m). Once more, the route turns south up a long, green valley, but now below craggy hills. Again a rocky defile leads over a low pass – Punta Yanajirca, at 14,900 feet (4,540m) – and suddenly the shapely, less savage peaks of the Cordillera Raura are ranged ahead. Above the twin Tactapata tarns, cradled among limestone hills, a grassy spur provides a superb campsite.

→

→ *On a clear morning, the east faces of the principal Huayhuash peaks are reflected in the still surface of Laguna Carhuacocha.*

VARIATION: 8 miles (13km). Laguna Carnichero to Laguna Viconga. Easy going southward via the undemanding Portachuelo de Huayhuash pass (15,580 feet/4,750m) to Laguna Viconga (10,700 feet/3,250m).

•DAY 8 9 miles (14.5km)

Laguna Tactapata to Laguna Niñococha

Below the lakes a river meanders across a large meadow. This is the Rio Lauricocha, the infant Marañon, main feeder of the mighty Amazon. Trekkers must wade across its outflow from Laguna Patarcocha, lowest of a chain of deep-set paternoster lakes. Passing isolated cottages, a good track climbs gradually above the eastern shore of the first two lakes before crossing the river by a waterfall and ascending the steep, western slopes above a third lake. Suddenly the trail becomes a dirt road, and at the lake head sprawls the mining settlement of Mina Raura, a fair-sized town with vehicles, mule trains, a market and comparative civilisation. The chain of lakes continues, however, smaller and smaller, until some way above the final buildings lies the highest – Niñococha or Lake of the Child – into which the glaciers flow. This is the source of the Amazon (15,600 feet/4,750m), and the final campsite lies close below.

VARIATION: 11 miles (17.5km). Laguna Viconga to Rio Huayllapa. A hard climb back over the watershed, via the Punta Cuyoc pass at 16,400 feet (5,000m).

extensions

Day 9 Rio Huayllapa to Punta Tapush The trail descends to Huayllapa village before another long climb, with fine close-up views. After 9 miles (14.5km) you will reach a camp just over a high pass.

Day 10 Punta Tapush to Huayan Downhill all the way for 14 miles (22.5km) to camp beside the Rio Achin.

Day 11 Huayan to Chiquian Another long day – 14 miles (22.5km) – following the river past Pacllon village to the Rio Pativilca, where it rejoins Day 1's route.

factfile

OVERVIEW

This is a seriously committing, high-altitude journey through rugged, remote, and glaciated country. These are the most savage mountains in Peru. The range is only sparsely inhabited, and complete self-sufficiency is essential. The route is around 73 miles (117km) and takes 8 days, while the better-known circuit back to Chiquian takes 11 days and totals some 112 miles (180 km).

Usual Start: Chiquian, some 19 miles (30km) off the Lima–Huaraz highway.

Finish: Mina Raura. The end of the regular circuit is Chiquian.

Difficulty & Altitude: While the landscape is fairly rugged, bare and arid, the trail is never really difficult or dangerous. However, there are frequent long, steep ascents and several quite lengthy days, so some experience and reasonable stamina are called for. The highest point is 15,400 feet (4,700m) or, on the popular circuit, 1,000 feet (300m) higher, while the lowest point is 13,300 feet (4,050m). This is continuously high country and adequate acclimatisation is essential. From Laguna Viconga, on the Amazon flank, escape is possible to the town of Cajatambo, some 13 miles (21km) south.

vicuna

ACCESS

Airports: Lima (Jorge Chavez International Airport). Anta is a small regional airport serving Huaraz and the Rio Santa Valley.

Transport: Chiquian is reached direct from Lima or the busy mountain resort of Huaraz. Mina Raura has a road link to the main coast highway, 50 miles (80km) north of Lima. Both the TUBSA and Landauro bus lines run a direct service from Lima, a 10-hour journey. From Huaraz there is no direct service, but after alighting from any south-bound bus at Conococha road junction it is possible to hitchhike the 20 miles (32 km) to Chiquian. From Mina Raura trucks and buses descend to Churin, from where buses can be taken to Lima.

Passport & Visas: Visas are required by all visitors except North Americans and Western Europeans. A Tourist Card issued by immigration on arrival must be retained and shown on demand and departure.

Permits & Restrictions: One is expected to register at the Chiquian Police Station on arrival.

LOCAL INFORMATION

Maps: The official 1:100,000 I.G.M. sheets 21-j Yanahuanca (covers both the Cordillera Huayhuash and the Raura) and 21-i Chiquian (covers the western approaches); South American Explorers Club publish an excellent 1:80,000 trekking map but it does not include the Cordillera Raura.

Guidebook: *Trails of the Cordilleras Blanca & Huayhuash of Peru*, Jim Bartle (Healdsburg) is the excellent definitive volume.

Background Reading: *Peru – A History*, H. Mellersh (Plata); *Backpacking and Trekking in Peru and Bolivia*, Hilary Bradt (Bradt Publications/Hunter); *Peru – A Travel Survival Kit*, Rob Rachowiecki (Lonely Planet).

Accommodation & Supplies: Only the very hardy will backpack this route. The hire of an *arriero* (donkey driver) with a few *burros* (donkeys) is relatively inexpensive. *Arrieros* belong to an official association and the Huascaran National Park, based in Huaraz, has set a proper scale of hire charges. Alternatively the expedition can be organised by one of several trekking outfitters based in Huaraz or Lima. Basic food-stuffs, meals and accommodation can be found in Chiquian and Llamac, and Pocpa, Huayllapa, and Pacllon on the circuit route. Mina Raura has a busy market, several cafés, and simple hotels.

Many trekking supplies can be obtained in Huaraz.

Currency & Language: Peruvian Inti of 100 centimos, but US dollars ($U.S) are also useful. Spanish is the national language, but Quechua, the language of the Incas, is the first language of the mountain peasants or *campesinos*. English is often understood by those involved in tourism.

Photography: No restrictions.

Area information: Parque Ginebra 28-G, Huaraz, Peru (tel.: 00 5144 721 811).

TIMING & SEASONALITY

Best Months to Visit: May until mid-September is the prime season for trekking.

Climate: This is mountain country and rain – or snow at the higher elevations – may fall at any time. In clear weather, temperatures can rise to over 80°F (27°C) and hard night frosts may be expected. In June, nights last some 11 hours with little or no twilight. The rainy season is December to March.

HEALTH & SAFETY

Vaccinations: Typhoid and tetanus are a matter of routine. A polio booster and gamma globulin against hepatitis are wise precautions.

General Health Risks: Hypothermia is a risk, so carry appropriate spare warm and weatherproof clothing. Dehydration is a potential hazard as the route includes several long, steep, shadeless ascents. Drink plenty and carry sterilised water. Diarrhoea is also a strong possibility, and likely to be contracted from eating locally prepared food. Proper precautions should be taken against sunburn and heat exhaustion. Rabies is common, so, to be safe, keep away from all dogs.

puma

Special Considerations: Altitude is likely to affect all who go high and the consequent initial headache can be treated with aspirin. To help avoid altitude sickness spend several days based at Huaraz (around 10,200 feet/3,100m), making gentle acclimatisation excursions. Both medical and baggage insurance are wise precautions.

Politics & Religion: Terrorism has now all but disappeared. Nevertheless, travellers should seek local advice before setting off into the hills. Essentially a Roman Catholic country, but Peruvian Indians often leaven Christianity with their ancient beliefs.

Crime Risks: The threat of theft must be taken seriously, especially in Lima or Huaraz; armed robbery, though possible, is rare. Illegal "tolls" have occasionally been demanded with menace from trekkers. Tact, the offer of a contribution towards "road repairs", and a working knowledge of the language can defuse the situation.

Food & Drink: Sweetcorn, potatoes, and tinned fish might be bought in the villages. Pasta and rice dishes are usually safe. The beer is passable and bottled drinks include Incacola. Do not drink tap water.

HIGHLIGHTS

Scenic: Savage, ice-fluted peaks –but always at a distance; wide valleys, bare and green; lakes in profusion; and hardly a tree from one end of the trip to the other; altogether a feeling of utter remoteness.

Wildlife & Flora: Though wild country, cattle, sheep, and goats are grazed within reach of habitation and often fall prey to pumas. Fleeting glimpses of the shy vicuna may be seen on the Amazon side of the watershed. Viscacha, a small, whistling, marmot-like creature, is common among old moraine areas. The lakes are rich in waterfowl, including giant coot. In the sky condors are often seen. In the drier western valleys lupins abound, while several kinds of cacti and other succulents are common in the hot and arid lower valleys.

← *The infant Rio Huayhuash flows through a series of natural bridges near Lupag on its journey to join the Rio Nupe and the Amazon.*

temperature and precipitation

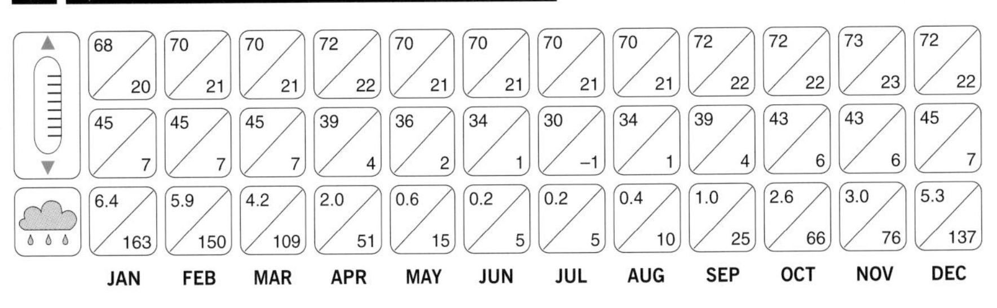

	JAN	FEB	MAR	APR	MAY	JUN	JUL	AUG	SEP	OCT	NOV	DEC
Max (°F / °C)	68 / 20	70 / 21	70 / 21	72 / 22	70 / 21	70 / 21	70 / 21	70 / 21	72 / 22	72 / 22	73 / 23	72 / 22
Min (°F / °C)	45 / 7	45 / 7	45 / 7	39 / 4	36 / 2	34 / 1	30 / −1	34 / 1	39 / 4	43 / 6	43 / 6	45 / 7
Precipitation (in / mm)	6.4 / 163	5.9 / 150	4.2 / 109	2.0 / 51	0.6 / 15	0.2 / 5	0.2 / 5	0.4 / 10	1.0 / 25	2.6 / 66	3.0 / 76	5.3 / 137

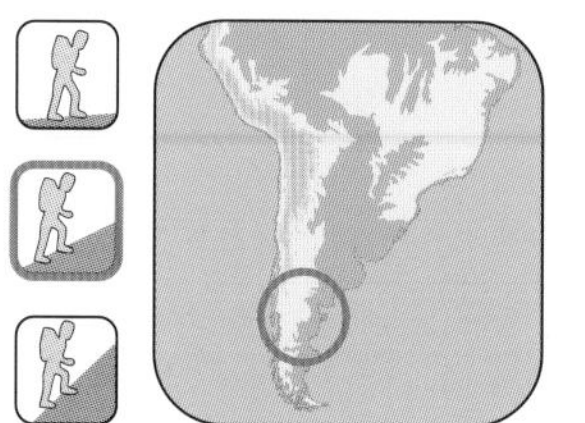

the TOWERS of PAINE

PATAGONIA, CHILE

Andrew Sheehan

THERE ARE ONLY A FEW PLACES ON THIS PLANET WHERE THE PRIMEVAL EARTH-SHAPING PROCESSES AND THE TITANIC STRUGGLE BETWEEN THE ELEMENTS ARE SO CLEARLY VISIBLE THAT THE LANDSCAPES ETCH THEMSELVES INTO MEMORY. PATAGONIA IS ONE OF THOSE PLACES, AND NOWHERE IS ONE CLOSER TO THIS DRAMA OF WIND, FIRE AND ICE THAN IN THE TOWERS OF PAINE NATIONAL PARK.

The centrepiece of the park is the Paine Massif, the remnant of a vast subterranean bubble of pink granite – a laccolith – capped by a resilient black shale, from which rise the *Torres* (towers) and *Cuernos* (horns) of Paine. The prosaic names for the Towers themselves, North Tower, Central Tower and South Tower, belie the first glimpse of these astonishing monoliths. The mind struggles to comprehend the scale of these cloud-wreathed spires, which soar in one vertical leap to almost 10,000 feet (3,050m).

First established in 1959 as Parque Nacional Lago Grey, the park was extended in 1962 and renamed. There are 15 peaks above 6,500 feet (2,000m), innumerable electric blue lakes, and, to the west, the vast Southern Ice Field (Campo de Hielo Sur), from which four large glaciers descend: Glaciers Grey, Dickson, Zapata and Tyndall.

•DAY 1 4½ miles (7.5km)

Guarderia Laguna Amarga (National Park entrance centre) to Hosteria Las Torres

Cross the Rio Paine below the Guarderia by the Puente Kusanovic, and follow the undulating gravel road to Hosteria Las Torres, where there is a hotel, refugio, campsite, and supplies kiosk. If visibility is good, expect excellent views of the Towers and Monte Almirante Nieto.

VARIATION: 9½ miles (15km). Guarderia Laguna Amarga to Campamento Seron. Begin the trek directly, by hiking straight to Campamento Seron. Cross the river (as above), then branch right, off the road, to follow the well-marked

trail north, parallel to the Rio Paine. The way follows river meadows for most of the way, crossing several tributaries, to the campsite.

•DAY 2 11¼ miles (18km)

Hosteria Las Torres to Torres del Paine viewpoint, and back

A steep climb from the suspension bridge over the Rio Ascensio leads to a descent into the Valle Ascensio, and its west bank is followed to the final ascent of the moraine that encloses the tiny, jade-green glacial lake below the Towers. From the top of the moraine, the Towers – if the view is clear – are a mesmerising sight.

Return to Hosteria Las Torres by the same route, or continue up the main valley for another hour to Campamento Japones before returning.

•DAY 3 8¾ miles (14km)

Hosteria Las Torres to Campamento Seron

From behind Hosteria Las Torres (on the north-eastern edge of the campsite), the path contours low, right round the ridge that falls from Cerro Paine. There are many shaded sections of woodland, with colossal fallen tree limbs, contorted, bleached and grey. Eventually, the path begins to descend steadily toward the Rio Paine, milky with glacial silt. Tributaries of the Rio Paine are crossed before the way continues across level fields, past more stretches of woodland, to the tranquil campsite of Campamento Seron.

•DAY 4 11¼ miles (18km)

Campamento Seron to Refugio Dickson

From Campamento Seron the path crosses meadows before tackling the lower slopes of the steep ridge, dropping north- →

↑ *A view that etches itself into memory: the Paine Massif, Chile's mountain showpiece, with the unique architecture of the* Cuernos *(horns), and the brooding bulk of the Cerro Paine Grande, its summits shrouded in cloud, to the right.*

east to Lago Paine. After contouring behind a tiny, jewel-like lake, the path climbs steeply to a col and some dramatic panoramas of Lago Paine and the sweeping, snow-plastered mountain walls of Cerro Paine Chico and Cerro Ohnet. As the path descends, distant mountains become mirrored in the blue of the lake. Just beyond the lake, Campamento Coiron is reached, and the descent continues to river meadows. The path continues to meander in the valley, until a short, sharp ascent gives a sudden glimpse of Lago Dickson and the campsite and refugio of Refugio Dickson (meals and supplies available). A short, steep descent leads to the edge of the grassy levels and the campsite, where icebergs of an otherworldly blue float sedately past.

•DAY 5 6¼ miles (10km)

Refugio Dickson to Campamento Los Perros

From the grassy levels of the campsite, the path climbs steeply through forest. An open area provides a welcome rest stop and excellent views back down the valley. A little farther, a substantial bridge crosses the Rio Cabeza del Indio. An interminable obstacle course begins; the path weaves round, under and over fallen trees. A gorge, where a waterfall thunders between rock walls, provides respite before returning to the fray. The maze continues until the trees thin out and the mountains appear. A narrow suspension bridge permits a crossing of the river and leads to the edge of the moraine slopes. From the moraine, one can look down into the lake, full of icebergs, and up to the glacier. The campsite, Campamento Los Perros, set in the edge of the forest on the banks of a stream, is only minutes away.

•DAY 6 3¾ miles (6km)

Campamento Los Perros to Campamento Paso

After crossing the stream on a tree-trunk bridge, begin a long, arduous and extremely boggy traverse across the hillside until you cross another stream, which gives access to open, drier slopes. Traversing above a small gorge, the path ascends scree and boulder slopes next to a cascading stream. The path climbs steadily up the scree, near a waterfall, until it begins to level out and enters the John Garner Pass. The Pass is short and, once crossed, opens out

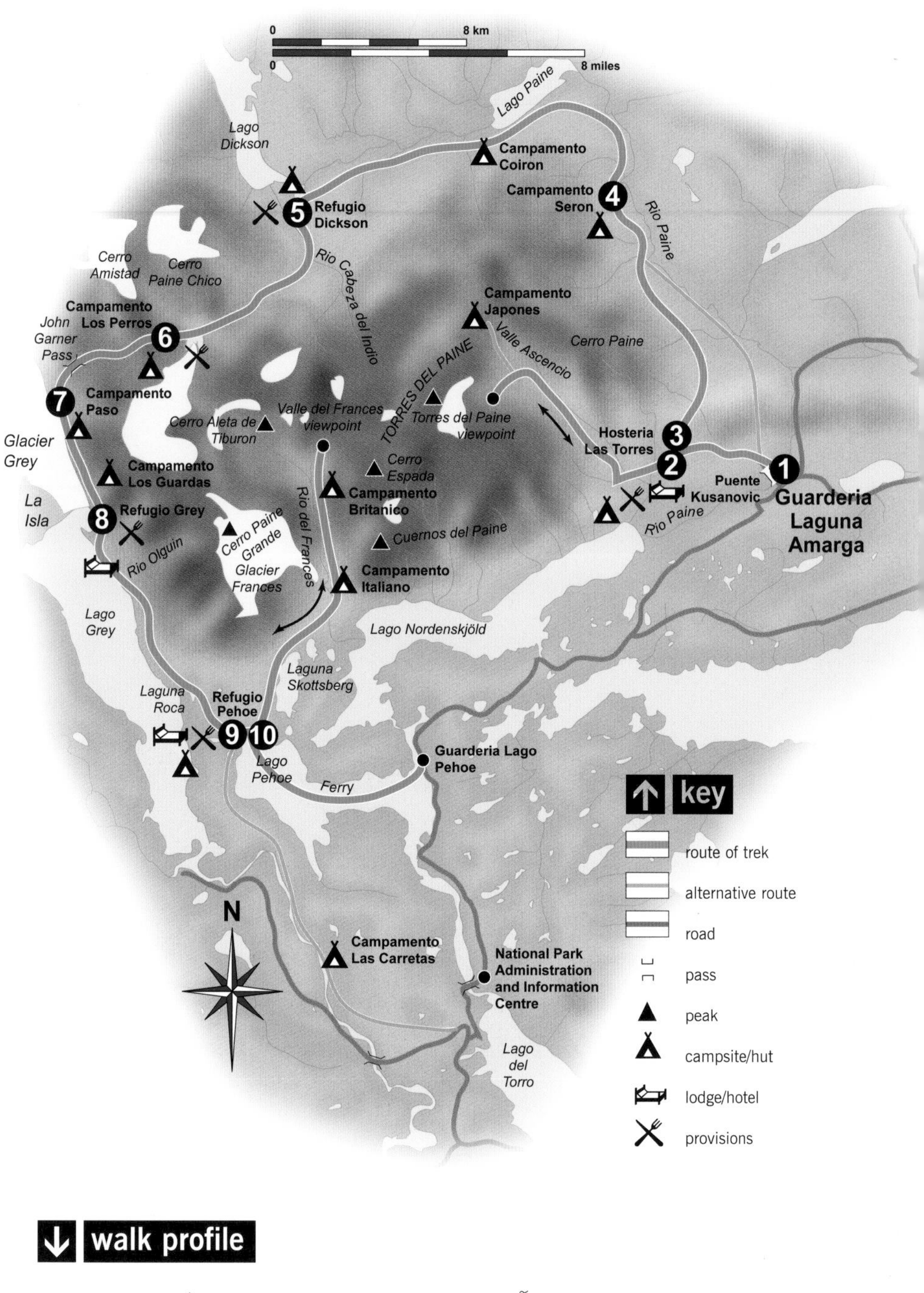

walk profile

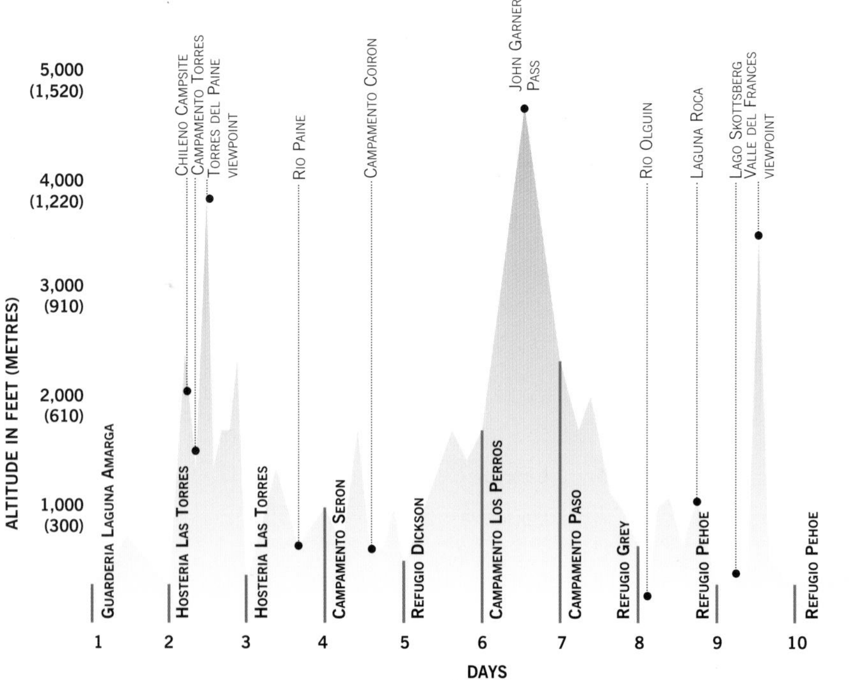

↑ *High in the Valle Frances, the aptly named Cerro Espada (Sword Mountain), and its neighbours, Cerro Hoja and Cerro Mascara, thrust themselves vertically from the sweeping curve of the valley side, to dominate the skyline.*

to one of the most inspiring panoramas of the whole route: far below is the colossal expanse of Glacier Grey, its surface scarred with crevasses. A gentle scree slope leads to the edge of the valley. From here it is a long, steep and sustained descent, progress frequently being made by swinging down tree roots and branches. The path leads directly into Campamento Paso.

•DAY 7 **4½ miles (7.5km)**

Campamento Paso to Refugio Grey

Another obstacle course over and round fallen trees leads to some more straightforward hiking, with the vast glacier a beautiful blue below and a craggy mountainside looming above. Some deep ravines, cut by raging torrents, are crossed with the help of flights of wooden steps. Passing Campamento Los Guardas, the path descends nearer to the glacier. A *nunatak* (island in the glacier), La Isla, marks the mouth of the glacier, where huge icebergs detach themselves with enormous roars. Beyond the end of the glacier, the path swings right and down, to the lakeside campsite and Refugio Grey.

•DAY 8 **6¼ miles (10km)**

Refugio Grey to Refugio Pehoe

Generally contouring through a mixture of open and wooded terrain, the path crosses the Rio Olguin. A section of steady ascent, then more contouring, follows. The path climbs to Laguna Roca, a small lake in a perfect setting, with smooth rock slabs rising out of its waters and southern beech along its shores. The path contours behind the lake, then crosses the isthmus between the two large lakes Grey and Pehoe. The path then enters a gorge-like valley and funnels out onto the shores of Lago Pehoe and the refugio and campsite. This place is very accessible and very popular, so accommodation and facilities may be booked up. Mice are a serious problem here.

→

From the Lago Pehoe campsite, looking north, towers the Cerro Paine Grande; to its right is the Valle del Frances.

•DAY 9 **16¼ miles (26km)**

Refugio Pehoe to Valle del Frances viewpoint, and back

This is a long day. From the campsite, the path crosses a ridge and then makes a long, winding descent almost to the shores of Lago Nordenskjöld. It contours above the lake, and then joins the Rio del Frances, crossing it by a large new suspension bridge. The route now follows the east bank of the river, first through trees, then over open sections of scree and rock outcrops left by the retreating glacier. Avalanches tumbling from Glacier Frances become louder and more visible as one ascends. More wooded sections and small valleys, then just beyond a marshy area lies Campamento Britanico. A little way beyond here, there are superb views of knife-edged peaks, with Cerro Espada, or Sword Mountain, prominent among them. After another 45 minutes, when one has passed below the main bulk of the north peak of Cerro Paine Grande, there are dramatic views of the mountains at the head of the valley, with the aptly named Cerro Aleta de Tiburon (Shark´s Fin Mountain) standing out above the rest. Return by the same route.

•DAY 10 **Ferry trip**

Campamento/Refugio Pehoe via ferry to Guarderia Lago Pehoe

Closing the circle, this is the last day of the circuit. The easiest option requires no more walking. Take the ferry from the landing stage on the lake, at the edge of the campsite, to Guarderia Lago Pehoe, on an inlet on the north-east side of the lake, which is on the main road out of the park. This ferry usually operates once a day on a first come, first served basis.

VARIATION: 11¼ miles (18km). Campamento/Refugio Pehoe to National Park administration and information centre. This option involves an easy hike to the park's main information and administration centre on the north-west shore of Lago del Toro. Climbing over a low rocky hill above the lake, the path follows easy meadows past a small campsite, Campamento Las Carretas, and on to the information centre. This is the bus terminus for services operating to and from Puerto Natales.

factfile

OVERVIEW

The two main options for long hikes within the park are the Circuit Trek and the Little Circuit, or "W". The route recommended here combines the Circuit Trek with the best elements of the "W", to give a 10-day outing. The Circuit Trek, circumnavigating the whole of the Paine Massif, is remarkable for its variety and contrasts, from strolling through valley-floor meadows to contouring steep mountainsides; from negotiating seemingly endless labyrinths of ancient fallen trees to scrambling up screes. Camping equipment and supplies for at least a week are necessary for the Circuit Trek itself. The "W" includes the ascent of two valleys that penetrate the heart of the Paine Massif, providing dramatic scenery. This walk is much shorter than the Circuit, and can be accomplished without tents by staying in the *refugios* or *hosterias*.

Start: Guarderia Laguna Amarga, Torres del Paine National Park.

Finish: Guarderia Lago Pehoe, Torres del Paine National Park.

Difficulty & Altitude: The Circuit is not difficult, although some sections are prone to extreme weather conditions, and the going at times is arduous. Paths are well marked throughout (though a map should be carried). Most bridges are in place, but are sometimes washed away. The John Garner Pass is the most infamous section of the Circuit; snowfall may render it difficult or impassable. The wardens have up-to-date information on the pass and other sections of the trails.

ACCESS

Airport: International airport at Santiago.

Transport: The jumping-off point for the Towers of Paine National Park is the tiny port town of Puerto Natales, 90 miles (145km) south-west of the park. To get to Puerto Natales from Santiago you can fly to Punta Arenas; from Punta Arenas, take a bus north 150 miles (250km) to Puerto Natales. Or, you could fly to Puerto Montt, then take the 4-day, 1,000-mile (1,600-km) trip south on the Navimag Company ship *MV Puerto Eden* to Puerto Natales. Long-distance buses are comfortable and cheap. However, although it is possible to take a bus all the way from Santiago to Puerto Natales, this involves several days on different buses, and a crossing of the Argentinian border. From Argentina, there are regular services between Puerto Natales and El Calafate (90 miles/150km to the north) and Rio Gallegos (on coast, 110 miles/180km to the east). Cars, including four-wheel drive, can be hired in Punta Arenas and Puerto Natales, but it is expensive.

caracara

Passport & Visas: A 90-day tourist visa is issued to citizens of the USA, Canada, Australia, and most western and eastern European countries upon presentation of a passport (valid for at least 6 months). New Zealand citizens need a consular visa.

Permits & Restrictions: All visitors must register at the C.O.N.A.F. (Corporacion Nacional Forestal) centre (Guarderia Laguna Amarga), and pay the entrance fee. You must specify how many days you intend to stay within the park; it is better to state a slightly longer time than necessary, rather than run out of time.

LOCAL INFORMATION

Maps: The locally produced 1:100,000 colour maps are adequate: Chileguide Guia de Viajes y Turismo Outdoors Map February 1999 Mapa de Excursionismo Parque Nacional Torres del Paine; Mapas Matassi no. 13 Torres del Paine and no. 18 Patagonia Sur–Tierra del Fuego.

Guidebook: *Trekking in the Patagonian Andes* (Lonely Planet) – includes a detailed description of the Circuit Trek. **Background Reading:** *The South American Handbook*, edited by Ben Box (Footprint) – updated and reissued each year; *South America on a Shoestring* (Lonely Planet) – updated and reissued each year; *Chile and Easter Island* (Lonely Planet); *Chile Handbook*, Charlie Nurse (Footprint).

Accommodation & Supplies: In Puerto Natales: accommodation from hostels to luxury hotels. *Hospedajes* (private houses offering rooms) are the cheapest option. There is a wide variety of restaurants, cafes, and bars. In the park, accommodation is limited but varies from five-star hotels to *refugios* and campsites. Purchase all food and other supplies before entering the park; what little is available there is extremely expensive. At manned campsites, kiosks sell a limited range of tinned goods, biscuits, chocolate and pasta. A full range of supplies is available in Punta Arenas and Puerto Natales. Camping equipment can be rented from several establishments.

Currency & Language: Chilean pesos; US dollars ($U.S) are accepted; traveler's checks are not useful, but a visa card is. Spanish; few people in Patagonia speak English so a few words of Spanish would be very useful.

Photography: No restrictions.

Area Information: The National Park is managed by C.O.N.A.F., Lago del Toro, Parque Nacional Torres del Paine, Patagonia (tel. 00 56 691 931).

These gaunt, grey skeletons of burnt trees are so common they could be an emblem of Patagonia.

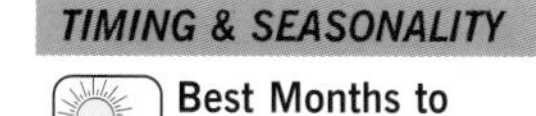

TIMING & SEASONALITY

Best Months to Visit: The Chilean summer (from December to the beginning of March) has the warmest weather and very long days. However, this is the peak period for tourists and, unless reservations have been made well in advance, it may be difficult or impossible to find flights or accommodation. Late March to early April can also have excellent weather, with the added bonus of autumncolour. Similarly, the months of October and November can be very pleasant. The park is open all year, but snowfall in winter (June to September) may make the Circuit Trek impassable, although other hikes are possible.

Climate: Patagonia is remarkable for the unpredictability of its weather, and notorious for its winds. At nights the temperature can drop to freezing point; during the daytime the sun can burn fiercely (highest factor sunblock essential). Snow can (and does) fall in mid-summer (February). Be prepared for all conditions.

HEALTH & SAFETY

Vaccinations: None required, although hepatitis A, typhoid, polio and tetanus are advisable.

General Health Risks: None.

Special Considerations: Mice are a great problem, gnawing holes through tents and rucksacks with ease. At night, suspend all food from branches, or wrap it in a plastic tarpaulin and place it securely on a picnic table. There is a slight risk of Hanta virus, which is transmitted in the urine or droppings of the mice, or by a bite. The C.O.N.A.F. centres have leaflets about the virus.

Politics & Religion: Chile has been a democracy since 1988, following Pinochet's military dictatorship, and many issues remain sensitive. However, the country is generally stable and visitors should experience no problems. Over 90 per cent of the population is Roman Catholic.

Crime Risks: Generally speaking, Chile is a remarkably safe country. In larger cities beware of pickpockets, and bag and jewellery snatchers. Be vigilant if you take the *MV Puerto Eden*.

Food & Drink: You are recommended to purify water in the park with iodine (or similar) tablets or drops, or by boiling.

HIGHLIGHTS

Scenic: In addition to the spectacular mountain scenery and deep iceberg-filled lakes, everywhere in the park you will see the grey, gaunt remains of trees, remnants of uncontrolled fires that have ravaged the park.

Wildlife & Flora: Supporting over 100 species of birds and 25 species of mammals, the Torres del Paine National Park has been declared a UNESCO biosphere reserve. The most famous residents of the park are the rarely sighted puma, the guanaco (a member of the camel family), and the condor. Not so well known, the huemul, a species of deer, is as elusive as the puma. There are also ñandus (lesser rheas), smaller versions of the ostrich, which may be seen grazing the plains. Upland geese are everywhere. Bright pink flamingos fringe the lakes, while green parrots scream noisily overhead. Perhaps the most distinctive bird, apart from the condor, is the Magellanic woodpecker, a large, jet-black bird – the male has an astonishingly brilliant red head and neck. Look out too for the caracara, a scavenger raptor. Around the campsites, tiny tree-runners and small finches are far from timid, while the southern lapwings, with their abrasive alarm calls, are a common sight and sound. The southern beech, *nothofagus*, exists in three varieties in Patagonia, from a low alpine shrub to a tall tree. In the autumn, the beech-covered mountainsides turn a distinctive red.

guanaco

temperature and precipitation

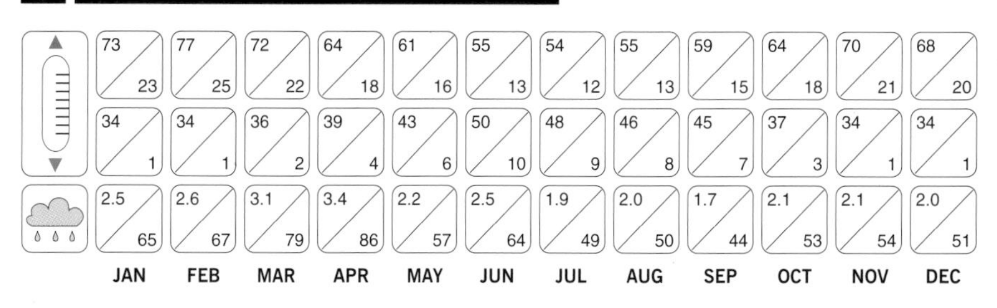

	JAN	FEB	MAR	APR	MAY	JUN	JUL	AUG	SEP	OCT	NOV	DEC
Max temp (°F / °C)	73 / 23	77 / 25	72 / 22	64 / 18	61 / 16	55 / 13	54 / 12	55 / 13	59 / 15	64 / 18	70 / 21	68 / 20
Min temp (°F / °C)	34 / 1	34 / 1	36 / 2	39 / 4	43 / 6	50 / 10	48 / 9	46 / 8	45 / 7	37 / 3	34 / 1	34 / 1
Precipitation (in / mm)	2.5 / 65	2.6 / 67	3.1 / 79	3.4 / 86	2.2 / 57	2.5 / 64	1.9 / 49	2.0 / 50	1.7 / 44	2.1 / 53	2.1 / 54	2.0 / 51

Europe

By way of the edelweiss of the high Alps, the wild herbs of the Mediterranean, and the daffodils of the English Lake District, the lands of Europe are rich in environmental, cultural, and linguistic diversity. Here you will find a living history that, in places, has remained intact since the stone age. Take the opportunity to explore the heartlands of Europe with these treks.

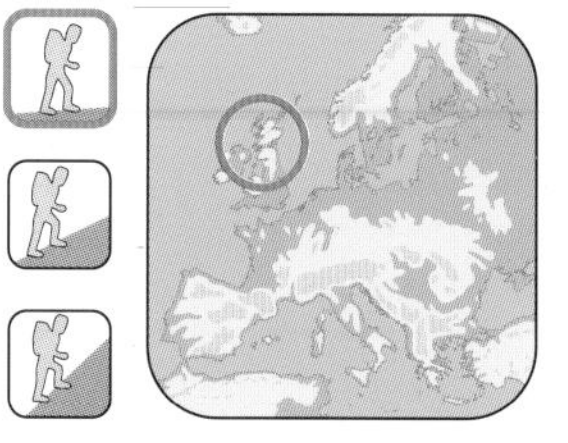

the WEST HIGHLAND WAY

SCOTLAND

Bill Birkett

THIS IS BY FAR THE MOST POPULAR LONG-DISTANCE TREK IN SCOTLAND, AND JUSTIFIABLY SO. FROM GLASGOW TO FORT WILLIAM, BY BONNIE LOCH LOMOND, OVER DESOLATE RANNOCH MOOR, TO THE FOOT OF BEN NEVIS; THIS WALK REACHES OUT FROM A GREAT URBAN CONURBATION TO EXPLORE THE UNTAMED WILDNESS AND GRANDEUR OF THE SCOTTISH HIGHLANDS.

There can be few more evocative sights than the sun rising beyond the desolate tract of wild, wet Rannoch Moor east of the Kingshouse Inn to colour blood-red the rocks of Buachaille Etive Mor, a mountain of perfect pyramidal proportions that stands guard at the head of Glen Etive and Glen Coe, and remains watchful as the route heads over the high pass of the Devil's Staircase to Kinlochleven.

The West Highland Way is a symbolic journey, from the pressures of the twenty-first century to the timeless freedom of the mountains and great open spaces. There is bed-and-breakfast or hotel accommodation available at each suggested stop. Alternatively, you can backpack the whole route, or use a combination of youth hostels and bunkhouses. On this route we take a stop on the edge of Rannoch Moor, at the Kingshouse Inn, so taking seven days for the whole route.

itinerary

•DAY 1 12 miles (19.5km)

Milngavie to Drymen

A large notice at the railway station directs you to the start. Relatively flat going provides a leisurely stroll by way of Mugdock Wood and lochs Craigallian and Carbeth to gain and follow the line of a dismantled railway. The final section is by way of road, past a Roman fort, to the town of Drymen.

•DAY 2 14 miles (22.5km)

Drymen to Rowardennan

After Garadhban Forest, take the high-level route by way of Conic Hill for the best views over Loch Lomond, the largest freshwater lake in Britain. Surrounded by sylvan splendor, this "Queen of Scottish Lochs", studded with 30 wooded islands, is some 22½ miles (36km) long and 623 feet (190m) deep. This is Rob Roy country – a legendary

↑

Ben Nevis – seen here rising above the salty waters of Loch Linnhe – is, at 4,409 feet (1,344m), the highest mountain in the United Kingdom. It towers above Fort William, and marks the end of the trek. If you have any energy left you may want to climb it – take care though, as its Gaelic name translates to Venomous Mountain.

protector of the poor akin to England's Robin Hood.

Alternating between shore and road, the way to Rowardennan is charming in good weather, but after a wet spell the woods remain damp long after the rain has gone.

•DAY 3 **19 miles (30.5km)**

Rowardennan to Crianlarich

Beyond Ptarmigan Lodge you have a choice of routes. Either follow the track or keep to the shoreline. The latter is more demanding, and there is a long way to go. It does, however, pass Rob Roy's Prison, a rift in the cliff, and takes in a view west over Tarbet, on the opposite shore of Loch Lomond, to the spectacular horn peaks of the Cobbler, one of the most distinctive mountains in Scotland. A footbridge crosses Snaid Burn to reveal the grand Inversnaid Hotel. Rob Roy's Cave, a natural cavern by the water's edge, also sheltered Robert the Bruce, the beleaguered Scottish king, after the battle of Dail Righ in 1306. The going by the loch side is now quite arduous, but in spring it's enlivened by a sea of yellow daffodils on Island I Vow. Beyond the head of the loch, the way rises by the flanks of Cnap Mor to follow Glen Falloch, traversing above the gorge that contains the Falls of Falloch. Climb next to Derrydaroch and the remnants of the ancient Caledonian Forest. The route crosses under the railway and over the road to gain the Old Military Road. Descent through the forest leads to Crianlarich.

•DAY 4 **13 miles (21km)**

Crianlarich to Bridge of Orchy

A climb leads back up through the forest before striking a line back to the road and along Strath Fillan. As it passes the ancient remains of St. Fillan's Church, the way runs between two railway lines and weaves a somewhat tortuous route, repeatedly crossing the main road. A little beyond this point, at the Holy Pool, the sick were dipped in the River Fillan to be healed, and lunatics and witches were bound and thrown in for cure or judgement. Tyndrum marks the start of the climb over the narrow pass between Beinn Bheag and Beinn Odhar. Ahead, the conical flanks of Beinn Dorain swoop unbroken for 3,000 feet (910m) to the Allt Kinglass river. This powerful scene marks the change to the wilder face of the Western Highlands. A little farther beneath these slopes lies Bridge of Orchy.

•DAY 5 **12 miles (19.5km)**

Bridge of Orchy to Kingshouse Inn

This is a splendidly wild section of the trek, pushing northwards between the heights of the Black Mount and the remote emptiness of Rannoch Moor. It ends at the Kingshouse Inn, beneath the majestic heights of Buachaille Etive Mor, which strategically commands the heads of Glen Etive and Glen Coe. This is the very doorstep of the Western Highlands. Despite the isolation and impressive surroundings, the going is relatively easy. Cross the bridge and follow the Old Military Road, rising to Mam Carraigh before dropping to cross by the head of Loch Tulla. From here the route follows the old drovers' track on to the moors. Continue across Ba Bridge, with the formidable expanse of Coireach a' Ba to the west and the myriad lochans and endless moss of Rannoch Moor to the east. Slow descent leads to Blackrock Cottage, and the welcoming portals of the Kingshouse Inn. →

→ *The pyramidal red rocks of Buachaille Etive Mor stand guard over the heads of Glen Coe and Glen Etive.*

•DAY 6 9 miles (14.5km)

Kingshouse Inn to Kinlochleven

The River Coupall winds beneath the Buachaille to pass Jacksonville, hut of the legendary Creagh Dhu climbing club, before the way snakes rightwards from the head of Glen Coe, zigzagging over the infamous Devil's Staircase to the highest point of the trek. It was in Glen Coe that, on February 12th, 1692, the Campbells treacherously massacred the MacDonalds – hence its alternative name, the Glen of Weeping. The Old Military Road leads to the pipeline that falls from Blackwater Reservoir to Kinlochleven. These pipes once supplied the water that powered the turbines of the aluminium works. Kinlochleven used to be a hive of industry and enterprise amid this wilderness setting.

•DAY 7 14 miles (22.5km)

Kinlochleven to Fort William

The last leg of the trek circumnavigates westwards beneath the Mamores, whose white quartzite tops give the appearance of permanent snow cover, to traverse within the black conifers of Nevis Forest, before emerging into the daylight of Glen Nevis. A short journey down this attractive valley, directly beneath the flanks of Ben Nevis, leads back to sea level and the many distractions of Fort William.

→ key

- route of trek
- road
- church
- peak
- lodge/hotel
- provisions

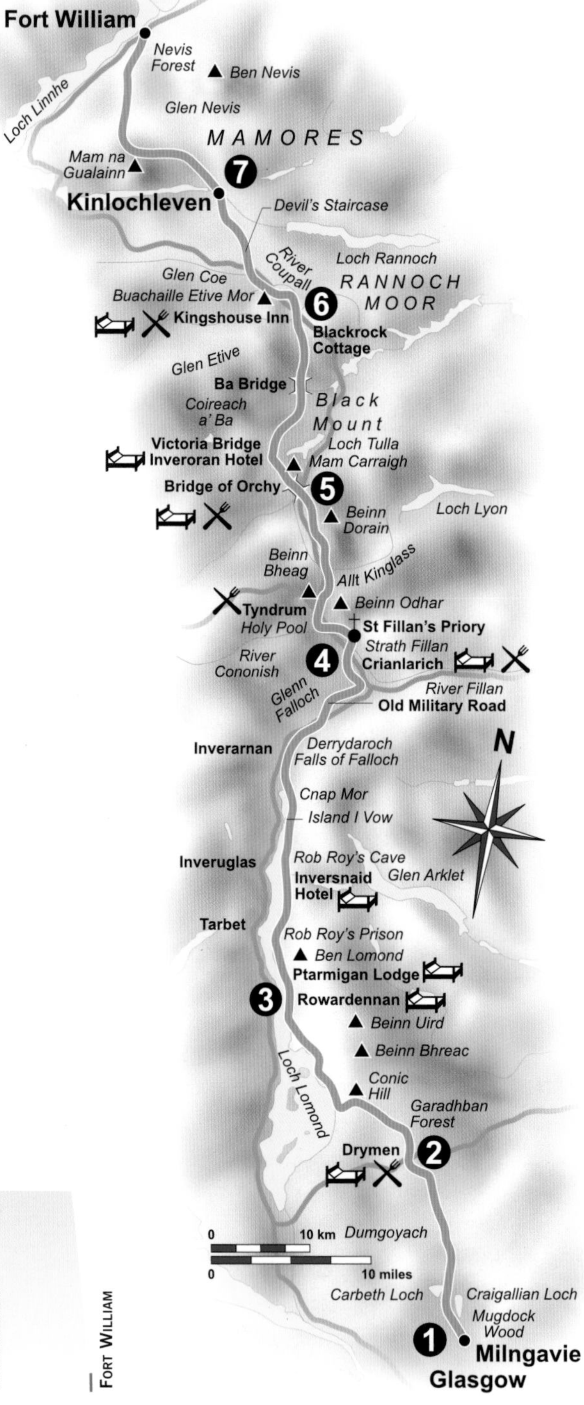

↓ walk profile

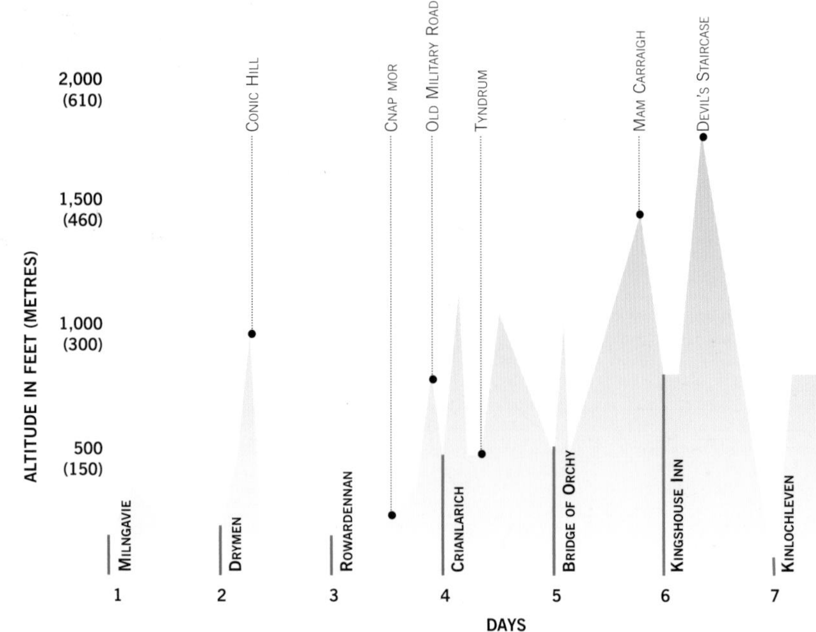

factfile

OVERVIEW

A walk from the suburbs of Glasgow through the dramatic landscape of the Western Highlands of Scotland. Explore the length of Loch Lomond; traverse the wilds of Rannoch Moor; cross the head of Glen Coe; and, finally, follow down Glen Nevis to finish at Fort William. The route is some 93 miles (150km) long and takes, on average, 7 days to complete.

Start: Milngavie.

Finish: Fort William.

Difficulty & Altitude: The way is well waymarked, and route-finding is straightforward. The route makes use of ancient trackways and the abandoned military roads of the Jacobite era. Nevertheless, the going can be both energetic and arduous. The main climb, at the head of Glen Coe, ascends the Devil's Staircase to an altitude of 1,800 feet (548m).

ACCESS

Airports: International airport at Glasgow.

Transport: The trek is served by good road and rail links.

Passport & Visas: Passport with tourist visa for all non-Europeans.

Permits & Restrictions: None.

LOCAL INFORMATION

Maps: Harveys Walker's Routes 1:40,000 West Highland Way; Ordnance Survey 1:25,000 Outdoor Leisure 39 Loch Lomond and 38 Ben Nevis & Glen Coe.

Guidebooks: *The West Highland Way*, Robert Aitken (HMSO); *The West Highland Way* (Footprint); *The West Highland Way*, Anthony Burton (Aurum Press); *A Guide to the West Highland Way*, Tom Hunter (Constable).

Background Reading: *Guide to the National Trails of Britain and Ireland*, Paddy Dillon (David & Charles).

Accommodation & Supplies: The varied accommodation along the route includes inns, bed-and-breakfast, youth hostels (at Rowardennan, Crianlarich, and Fort William only) and bunkhouses (sleeping bags required). Discerning backpackers will generally find an informal site to pitch a tent, though official campsites are in limited supply en route. Beyond Balmaha, on the shore of Loch Lomond, availability of supplies is limited to Crianlarich and Tyndrum, so be prepared.

Currency & Language: Pounds sterling (£). English (but the Scottish accent may prove indecipherable to many).

Photography: No restrictions.

Area Information: Tourist information centres at Glasgow, Drymen, Tyndrum, Kinlochleven, and Fort William.

rowan tree

TIMING & SEASONALITY

Best Months to Visit: May to September.

Climate: Officially Britain has a temperate climate, mild and wet. It can be incredibly varied; in general terms, however: spring (late March to May) can provide sunshine or rain with a lingering of snow on high ground; summer (June to August) can be hot and sunny or, equally, very wet and windy; autumn (September to October) is cooler, but it can provide the most stable period of weather; winter (November to February) is colder, and snow and ice may be expected on high ground at any time. Temperatures can range between the extremes of -4 to 82°F (-20 to 28°C), though a daily average summer value could be around 65°F (18°C). Particularly during the winter months, extreme mountain weather conditions of cold, heavy snowfall and blizzard can prevail at any time.

HEALTH & SAFETY

Vaccinations: None required.

General Health Risks: No special precautions are necessary. Britain has a good health service.

Special Considerations: The voracious midges – tiny flies, swarms of which may become mobile in calm, wind-free conditions between July and August – can become unbearably irritating and present a serious threat to enjoyment of this route. Midge repellent is readily available in out-door shops and is an essential requirement. Keep clear of red deer stags rutting in September; they are a danger only if provoked.

Politics & Religion: Scotland is a stable country. Christianity is the major religion.

Crime Risks: Low.

Food & Drink: It is best to carry sufficient drink with you, though with discretion water from mountain streams should be safe for drinking if you ensure no dwelling places are located above and that there are no obvious sources of pollution.

HIGHLIGHTS

Scenic: High, misty mountains, great tracts of wildness and wet, rugged grandeur all contrast markedly with the sylvan splendor of the glens, fresh-water lochs, and deeply penetrating sea lochs. You may well encounter wild-looking, shaggy, red-haired Highland cattle with their considerable, curving, pointed horns. In the summer months, steam trains may be seen running the nearby West Highland Line.

Wildlife & Flora: A great variety of wildlife, particularly birdlife, may be observed on this trek. In the mountains you may see the golden eagle, and the high ground and cliffs are occupied by ravens and peregrine falcons. Look out for pine marten in the trees, and on open ground, particularly the high moors, red deer, hare and stoat are common sights. Wildcats have been reported within the area, though you would be exceedingly fortunate to see one. There are many varieties of pine and broadleaved trees. The bright red berries of the rowan tree and the purple heather add splendid colour in autumn.

The isolated Kingshouse Inn, a traditional coaching inn, stands on the edge of the great wilderness of Rannoch Moor.

temperature and precipitation

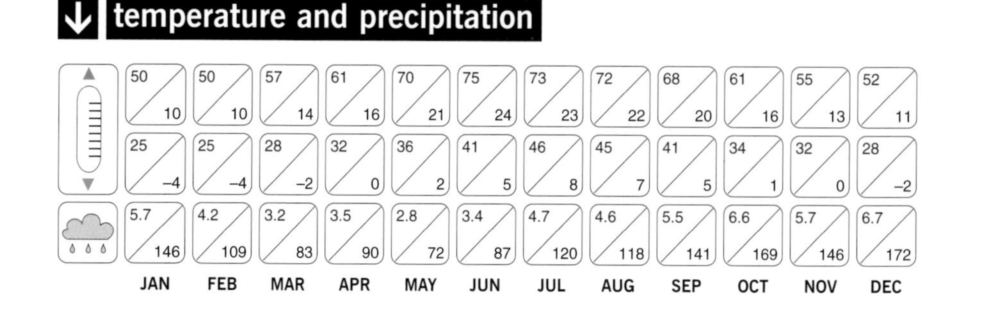

	JAN	FEB	MAR	APR	MAY	JUN	JUL	AUG	SEP	OCT	NOV	DEC
Max temp (°F / °C)	50 / 10	50 / 10	57 / 14	61 / 16	70 / 21	75 / 24	73 / 23	72 / 22	68 / 20	61 / 16	55 / 13	52 / 11
Min temp (°F / °C)	25 / –4	25 / –4	28 / –2	32 / 0	36 / 2	41 / 5	46 / 8	45 / 7	41 / 5	34 / 1	32 / 0	28 / –2
Precipitation (in / mm)	5.7 / 146	4.2 / 109	3.2 / 83	3.5 / 90	2.8 / 72	3.4 / 87	4.7 / 120	4.6 / 118	5.5 / 141	6.6 / 169	5.7 / 146	6.7 / 172

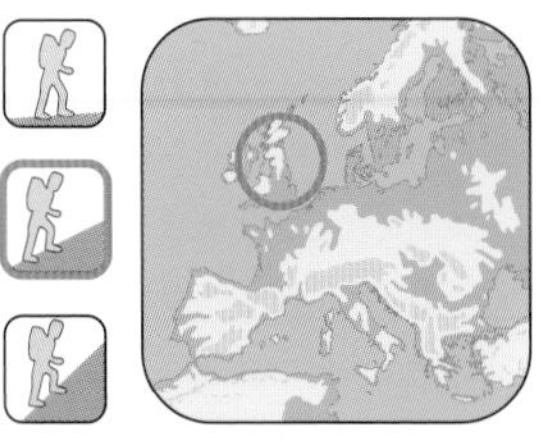

a COAST to COAST WALK

ENGLAND

Bill Birkett

FROM THE IRISH SEA TO THE NORTH SEA, THIS ROUTE FOLLOWS A LINE WEST TO EAST, TRACING THE NATURAL FEATURES OF THE LANDSCAPE OF NORTHERN ENGLAND; OVER MOUNTAIN PASSES, THROUGH ICE-CARVED DALES, AND ALONG RIVER VALLEYS, TO CROSS THE HIGH PENNINE HILLS, "THE BACKBONE OF ENGLAND".

Spanning some 190 miles (300km), the Coast to Coast was first pioneered by the North Country's legendary walker A.W. Wainwright, and featured in his guidebook in 1972. Passing through three National Parks, the walk contrasts the sublime and dramatically picturesque landscape of the English Lake District with the sinuous valleys of the Yorkshire Dales and the stark banks and open spaces of the North York Moors. A journey through history reveals burial mounds and long barrows of prehistory, the mountain Roman Road of High Street, the Nine Standards of the Viking raiders, and the classic Norman town of Richmond. From the hill-farming idyll of the Lake District to the industrial heritage of the Yorkshire Dales' lead mines and the North York Moors' Rosedale Ironstone Railway, the trek unfolds a colourful tapestry of man's ingenuity and resourcefulness.

itinerary

•DAY 1 14 miles (22.5km)

St. Bees to Ennerdale Bridge

Begin on the red sandstone cliffs of St. Bees; during the nesting season (April to June), these are thronged with important seabird colonies of guillemots and razorbills. The route heads first west then north, before turning east to pass through Sandwith and the narrow industrialised belt of West Cumbria. An ascent of Dent leads into the Lake District National Park, and the road is followed passing Kinniside Stone Circle and down to the village of Ennerdale Bridge. The Lake District is a breathtakingly beautiful corner of England. The neat, stone farms and whitewashed cottages nestle on the sides of hills or deep in the dales in perfect proportion to tree, lake and fell. The mountains, or fells as they are known locally, may only occasionally lift their heads above the 3,000-foot (910-m) mark, yet they rise with an appeal and beauty rarely matched by higher hills.

Lovely Grasmere village stands at the head of Grasmere lake, with Helm Crag to the left and Tongue Gill rising to Grisedale Pass to the right. The village is full of interest, and is a popular place to stop along the trek. In springtime, Grasmere is noted for its bloom of yellow daffodils, immortalised by the poetry of its most famous resident, the Romantic poet William Wordsworth.

On England's west coast, where the waters of the Irish Sea enter the Solway Firth to divide England from Scotland, the soft red sandstone of St. Bees Head marks the start of the Coast to Coast walk.

•DAY 2 14½ miles (23km)

Ennerdale Bridge to Rosthwaite

Beneath rocky Angler's Crag the path traverses the shore of Ennerdale Water to enter the wilds of upper Ennerdale, before rising to cross the fells. The high-level and particularly recommendable alternative route climbs the Scarth Gap Pass to traverse wonderful Haystacks and visit Innominate Tarn. Honister Pass leads down to the hamlets of Seatoller, Rosthwaite and Stonethwaite in sylvan Borrowdale. Borrowdale, with its blue-grey slate rocks, green fields, purple heathers, luxuriant bracken, crystal clear waters and craggy heights, is a prize to behold.

•DAY 3 7 miles (11.5km)

Rosthwaite to Grasmere

The path by Stonethwaite Beck leads to Greenup Gill beneath the imposing bulk of Eagle Crag. Once out on the tops, Greenup Edge leads either to a high traverse of the ridge over Gibson Knott to Helm Crag – noted for its "Lion and Lamb" rock pinnacle – or down the valley of Easedale. Both routes lead to Grasmere, once the home of William Wordsworth (1770–1850), the English Romantic poet who celebrated the daffodils that bloom there in spring.

•DAY 4 8½ miles (13.5km)

Grasmere to Patterdale

Tongue Gill rises to Grisedale Pass and another choice of routes. The high-level path to Helvellyn should not be underestimated, particularly when cloaked in snow or ice; despite its great popularity, exposed Striding Edge requires a head for heights and, in winter, a fair degree of mountain expertise. An easier track leads down the Grisedale Valley under ancient copper mines. Both routes lead to Patterdale.

•DAY 5 16 miles (25.5km)

Patterdale to Shap

The high crossing of High Street Roman Road and Kidsty Pike, at an altitude of 2,559 feet (780m), gives access to the drowned valley of Mardale and Haweswater Reservoir, as the route nears the eastern edge of the Lake District National Park. Although fairly easy going, this is wild

From beautiful sylvan Borrowdale, the path rises steeply above Greenup Gill, and continues to make the ascent of Greenup Edge.

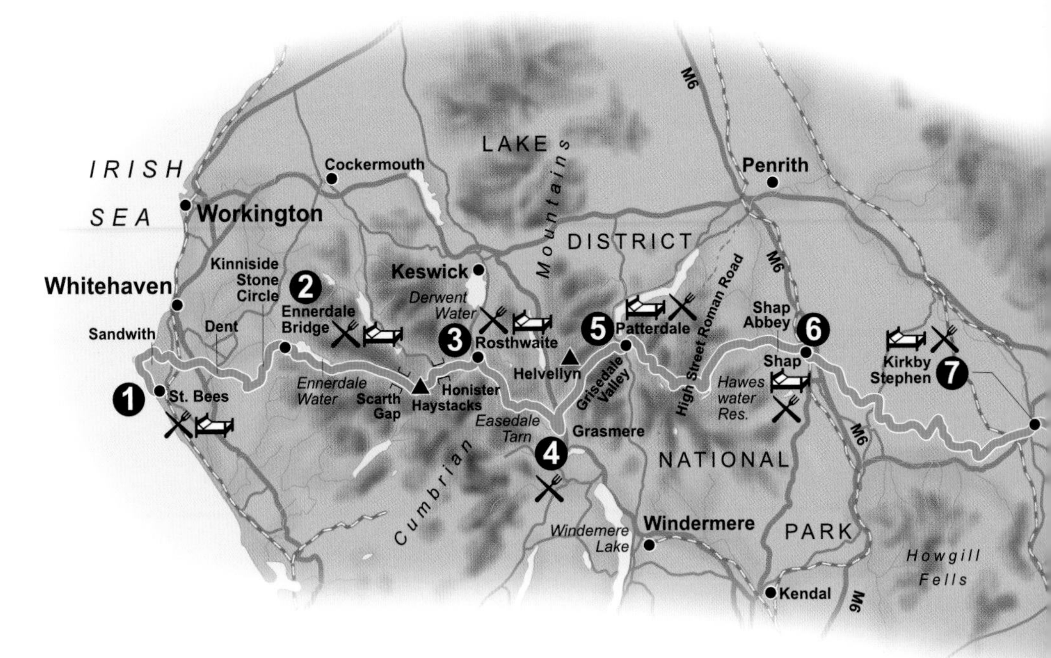

countryside, where the golden eagle, red deer, red fox and Lakeland fell pony may be seen. Beyond the remains of Shap Abbey, a short climb leads to Shap village.

•DAY 6 **20 miles (32km)**

Shap to Kirkby Stephen

A footbridge crosses the M6 motorway, and the striking volcanic features of the Lake District are now replaced with the clint and gryke of the limestone plateau of "Westmorland" (the original name of the county before it became Cumbria). Long barrows, known locally as Giant's Graves, cairns, earthworks and relics of prehistory abound before the market town of Kirkby Stephen signals the near end of Cumbria.

•DAY 7 **13 miles (21km)**

Kirkby Stephen to Keld

Above Kirkby Stephen stands the hill of Nine Standards Rigg, part of the Pennines. Nine substantial circular stone cairns stand on the summit ridge and are said to have been built by Viking raiders hoping to intimidate the residents of Kirkby Stephen before making their raid.

The Pennines, a long, dividing chain of hills, form the watershed between western and eastern England. Beyond lie the Yorkshire Dales. Whitsun Dale leads to Swaledale (pronounced "swaddle" locally) to pass a series of little waterfalls ("forces") where the dark gold, peaty waters of the River Swale tumble over resilient bands of limestone. The little-changed hamlet of Keld marks the trek's halfway point.

•DAY 8 **11 miles (17.5km)**

Keld to Reeth

Keld and Swaledale are soon left behind for the starker world of high moor. This is a bleak landscape, for centuries

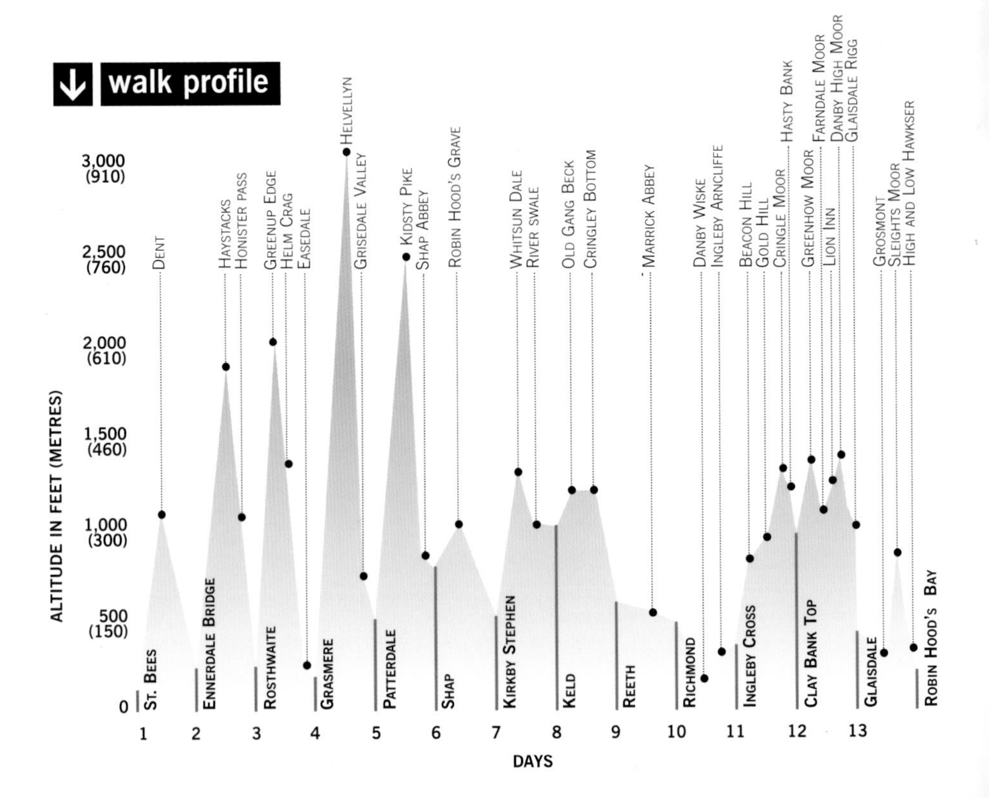

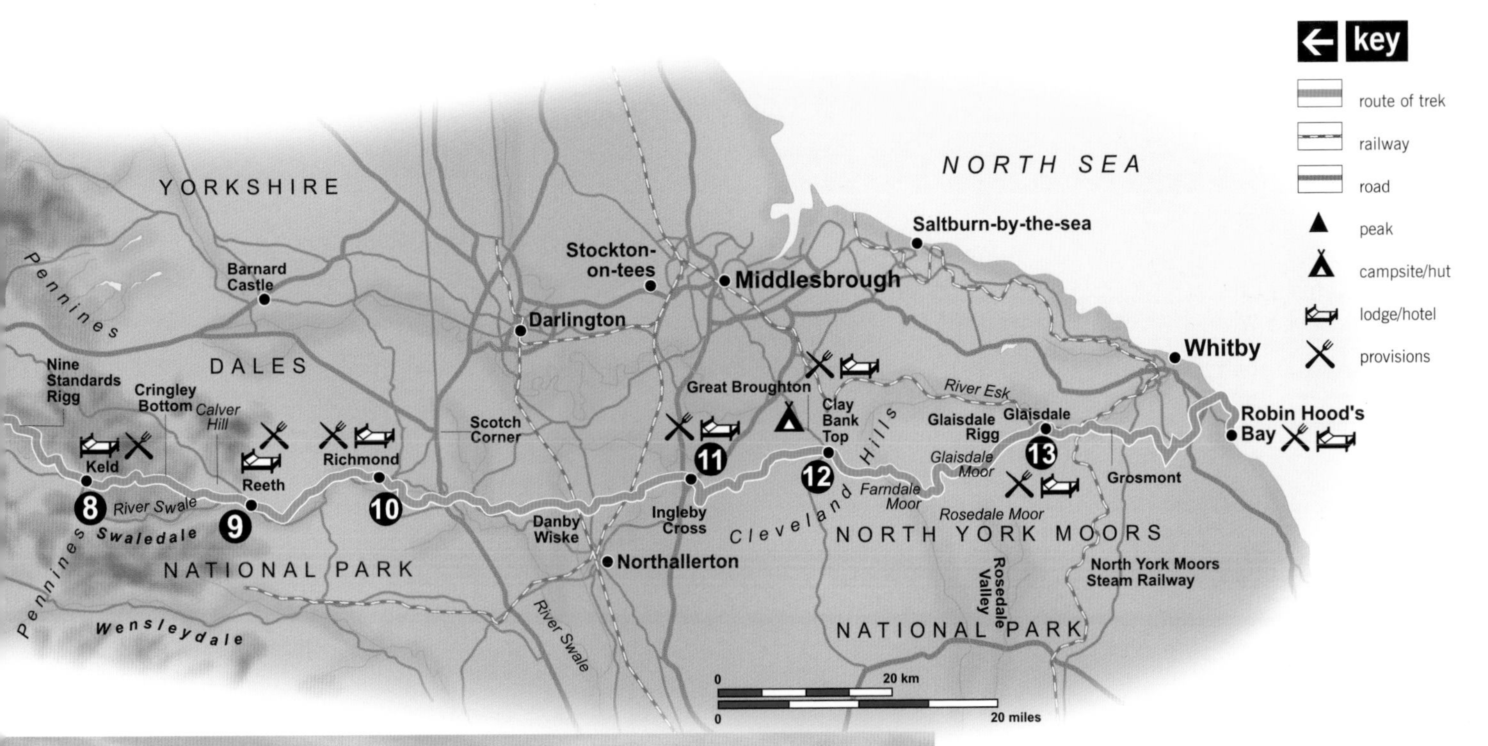

The stone cairns of the Nine Standards are said to have been built by Viking raiders to intimidate the residents of Kirkby Stephen below.

ravaged for its mineral wealth. Ingots of lead with Roman inscriptions have been found and the area may even have been mined in prehistoric times. Old mine buildings litter the area around Old Gang Beck., cross Cringley Bottom and traverse the flanks of Calver Hill, before a steep descent down Skelgate Lane to Reeth.

•DAY 9 **11 miles (17.5km)**

Reeth to Richmond

The walk rejoins briefly the banks of the River Swale. Pass the ruins of the twelfth-century Marrick Abbey before leaving the Dales National Park and making your way to Richmond. With its Norman castle, arched stone bridge, cobbled streets and historic inns, Richmond is the archetypal English town – take the time to enjoy it.

•DAY 10 **23 miles (37km)**

Richmond to Ingleby Cross

Known locally as the Vale of Mowbray, though more generally identified as the northen end of the Vale of York, this connecting valley to the foot of the North York Moors is relatively flat and provides easy going. The walk follows the banks of the River Swale beneath the busy A1 road, before arable fields of wheat and barley lead on to pass Danby Wiske. In a mile (1.5km) you reach the battlefield of The Standard where, in 1138, the English repelled the raiding Scots army. Ingleby Arncliffe leads to Ingleby Cross.

•DAY 11 **12 miles (19.5km)**

Ingleby Cross to Clay Bank Top

Imagine a basin standing above a flat table, with its northern rim tipped higher than the rest: this is the North York Moors. A series of river valleys flow due south from the rim. To the north, stand the flat industrialised plains of Teesside. This route follows the high rim until it reaches the Eskdale Valley, which falls east to Whitby and the North Sea. The rim comprises Beacon Hill, a drop to Scugdale, a rise to

Light and shade emphasize the exposed nature of Striding Edge. This ridge provides a popular high-level route from Helvellyn to Patterdale.

Gold Hill, Carlton Bank, Cringle Moor, to climb past the sandstone pinnacles of the Wainstones to a grand finale (for the day) along Hasty Bank, and finally dropping to the road at Clay Bank Top. The muddy track circumnavigating Hasty Bank to the north is the famous Lyke Wake route taken by the monks of the region who carried their dead 40 miles (64km) to be buried at Ravenscar, south of Robin Hood's Bay. Your nearest overnight accommodation will be found 2 miles (3km) north down the road at Great Broughton.

•DAY 12 19 miles (30.5km)

Clay Bank Top to Glaisdale

On this section wild moor contrasts markedly with the region's rich industrial heritage. Mysterious standing stones and ancient stone crosses heighten the already heady atmosphere. Urra and Greenhow moors lead to Farndale Moor and the deserted Rosedale Ironstone Railway. Constructed in 1861, this remarkable railway carried high-grade iron ore from the Rosedale Valley all along the heights to the iron masters of Cleveland, on the flat plains to the north. Today its route provides easy, level going. On Blakely Moor, the welcoming sight of the Lion Inn, dating from 1553, stands in isolated splendour at an altitude of 1,310 feet (400m). Traverse Danby High Moor and Glaisdale Moor before descending gradually to Glaisdale by way of Glaisdale Rigg.

•DAY 13 19 miles (30.5km)

Glaisdale to Robin Hood's Bay

Some may appreciate the hustle and bustle down by the River Esk to Grosmont. Here the mainline railway (Whitby to Middlesbrough) is joined by the preserved North York Moors Steam Railway. The latter, famous for its working steam engines, follows the glacial overflow valley down to Pickering. Onwards our route crosses Sleights Moor to cross Little Beck. Finally, posts mark the way over Graystone Hills before passing the Hawksers, High and Low, to gain the shale cliff edge, which traverses above the North Sea and down to the sheltered haven of Robin Hood's Bay.

This is more than just a walk through an attractive landscape; it is a journey through time, and a reflection on the gritty, independent character of the north of England.

factfile

OVERVIEW

A walk across the width of England from west to east, from the Cumbrian to the North Yorkshire coast, from the Irish to the North Sea, via three national parks – the Lake District, the Yorkshire Dales, and the North York Moors. The route is 188 miles (302km) long and takes, on average, 13 days to complete.

Start: St. Bees, Cumbria

Finish: Robin Hood's Bay, North Yorkshire.

Difficulty & Altitude: Although mountain and moor predominate, paths are well defined. The going is energetic and can be particulary arduous in inclement weather. In winter, particularly if the high-level variations are taken, basic hill-walking skills are vital. The low-level route (crossing the Lake District's High Street) reaches 2,500 feet (760m), and the highest high-level alternative (over the summit of the Lake District's Helvellyn) gains an altitude of 3,118 feet (950m). Escape to shelter and habitation is never more than a few hours' distant, and all the high-level sections are served by mountain rescue teams.

ACCESS

Airports: The nearest international airport is Manchester (75 miles/120km south), which has good rail and road connections.

Transport: The start and finish are served by good road and rail links. St. Bees has a railway station; from Robin Hood's Bay a bus must be taken to the railway station at Scarborough or Whitby.

Passport & Visas: Passport with tourist visa for all non-Europeans.

Permits & Access: No restrictions.

LOCAL INFORMATION

Maps: Two excellent Ordnance Survey 1:25,000 scale maps: Outdoor Leisure Maps No.33 Coast to Coast Walk St. Bees to Keld, No.34 Coast to Coast Walk Keld to Robin Hood's Bay.

Guidebooks: *Coast to Coast,*

daffodil

Ronald Turnbull (Dalesman); *Coast to Coast Walk*, Paul Hammond (Hillside).

Background Reading: *A Coast to Coast Walk*, A.W. Wainwright with photography by Derry Brabbs (Michael Joseph).

Accommodation & Supplies: There is plentiful and varied accommodation along the whole route, including hotels, inns, bed-and-breakfast and youth hostels. Stilwell's *National Trail Companion* details accommodation en route. *The Coast to Coast Bed-and-Breakfast Accommodation Guide* is available from Mrs Doreen Whitehead, Butt House, Keld, near Richmond, North Yorkshire DL11 6LJ, England. Camping is not well catered for and will be found to be impracticable en route. At each stop there are shops and supplies.

Currency & Language: UK pounds sterling (£). English; some may find the rich North Country dialects hard to understand.

Photography: No restrictions.

Area Information: Lake District National Park Visitor Centre, Brockhole, Windermere, Cumbria LA23 1LJ, England (tel.: 00 44 15394 46601). Yorkshire Dales National Park Information Services, Colvend, Hebden Road, Grassington via Skipton, North Yorkshire BD23 5LB, England (tel.: 00 44 1756 752748). North York Moors National Park Authority, The Old Vicarage, Bondgate, Helmsley, York YO62 5BP, England (tel.: 00 44 1439 770657).

TIMING & SEASONALITY

Best Months to Visit: April to October.

Climate: England has a temperate climate, mild and wet, but the weather can be incredibly varied and each day can be quite different from the next in any season. In general terms: spring (late March to May) brings sunshine or showers with a lingering of snow on high ground above 1,000 feet (300m); summer (June to August) can be either hot and sunny or wet and windy; autumn (September to October) is cooler but may provide the most stable period of weather; winter (November to February) is colder again, and snow and ice can be expected on high ground at any time. Temperatures can range between the extremes of 4 to 82°F (-20°C to 28°C), but a daily average summer value could be around 65°F (18°C).

HEALTH AND SAFETY

Vaccinations: None required.

General Health Risks: No special precautions are necessary. Britain has an excellent National Health Service which is free.

Special Considerations: Voracious midges (swarms of tiny flies) can be become unbearably irritating on summer evenings; midge repellent is available in shops specialising in outdoor activities. The adder is Britain's only venomous snake and – if you are lucky, for it is exceedingly shy – you may encounter it on this trek. Although its bite may be painful, it rarely proves fatal, although medical assistance should be sought.

Politics & Religion: England is a politically stable country. Christianity, in many forms, is the major religion, but it is a multiracial society tolerant of many different beliefs.

Crime Risks: Low.

Food & Drink: Most types of Western food are readily available. It is best to carry sufficient drink with you, although mountain streamwater en route should be safe for drinking if you ensure that no dwelling places are located above you and that there are no obvious sources of pollution.

peregrine falcon

HIGHLIGHTS

Scenic: These northern regions of England are noted for their wide expanses of untouched landscape. The glistening lakes of the west, the cosy dales of Yorkshire, the open moorland where ponies roam, and the rough seas off both the east and west coasts, are just some of the many sides to England that you will see along this walk.

Wildlife & Flora: Look out for the golden eagle in the Lake District, and the raven and peregrine falcon on high ground throughout. The red squirrel is a particularly attractive inhabitant of the Lakeland dales and woods. Many varieties of broadleaved trees, colouring beautifully in autumn, are a feature. Plants vary from mountain alpines to bog orchids.

→ *The North Sea and Robin Hood's Bay mark the end of this long-distance walk across the breadth of England.*

temperature and precipitation

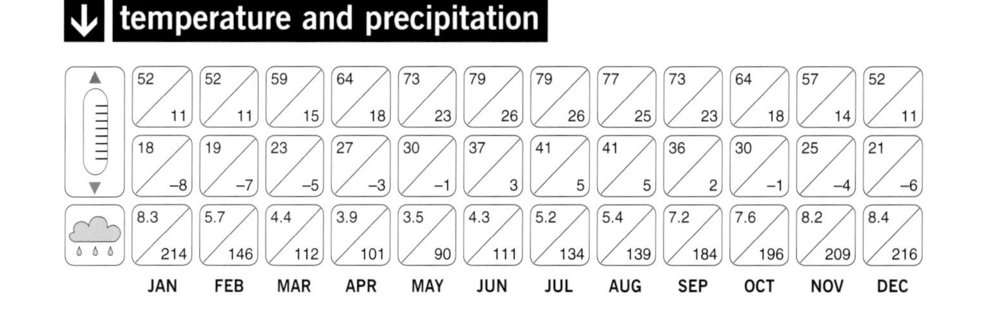

	JAN	FEB	MAR	APR	MAY	JUN	JUL	AUG	SEP	OCT	NOV	DEC
Max temp (°F / °C)	52 / 11	52 / 11	59 / 15	64 / 18	73 / 23	79 / 26	79 / 26	77 / 25	73 / 23	64 / 18	57 / 14	52 / 11
Min temp (°F / °C)	18 / –8	19 / –7	23 / –5	27 / –3	30 / –1	37 / 3	41 / 5	41 / 5	36 / 2	30 / –1	25 / –4	21 / –6
Precipitation	8.3 / 214	5.7 / 146	4.4 / 112	3.9 / 101	3.5 / 90	4.3 / 111	5.2 / 134	5.4 / 139	7.2 / 184	7.6 / 196	8.2 / 209	8.4 / 216

NORTHERN KUNGSLEDEN

LAPLAND, SWEDEN

Chris Townsend

SOUTHERN SWEDEN IS A LAND OF FORESTS AND LAKES, BUT IN THE NORTH, IN LAPLAND, THERE ARE MOUNTAINS – 600 MILES (1,000KM) OF THEM, STRADDLING THE ARCTIC CIRCLE, AND RISING TO 6,944 FEET (2,117M) ON KEBNEKAISE, THE HIGHEST MOUNTAIN IN SWEDEN AND ARCTIC SCANDINAVIA. THE KUNGSLEDEN RUNS SOUTH TO NORTH THROUGH THE HEART OF THESE MOUNTAINS.

On its way through the mountains, the Kungsleden traverses three national parks: Sarek, Stora Sjöfallets and Abisko – and passes close by Kebnekaise, which can be climbed as a side trip. As well as snow- and glacier-clad peaks, Swedish Lapland has huge lakes, powerful rivers, thundering waterfalls, wide sweeps of Arctic tundra and dense forests of birch and pine, making up a vast, unspoilt wilderness. In this far northern land there is a special quality about the light, a subtle play of sun and shadow, that is unique to the Arctic. At mid-summer it is always light and even later in the season daylight hours are long. The twilight can seem endless, a long, slow fading of the light, with the colours of the sky and the mountains slowly changing as the sun sets. Dawn, by contrast, comes quickly, the greyness of first light suddenly sparkling as the sun appears.

itinerary

•DAY 1 **10 miles (16km)**

Kvikkjokk to Pårte

Kvikkjokk is a tiny hamlet situated at 1,150 feet (350m) at the end of a very long one-way road. It lies on the wild Kamajåkkå river, not far from the borders of Sarek National Park. The Kungsleden starts gently, meandering north-east through mixed forest, past Stuor-Tata lake to Stabtjakjaure lake and Pårte, a self-service lodge.

•DAY 2 **15 miles (24km)**

Pårte to Aktse

Beyond Pårte the trail climbs over increasingly rocky terrain to pass between the craggy knolls of Faunåive and Huornatj. The views are wide-ranging, across lakes, and east and west to distant peaks. The trail traverses the hill-side before descending back into open forest. To the north rises a very steep, rocky mountain wall called Tjakkeli. On reaching Aktse lake you can either wait for the ferry or row the 2 miles (3km) to the lodge. There is a small hut called Laitaure to shelter in while you wait.

•DAY 3 **8 miles (13km)**

Aktse to Sitojaure

Aktse lies at the mouth of Rapadalen, a long, impressive valley that stretches into the heart of Sarek. As you climb to cross the mountain shoulder of Njunjes, there are good views up Rapadalen. If you have time, you could make a diversion north-westwards up Skierfe (3,867 feet/1,179m), from where you can look right up marsh- and pool-filled Rapadalen to glacier-covered peaks. Once across Njunjes, the trail descends to the trees and the shores of Kabtåjaure lake. Again, you can wait for the ferry or row to the lodge.

•DAY 4 **12 miles (19.5km)**

Sitojaure to Saltoluokta

Climbing again, the trail reaches long wide Autsutjvagge, a wild upland valley with a stream running through it and the

↑
The mountain lodge of Singi in Tjäktjavagge is a good base for an ascent of Sweden's highest mountain, Kebnekaise.

rock walls of Sjäksjo rising above. After 7 miles (11.5km) in Autsutjvagge, the trail cuts over a small hill called Kåinutålke and descends to the Saltoluokta fjällstation on the shores of Langas lake.

•DAY 5 **10 miles (16km)**

Saltoluokta to Teusajaure

Across the lake from Saltoluokta lies a main road. You could follow this for 16 miles (25.5km) to where the Kungsleden continues; but it's better to take a ferry to Kebnats, and then catch a bus to Vakkotavare lodge on the shores of Suorvajaure lake. The character of the trail changes as it turns westwards – from now on you walk through the heart of the mountains rather than skirting their edges.

The trail resumes with a steep climb through trees, then traverses between high mountain peaks, with the wedge of Kappetjåkkå standing out, before a quick descent into birch woods and Teusajaure, a lovely lake surrounded by fine mountains. You have to row to the lodge on the far side.

←
Views of the snowclad peaks of Sarek National Park greet you early on in the trek at the whitewater of the wild Kamajåkkå river.

•DAY 6 **11 miles (17.5km)**

Teusajaure to Singi

Another steep climb leads to the crossing of a broad mountain ridge by way of the Muorki Valley. A descent follows, down to the wild rush of the Kaitumjåkka river and beautiful Kaitumjaure lake. The trail then climbs into the spectacular 20-mile (32-km) long Tjäktjavagge that runs past Kebnekaise and through the centre of a mass of rugged mountains. The often snowcapped summits of these mountains are steep-sided, with long, flat ridges, and glaciers on the north and eastern faces. The next lodge, Singi, can be used as a base for an ascent of Kebnekaise, which lies to the north-east. The climb is steep but not difficult, and, in good weather, the views from the summit are superb.

•DAY 7 **15 miles (24km)**

Singi to Tjäktja

The walk up Tjäktjavagge is probably the most scenic part of the Kungsleden. Mountains of every shape and size soar on either side. Kebnekaise can be seen at several points, and there is a wonderful view up the side valley of Kuopervagge to the cluster of glaciers and summits of Kaskasatjåkka. Eventually the head of the Tjäktjavagge is reached and the path climbs to the Tjäktja Pass, at 3,740 feet (1,140m) the highest point on the Kungsleden. Ahead lies Alisvággi valley with Tjäktja lodge at its head.

→

•DAY 8 **8 miles (13km)**

Tjäktja to Alesjaure

The trail winds on slowly, down widening Alisvaggi. The Alesgätno river meanders through the valley, sometimes close to the trail, and with many oxbow lakes, loops and broad sections dotted with small islands. Alesjaure lodge lies across the river just before it enters Alisjávri lake.

•DAY 9 **12 miles (19.5km)**

Alesjaure to Abiskojaure

After following the shores of three scenic lakes, the trail descends into Gárddenvággi, a narrow valley hemmed in between the steep walls of Kartinvare and Siellanjunnji. The high mountains and their valleys are being left behind, and soon the path drops more steeply into trees and the Abisko valley. Abiskojaure lies on the shores of Abiskujávri lake.

•DAY 10 **9 miles (14.5km)**

Abiskojaure to Abisko

The last stage of the Kungsleden runs down the beautiful Abisko valley beside the wide Abiskojåkka river to Torneträsk, a huge lake. Mountains still rise all around but they are farther away now and the walking is in forest not over tundra – a relaxing end to a superb hike.

↑ *Practise your rowing skills before embarking on this trek, as Aktse lake is one of three lakes on the route where rowing boats are made available to trekkers needing to reach lodges on the far side of the lakes.*

→ key

- route of trek
- recreational area boundary
- minor track
- road
- pass
- peak
- campsite/hut
- lodge/hotel
- provisions

Torneträsk
ABISKO NATIONAL PARK
Abisko
Abiskojåkka
Abiskojávri
10
Abiskojaure
Sielanjunnji
Kartinvare
Miesakjávri
Ahpparjávri
Rádujávri
Alisjávri
Alesjaure
9
N

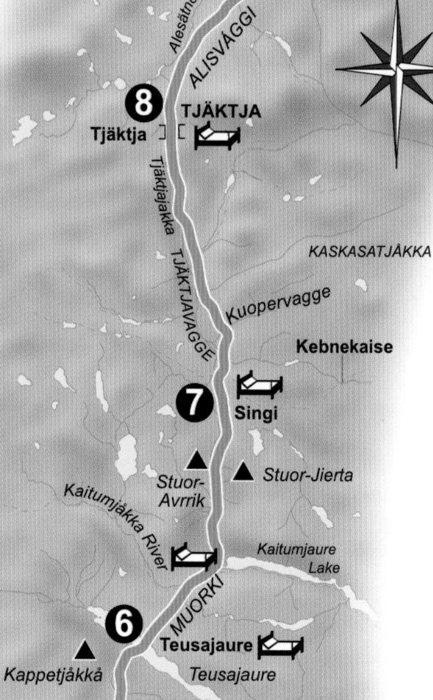

STORA SJÖFALLETS NATIONAL PARK
Suorvajaure
Vakkotavare
Langas
Kebnats
Lulep Kierkau
5
Saltoluokta
Kånutälke
Sjäksjo
AUTSUT-JVAGGE

↓ walk profile

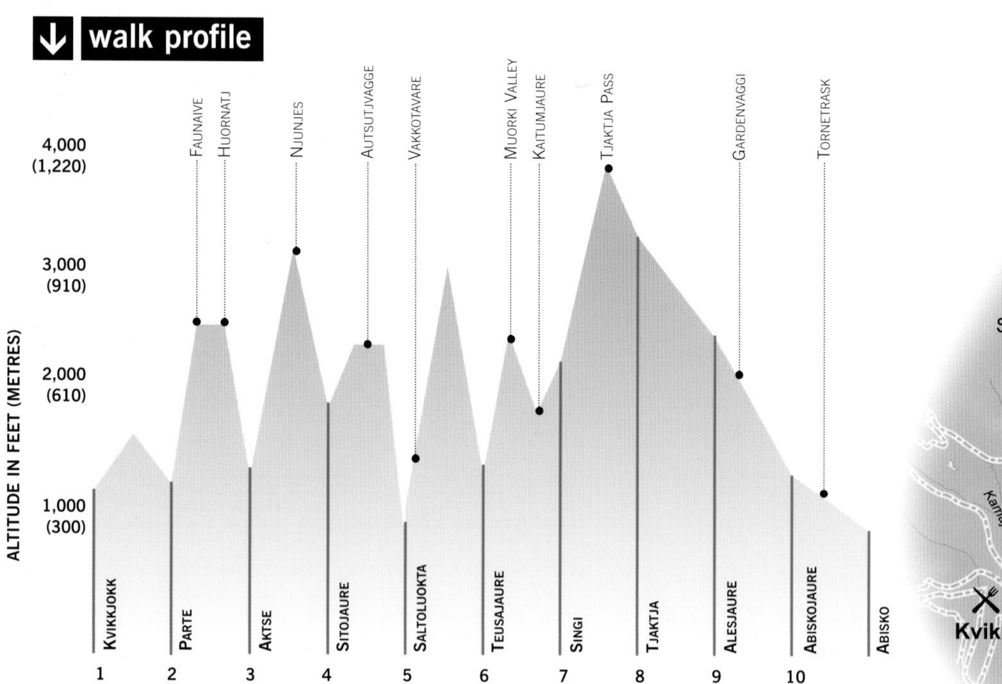

factfile

OVERVIEW

The Kungsleden, or King's Way, is a long-distance trail running through the mountains of Arctic Sweden. The whole Kungsleden is 280 miles (450km) long. The most scenic northern section, described here, is 110 miles (177km) long and can be walked in 1 to 2 weeks.

Start: Kvikkjokk.

Finish: Abisko.

Difficulty & Altitude: The trail is well marked and quite easy to follow. It can be wet and muddy in places, but the worst areas have wooden walkways. There are a few steep climbs, but they are short. The walking is mostly on fairly level terrain. The altitude is low, the highest point being 3,740 feet (1,140m). The many rivers on the route are bridged, but four long lakes have to be crossed by ferry or rowing boat. There must always be at least one boat on each shore so if you find just one boat you should row across, row back towing another boat, then row back again.

ACCESS

Airports: Stockholm, from where there are flights to Kiruna and Gullivare.

Transport: Train (Swedish State Railways) from Stockholm, Gullivare, or Kiruna to Murjek, from where there are buses to Kvikkjokk. Abisko is on the railway line.

Passport & Visas: Passports required by all visitors.

Permits & Restrictions: None.

reindeer

LOCAL INFORMATION

Maps: Fjällkarten 1:100,000 topographic maps BD10 Sareks National Park, BD8 Kebnekaise–Saltoluokta, BD6 Abisko–Kebnekaise.

Guidebooks: None in English.

Background Reading: *Scandinavian Mountains*, Peter Lennon (West Col) – a useful general guide to the mountains of Sweden and Norway.

Accommodation & Supplies: Numerous mountain lodges mean there is no need to carry a tent, though there are many splendid campsites. Sleeping bags should be carried in case there is a shortage of bedding in a lodge, of which there are two types. Self-service lodges (12 on the route) are wardened, and provide bedding, heating and cooking facilities; some sell food supplies. Fjällstations (three on the route) are much larger; they are like hotels, with restaurants, hot water and showers. Meal times are fixed but they also have self-catering facilities and stores. Small huts and wind shelters on the trail are for protection from the weather rather than overnight stays, but could be used for the latter in emergency. A good choice of provisions, and cheaper than in the lodges, can be purchased in Kvikkjokk and Abisko.

Currency & Language: Swedish Kroner. Swedish; English is widely spoken.

Photography: No restrictions.

Area Information: Swedish Tourist Office: Svenska Turistforeningen, Box 25, 101 20 Stockholm, Sweden. Websites: www.hejoly.demon.nl/countries/sweden.html; www.lahn.de/homes/sirius/trails/k.htm www.merausverige.nu

TIMING & SEASONALITY

Best Months to Visit: The trail is open from around June 20 to September 20, after which the ferries are taken out before the autumn freeze. Late August and September is perhaps the best time as the mosquitoes have gone and the fall colours are spectacular.

Climate: Although this is the Arctic, summers can be quite hot. However, the weather is very changeable, and rain and low cloud can blow in very quickly. Be prepared for anything!

HEALTH & SAFETY

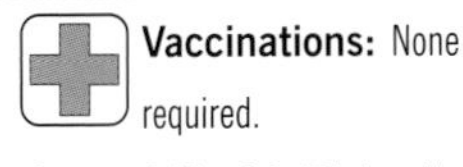

Vaccinations: None required.

General Health Risks: No special risks. Health insurance should be taken out.

Special Considerations: None.

Politics & Religion: No concerns.

Crime Risks: Low.

Food & Drink: Water in the wild is considered safe.

HIGHLIGHTS

Scenic: Arctic Scandinavia contains the largest unspoilt, wild country left in western Europe. The landscape is magnificent throughout, but the lakes stand out for their stark beauty, while there are magnificent views of the Sarek mountains and the summits and glaciers around Kebnekaise throughout the walk.

Wildlife & Flora: Reindeer will almost certainly be seen during the walk. These are not wild, but are owned by the local Sami people, and shouldn't be disturbed as this can make rounding them up difficult. Mountain and tundra birds likely to be seen include willow grouse and ptarmigan. Golden eagles and rough-legged buzzards soar above the mountains. The forests are of pine, spruce and birch. Early in the summer wildflowers decorate the forest and tundra, while later in the year there are bilberries and cloudberries.

In autumn, the forests of the Kungsleden turn into a spectacular sea of warm hues.

temperature and precipitation

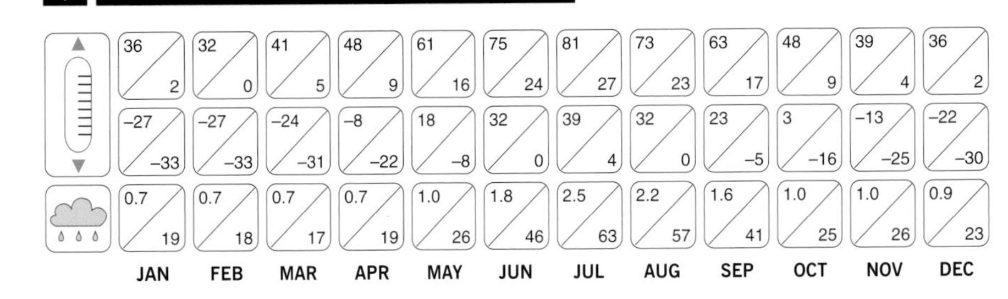

	JAN	FEB	MAR	APR	MAY	JUN	JUL	AUG	SEP	OCT	NOV	DEC
Max temp (°F / °C)	36 / 2	32 / 0	41 / 5	48 / 9	61 / 16	75 / 24	81 / 27	73 / 23	63 / 17	48 / 9	39 / 4	36 / 2
Min temp (°F / °C)	−27 / −33	−27 / −33	−24 / −31	−8 / −22	18 / −8	32 / 0	39 / 4	32 / 0	23 / −5	3 / −16	−13 / −25	−22 / −30
Precipitation (in / mm)	0.7 / 19	0.7 / 18	0.7 / 17	0.7 / 19	1.0 / 26	1.8 / 46	2.5 / 63	2.2 / 57	1.6 / 41	1.0 / 25	1.0 / 26	0.9 / 23

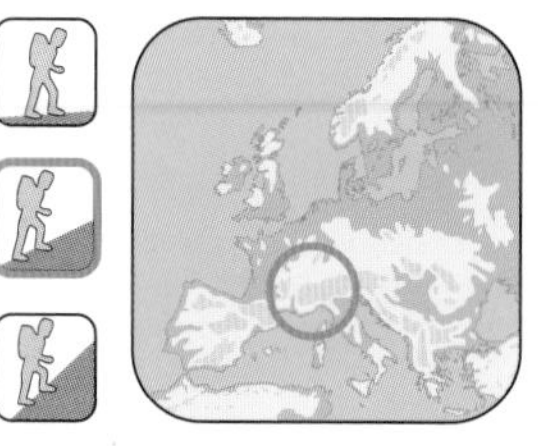

TOUR of MONT BLANC

FRANCE/ITALY/SWITZERLAND

Walt Unsworth

THE WALK ROUND MONT BLANC AND ITS SATELLITES HAS BEEN ONE OF THE MOST POPULAR LONG-DISTANCE TRAILS IN THE ALPS SINCE THE EARLIEST DAYS OF ALPINE EXPLORATION. THE VICTORIANS WOULD DO IT IN FIVE DAYS, OR EVEN LESS – BUT THEY WERE PRODIGIOUS WALKERS!

Like many such trails, the T.M.B. has been treated as a racecourse by some, but this misses the essential point of doing it in the first place, which is to enjoy the outstanding scenery.

The walk basically involves visiting in turn the seven valleys around Mont Blanc, hopping from one to another by a series of high passes. The valleys are (counter-clockwise from Contamines): Vallée de Montjoie, Vallée des Glaciers, Val Veni, Italian Val Ferret, Swiss Val Ferret, Vallée du Trient, and the valley of the River Arve – better known as the Chamonix Valley – where the walk starts and ends for most people. In general, the paths are good and easy to follow, but if snow is lying on the high passes care is needed.

The following is a valley to valley description of the trek. To get the best out of the tour, it is recommended that days out be taken at Chamonix and Courmayeur for cable-car visits to the Aiguille du Midi and Col du Géant, for their tremendous views.

itinerary

• DAY 1 **11½ miles (18.5km)**

Les Houches to les Contamines

From the main street in les Houches, a path climbs steeply up the wooded hillside to the Col de Voza, offering fine backward views of the Chamonix Valley. The gaunt hotel at the col is reached in just under 3 hours. From here you leave the original T.M.B. route and take the more adventurous one toward the vast Bionnassay Glacier.

At the well-named Bellevue, the eye sweeps round the peaks from the Dôme du Goûter to Mont Blanc and the Aiguilles, the Brévent and the Aiguilles Rouges. The path becomes rough and less distinguishable as it descends to a bridge, Pont des Places, over the Bionnassay torrent, and then via a metal ladder to the foot of the Col de Tricot, the highest point of the day's walk at 6,955 feet (2,120m). The path then descends very steeply to the Chalets de Miage and the outflow of another glacier before climbing again to a shoulder of Mont Truc to reach the chalets of the same name. It's then down, through the woods, to les Contamines.

This makes for a hard first day, but there is accommodation en route at the various hamlets if you don't wish to do it all in one go.

VARIATION: If the weather is poor, it is better to take the original T.M.B. route from the hotel at the Col de Voza steeply down to Bionnassay hamlet and beyond to the main valley floor, where a road leads to les Contamines. The road is tedious, but the valley is quite fine, and the walk is much shorter and less strenuous than the alternative. In early summer, the valley is full of flowers.

• DAY 2 **12½ miles (20km)**

Les Contamines to les Chapieux

Anyone backpacking needs to stock up in les Contamines because there is not much else until Courmayeur is reached a few days later. This is a delightful undeveloped village, and quite a contrast to nearby Chamonix. From the village

↑ *The Brenva face of Mont Blanc, one of the most challenging mountains in Europe, as seen from the Col du Géant.*

the T.M.B. continues along the valley until it can quit the tarmac and follow a pleasant path by the River Bonnant to the ancient chapel of Notre Dame de la Gorge, whose white walls contrast sharply with the dark forest around it.

This is one of the ways the Roman legions came into Gaul, and the T.M.B. follows a paved Roman way steeply up from the chapel until it ends at a bridge over a torrent. The path continues past the Nant Borrant hotel to the higher restaurant at La Balme. It takes about 4 hours to reach here, so stop for a rest and a coffee while enjoying the wide panorama of the Montjoie Valley and the great peaks of the Miage.

The way ahead seems steeply formidable. You will have to toil up great swellings of shale-like slopes until a huge cairn is reached. This is the Plan des Dames. The cairn is said to be a memorial to two women who perished in a storm in this lonely place. The Col du Bonhomme (7,641 feet/2,329m) lies above, and beyond it there is a traverse across the hillside below a curious rocky tower, called the Roche du Bonhomme, to the higher Col de la Croix du Bonhomme (8,123 feet/2,476m). If snow is left over from the previous winter, the ascent of the cols, and especially the traverse between them, can be a bit tricky; an ice-axe or trekking poles give a better feeling of security.

Beyond the Col de la Croix is the steep descent to the Val des Glaciers, past an alpine refuge, with the perfect snow cone of Mont Pourri floating in the distant haze. The descent ends at the hamlet of les Chapieux, where accommodation and meals are available.

← *From the Col Chécrouit there is an idyllic view down to the pretty, unspoilt village of Courmayeur in Italy.*

•DAY 3 **9¾ miles (15.5km)**

Les Chapieux to Elisabetta Hut

Ahead lies a long, tedious metalled road crawling up the valley to the café-refuge of les Mottets, at the foot of the Col de la Seigne. The walk up the road is dominated by the spendid Aiguille des Glaciers, which rises at the valley head from the curiously named Glacier des Glaciers.

From les Mottets, the path climbs the Col de la Seigne (8,255 feet/2,516m), beyond which lies Italy. The path is surprisingly good to the col, and the ascent easy (less than →

2 hours). The view down the other side is amazing as it sweeps down the long straight Val Veni, then up the opposing Val Ferret to the Swiss frontier – a glance which takes you entirely across Italy! Late snow can lie on the col and care is needed, but in a little over an hour you can reach the splendid Elisabetta Hut, overlooking the Lex Blanche glacier.

VARIATION: There is an alternative way from the Col de la Croix to les Mottets, and that is to climb to the Col des Fours (8,743 feet/2,665m), which crosses the Fours ridge and cuts out the descent to les Chapieux and the road walk. However, the scenery is dreary and there is some danger on a tricky descent, especially if conditions are less than perfect.

•DAY 4 **13½ miles (21.5km)**

Elisabetta Hut to Courmayeur

Below the hut, the path goes down the Val Veni to the Lac Combal, a swamp-like place with braided streams, where you meet the first real traffic for some time. The road from Courmayeur comes up to this popular spot, so all the peace you have enjoyed over the last couple of days dissolves in an instant.

The original T.M.B. route continued down the Val Veni to Courmayeur, but most walkers now leave the valley at the Combal bridge and climb steeply up the hillside to the right, past the ruined chalets of L'Arp Vielle, to a spur of Mont Favre. You will already have been conscious of the view across the valley, but from this place it is simply stunning. Seldom will you see such an array of mountain savagery. The long and stony Miage Glacier with little Lac Combal locked in its moraine arms, the huge face of Mont Blanc from which half a dozen other glaciers tumble, and, perhaps more impressive than the rest, the savage spire of Aiguille Noire de Peuterey, and the spiky pinnacles known as the Dames Anglaises, a French joke at the expense of Victorian English ladies. Just to show there is no ill-feeling, however, the wide hollow on the face of the mountain is known as the Fauteuil des Allemands, a joke on the width of German bottoms!

Just a little way along there is a favourite photo-spot

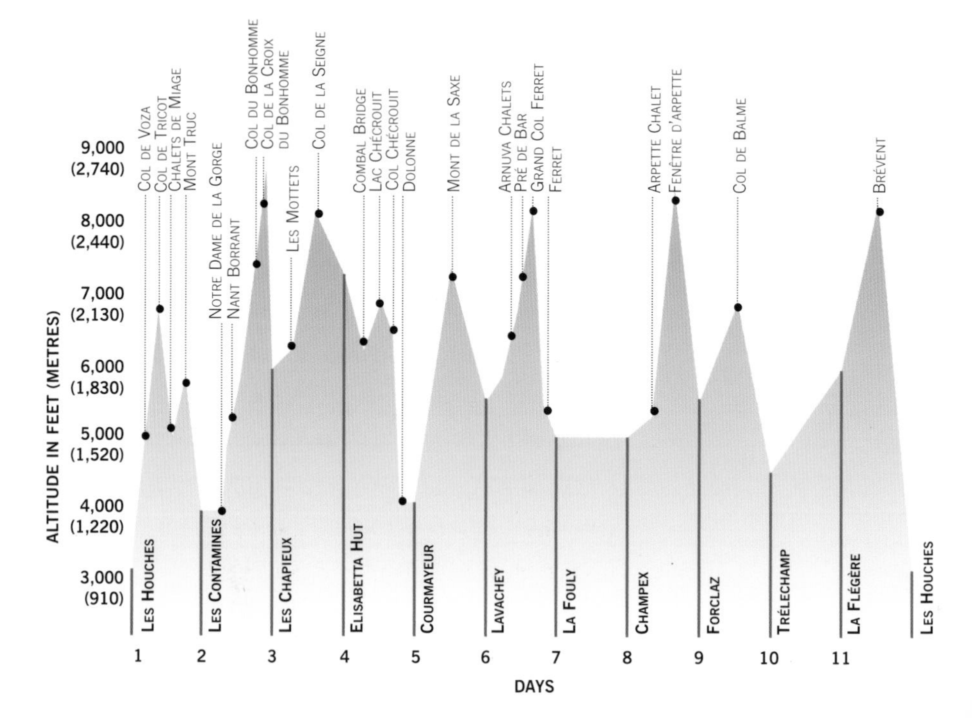

This dramatic pinnacle rises near the path to Flégère. It is a favourite with rock climbers, who are spoilt for choice in this region of the world.

at the little Lac Chécrouit, in whose waters the mountains are picturesquely reflected. Beyond the lake is the Col Chécrouit, from where a zigzag path zips down to Dolonne and Courmayeur. This path is less than good these days, so you can cheat by going straight down in the cable car, or take a long detour, much of it on roads, round Mont Chetif.

Courmayeur is a much changed place since the opening of the Mont Blanc Tunnel, the road from which goes through the town's outskirts. The constant roar of heavy traffic from Italy's industrial north and the swarm of tourists on day trips from Chamonix mean it is no longer the isolated village known to the pioneers. Nevertheless, it is worth spending some time here, if only to take the cable car up from la Palud, at the head of the valley, to the Torino Hut, from which it is just a short walk over snow to the Col du Géant. The prospect from here over the Vallée Blanche encompasses peak after famous peak, from the Brenva Face of Mont Blanc, over the Aiguilles, and round to the Géant. It is one of the world's great mountain prospects.

•DAY 5 **11¼ miles (18km)**

Courmayeur to Lavachey

From Courmayeur, the T.M.B. travels up the Italian Val Ferret to cross the Grand Col Ferret into the long Swiss valley of the same name. If the weather is fine and there is no residual snow from winter, it is worth the effort to climb up to the Montagne de la Saxe ridge from Courmayeur and follow it high above the Val Ferret. It gives some of the widest and grandest panoramas of the tour. The eye can travel back up the Val Veni to the Col de la Seigne and the whole of the Mont Blanc massif, huge and magnificent; nearer to hand is a long line of rocky peaks – Géant, Grandes Jorasses, Leschaux, Triolet, and Mont Dolent – which form a veritable wall, formidable in aspect.

In poorer weather there's a walk along the valley floor, where the prospect is not so wide but still impressive. The two ways meet at Lavachey (if the weather is really bad, you can catch a bus to here from Courmayeur).

•DAY 6 **11½ miles (18.5km)**

Lavachey to la Fouly

After Lavachey come the Arnuva chalets, where the road ends. A good path climbs to the grassy alp of Pré de Bar, where there is a fine mountain refuge. Once again, the backward view extends to the distant French frontier at Col de la Seigne; but nearer to hand is the curious Pré de Bar glacier, squeezed between rock walls like a dollop of toothpaste.

The path now mounts steadily to the Grand Col Ferret (8,323 feet/2,537m), a desolate spot if the mist is down, but easy to cross with a good path leading down into Switzerland. Eventually, a surfaced road appears and, a few →

To spend some time at the peaceful, lakeside resort of Champex is like stepping back in time to the Edwardian era, in terms of style and elegance.

minutes later, the village of Ferret. The Heidi-like chalets confirm that this is indeed Switzerland. Soon, the village of la Fouly is reached.

•DAY 7 **10½ miles (17 km)**

La Fouly to Champex

The Swiss Val Ferret is pretty but has none of the drama of its Italian counterpart. The mountains lie back, hidden. The walk travels down the valley to Issert, and from there climbs through the woods to the old-style holiday resort of Champex, complete with lake and paddleboats!

•DAY 8 **10½ miles (17km)**

Champex to Forclaz

Between Champex and the Chamonix Valley lies the valley of Trient, but nowadays the tour only skirts the valley's fringe. There are two ways of leaving Champex, and both are popular. The macho way is to go by the Arpette chalet up the long valley of the same name and cross the high notch in the ridge above, called the Fenêtre d'Arpette (8,743 feet/2,665m), the highest point on the tour. It is a splendidly wild journey into the heart of the mountains, and the descent on the other side, above the Trient Glacier, is quite spectacular. It goes alongside a splendid bisse, or watercourse, to the hotel at Forclaz.

VARIATION: If the weather is poor, or if there is a lot of snow about, the choice then must be the gentler Bovine Alp route, which, though much less rugged, offers some fine views over the Rhône Valley.

•DAY 9 **11¾ miles (19km)**

Forclaz to Trélechamp

Originally, the route wound down from the Bovine Alp route into the valley deep below, past the village of Trient, before climbing up to the Col de Balme on the French frontier, but now it is more usual to follow the narrow path to Les Grands, alongside yesterday's bisse, where there is a unique rock staircase, complete with handrails, leading up a cliff face to a thin path high above the Trient vale. The path leads straight across to the Col de Balme without losing height. If icy snow is on the ground, however, this traverse can be difficult, and an ice-axe or poles are useful.

There is a mountain hut at the Col de Balme and a view right down the Chamonix Valley to Mont Blanc, which seems to float like a cloud above it. From the hut, a good path leads off to the col Posettes, with more superb views of Mont Blanc, and then drops to the Col des Montets and Trélechamp.

•DAY 10 **5¼ miles (8.5km)**

Trélechamp to la Flégère

The last stage of the walk is a traverse of the Aiguilles Rouges. It includes an ascent of the Brévent (8,287 feet/2,526m), and will take a couple of days. The ridge makes a superb viewing platform for the whole of the Mont Blanc range, not so rugged or intimate as on the Italian side, perhaps, but very spectacular all the same. Chamonix lies in the valley below the ridge. With plenty of cable cars in sight, the paths are hardly lonely.

•DAY 11 **11½ miles (18.5km)**

La Flégère to les Houches

The path continues along the ridge to the col du Brévent, and then climbs to the summit of the Brévent itself, a good viewing platform for Mont Blanc and popular with tourists who come up by cable car. After the Brévent, it is all downhill back to les Houches.

factfile

OVERVIEW

A circular walk encompassing Western Europe's highest summit, Mont Blanc (15,770 feet/4,807m), and its satellite peaks. Usually referred to as the "T.M.B.", it is some 120 miles (190km) by the standard route, and involves 33,000 feet (10,000m) of ascent and the same of descent. The guidebook allows 11 days, but it can be done in much less time. The scenery throughout is quite staggering, which accounts for its tremendous popularity. The walk passes through France, Italy, and Switzerland in turn.

Start/Finish: This is a circular walk and can be joined at any point en route. It is usual, however, to start at les Houches and go round counter-clockwise.

Difficulty & Altitude: The route is waymarked with red and white stripes on stones, trees or buildings. If the stripes bend, the path bends in that direction too. Crossed stripes means you have gone the wrong way and should retrace your steps. Seven valleys radiate from the mountain core and each is visited in turn, crossing high cols. There are several such cols of 6,500 feet (2,000m) or more – the highest being the Fenêtre d'Arpette at 8,743 feet (2,665m) on the standard route or, on the alternative route over the Bovine Alp, the Grand Col Ferret at 8,323 feet (2,537m).

ACCESS

Airport: International airport at Geneva.

Transport: Taxi to Eaux-Vives Station (public transport is available, but complicated), then train to St. Gervais and narrow gauge from there to les Houches. St. Gervais and Martigny are also served by train from Paris.

Passport & Visas: Passport required for all visitors. Tourist visa required by Australasians.

Permits & Restrictions: None.

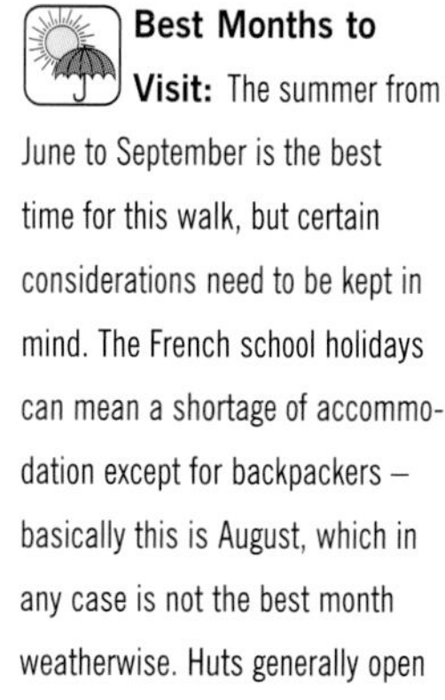
great yellow gentian

LOCAL INFORMATION

Maps: IGN 3630 O and 3531 E 1:25,000 or Didier & Richard No.8 Mt. Blanc and Beaufortain 1:50,000. Because the path is well-marked, the smaller scale map should suffice.

Guidebook: *Tour of Mont Blanc*, Andrew Harper (Cicerone). As well as the route directions, it is an invaluable source of information and addresses.

Background Reading: This is one of the world's most popular long-distance trails, and consequently it features in the brochures of many adventure travel companies. *Mont Blanc and the Seven Valleys*, R. Frison-Roche (Nicholas Kaye); *La Grande Ronde autour du Mont Blanc*, Samivel (J. Glénat) – French text.

Accommodation & Supplies: There is every sort of accommodation on this tour, from hotels to *dortoirs* (dormitories), and mountain huts. Backpacking is possible, as there are plenty of camping spots and places to re-stock food every couple of days.

Currency & Language: French and Swiss francs, Italian lira. Depending on whether you stay in Courmayeur for a break or not, reckon on lira for 4 days, Swiss francs for 3 days and French francs for the rest. The common language is French, which is generally understood even in Courmayeur.

Photography: No restrictions.

Area Information: Azienda Promozione Turistica Monte Bianco, Piazzale Monte Bianco, Courmayeur (tel.: 00 39 165 842060, fax: 00 39 165 842072).

TIMING & SEASONALITY

Best Months to Visit: The summer from June to September is the best time for this walk, but certain considerations need to be kept in mind. The French school holidays can mean a shortage of accommodation except for backpackers – basically this is August, which in any case is not the best month weatherwise. Huts generally open in June and close in September, but the precise time depends on the weather and the guardian.

Climate: If the preceding winter has been very snowy, the high passes can be snow-covered well into July, in which case an ice-axe is useful as a precaution. The following places need particular care at such time: traverse from Col du Bonhomme to Col de la Croix, Col des Fours, Col de la Seigne, Lac Chécrouit traverse, Fenêtre d'Arpette, and traverse Les Grandes to Col de Balme. Where possible, take the easier alternative.

marmot

HEALTH & SAFETY

Vaccinations: None required.

General Health Risks: No concerns.

Special Considerations: None.

Politics & Religion: No concerns.

Crime Risks: Low.

Food & Drink: Tap water is safe to drink.

HIGHLIGHTS

Scenic: The scenery is staggering throughout, but special highlights include the Aiguille Noire as seen from the Chécrouit alp, and Mont Blanc seen from Brévent.

Wildlife & Flora: All round the walk marmots whistle their warnings at the approach of strangers. In spring, the lower meadows, especially near les Contamines, are full of gentians and other alpine flowers.

From Forclaz the route climbs through dramatic scenery at les Grands; this avoids the old, steep route down to Trient.

temperature and precipitation

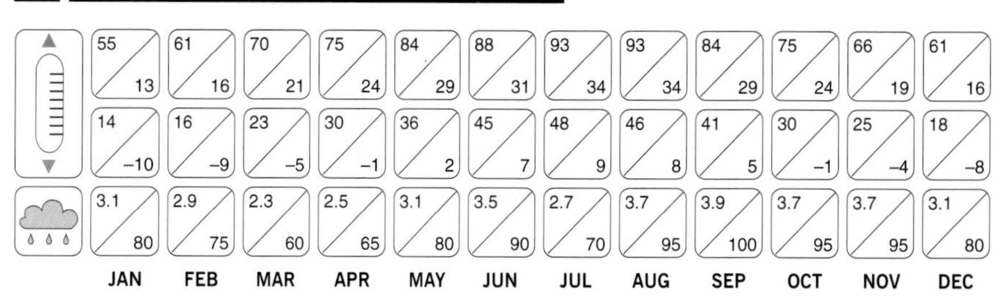

	JAN	FEB	MAR	APR	MAY	JUN	JUL	AUG	SEP	OCT	NOV	DEC
Max temp (°F / °C)	55 / 13	61 / 16	70 / 21	75 / 24	84 / 29	88 / 31	93 / 34	93 / 34	84 / 29	75 / 24	66 / 19	61 / 16
Min temp (°F / °C)	14 / –10	16 / –9	23 / –5	30 / –1	36 / 2	45 / 7	48 / 9	46 / 8	41 / 5	30 / –1	25 / –4	18 / –8
Precipitation (in / mm)	3.1 / 80	2.9 / 75	2.3 / 60	2.5 / 65	3.1 / 80	3.5 / 90	2.7 / 70	3.7 / 95	3.9 / 100	3.7 / 95	3.7 / 95	3.1 / 80

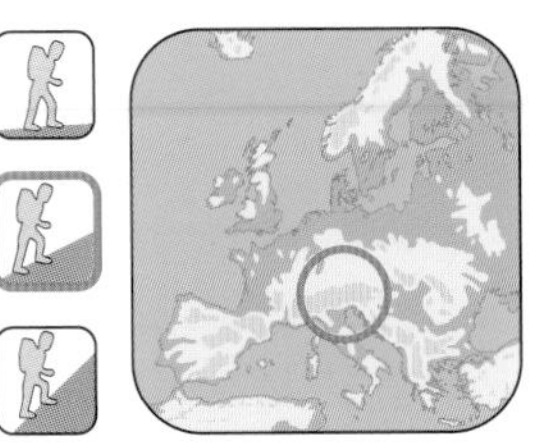

the STUBAI HORSESHOE

AUSTRIA

Richard Gilbert

THIS OUTSTANDING TREK CROSSES THE STUBAI AREA OF THE AUSTRIAN TYROL IN A WIDE HORSESHOE. IT IS SUITABLE FOR FAMILY VACATIONS, AND IS PARTICULARLY POPULAR WITH AUSTRIANS AND GERMANS. IT IS CUSTOMARY TO EXCHANGE THE GREETING "GRUSS GOTT" WITH FELLOW WALKERS WHOM YOU MEET ON THE PATH.

In summer the Stubai is breathtakingly beautiful, featuring a profusion of alpine flowers and glaciated peaks set against a blue sky. Lunch on a col can provide extensive views, stretching from the giants of the Swiss Alps to the rock towers of the Dolomites. The paths are clearly marked and easy to follow, and every night is spent in a comfortable mountain hut with meals provided. The longest day involves 6 hours' walking, thus the basic trek is undemanding, if not relaxing. Keen climbers, however, can include the ascent of several peaks, either on a hut to hut route or by spending an extra night at a hut and allowing a full day for the climb.

↓ itinerary

•DAY 1 5½ miles (9km)

Obertal to Bremer Hut

The post bus deposits you at the road end at Obertal. From here, continue along the jeep track up the valley to the restaurant at Laponis (2 miles/3km). Here you will see the terminus of a cable hoist, used to carry supplies up to the Bremer Hut. For a small fee they will take your rucksack up to the hut; the adjacent telephone enables you to contact the hut warden and tell him or her that your pack is on the hoist. It is well worth saving energy at this point in the trek to help the vitally important acclimatisation process.

↑ *Clouds clear from the Stubai mountains after a brief rain shower to reveal spectacular panoramic views from the Neue Regensberger Hut. The huts in the Stubai offer a warm welcome, comfortable beds and a varied menu of delicious meals.*

← *Walking hut to hut in the Stubai mountains is both relaxing and exhilarating. The paths are clearly marked by cairns and painted rocks, and route-finding is never difficult.*

The path winds up the steep hillside, firstly through pine woods and then across open, grassy alps and finally boulder fields. The Bremer is an old wooden hut perched on the Mitteregg spur at the foot of the east ridge of the Innere Wetterspitze at 7,917 feet (2,413m); it overlooks the Simming glacier.

•DAY 2 3 miles (5km)

Bremer Hut to Nurnberger Hut

The path, marked R102, leaves from behind the Bremer Hut and climbs steadily across grassy and then boulder-strewn slopes, speckled with snow gentians, to the Simmingjochl at 9,068 feet (2,764m). Fixed cables assist the final pull up to the col, where there is an old custom house.

Climbers may wish to leave their rucksacks beside the custom house and continue along the ridge to the Ostl Feuerstein (10,722 feet/3,268m). This climb, with a return to the packs, will take an extra 5 hours.

The far side of the Simmingjochl is very steep, and more security cables will be found. Eventually, having crossed patches of old snow, you will reach a ring of glacial debris with boulders strewn everywhere and you must pick your way down carefully. The Nurnberger hut can be seen across the cirque, but first you have to cross the foaming Langtal River by a plank bridge. The Nurnberger is a substantial four-storey, stone building, run very efficiently by the warden and a team of assistants. It is the base for the popular ascent of the Wilder Freiger (7 hours), a thoroughly recommended climb.

→

•DAY 3 **3 miles (5km)**

Nurnberger Hut to Sulzenau Hut

The path climbs quite steeply behind the hut for 1,148 feet (350m) to reach the Niederl col under the rocky peak of the Maier Spitz. Fixed cables are provided where necessary.

Beyond the Niederl, the path is extremely narrow as it zigzags down a series of ledges to reach two small lakes. Continuing down to the much larger turquoise lake, the Grunau See, you can enjoy spectacular views across the valley to the Wilder Freiger. Moraines and then grassy slopes lead down to the Sulzenau Hut, a modern, wedge-shaped design built in 1975 on the site of the old hut, which was demolished by a freak avalanche.

•DAY 4 **3 miles (5km)**

Sulzenau Hut to Dresdner Hut

The path goes up the valley, first over grassy alps and then along the lateral moraine on the north side of the Sulzenau glacier. It crosses a rocky spur with the help of iron stanchions and cables, and then climbs steeply to the Peil Joch, where there is an extraordinary collection of cairns.

Stop at the Peil Joch to marvel at the icefall and the gaping crevasses of the Sulzenau glacier, and at the huge glacier system that runs right up to the Zuckerhutl and the

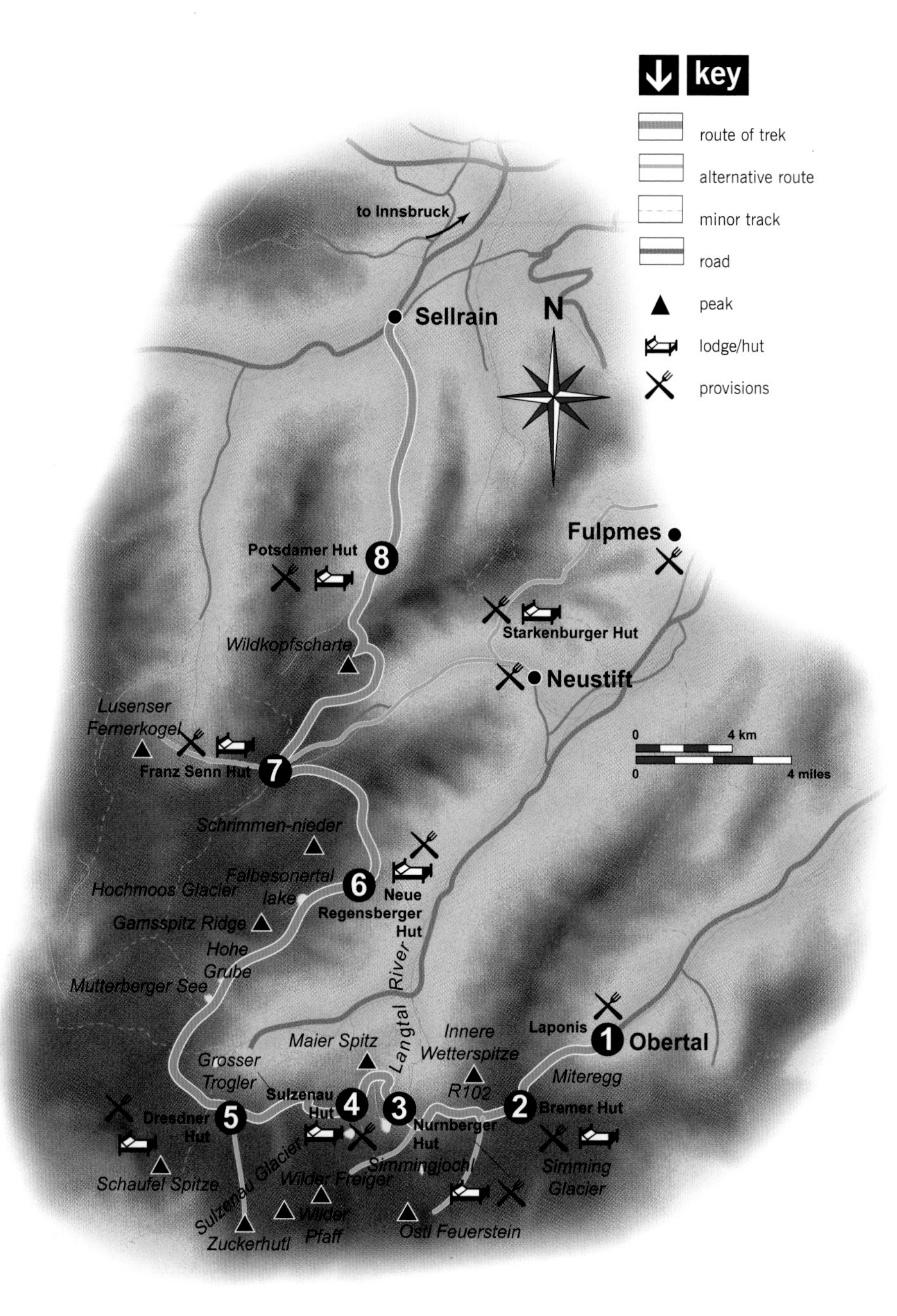

walk profile

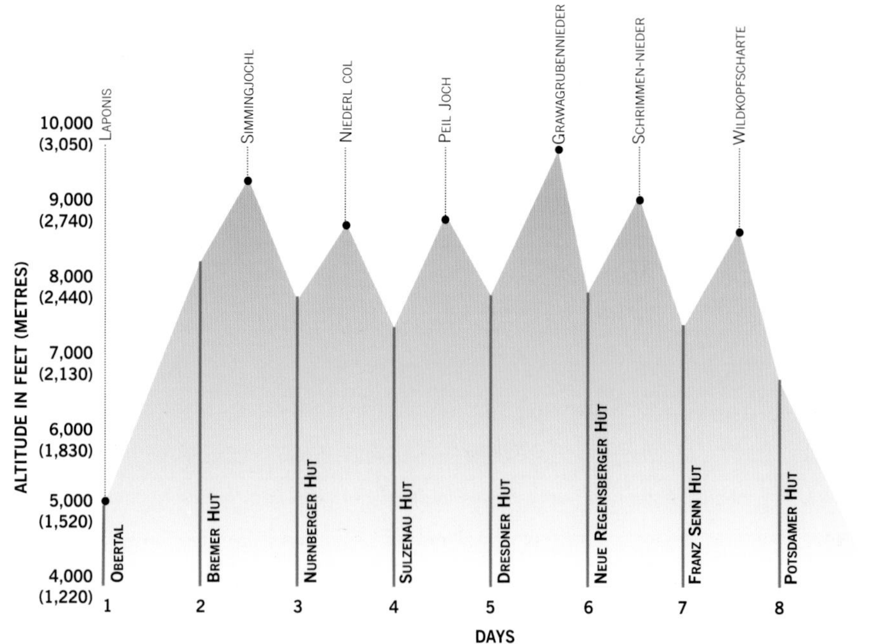

Wilder Pfaff. A rather loose descent down the rocks of the Grosser Trogler takes you to the Dresdner Hut, which is set amid pylons and cable-car stations. The hut is large and modern, and is frequented by summer skiers during the day; it provides wonderful views of the Schaufel Spitze at the head of the valley wall.

Climbers might like to spend an extra night at the Dresdner in order to climb the Zuckerhutl, at 11,500 feet/3,450m (7 hours), the highest peak in the Stubai Alps.

•DAY 5 **6¾ miles (11km)**

Dresdner Hut to Neue Regensberger Hut

This is the longest and most arduous day of the trek, and an early start is advised. Take the path westwards beside

As you cross the Niederl col between the Nurnberger and Sulzenau huts, the Wilder Frieger and its accompanying glaciers burst into view. Startling contrasts between glistening ice fields and green meadows characterise the Stubai Horseshoe trek.

the stream, and at a signpost turn off right along another path marked Mutterberger See. This path crosses the west ridge of the Egesengral and then descends to the lake of Mutterberger See.

A second lake is passed at Hohe Grube, and then the traverse path switchbacks over the Gamsspitz ridge and zigzags up to the Grawagrubennieder col at 9,449 feet (2,880m). Descending north, you reach the Hochmoos glacier, which usually has a well-beaten and cairned path down it in summer. After passing the Falbesonertal lake, the path wanders through grassy meadows, grazed by sheep with bells, and arrives at the Neue Regensberger Hut, perched on the lip of the valley.

•DAY 6 4 miles (6.5km)

Neue Regensberger Hut to Franz Senn Hut

This well-marked path (no. 133) traverses the grassy slopes in a north-easterly direction before turning north and climbing steeply to the Schrimmen-nieder, (8,878 feet/2,706m). After a rough and stony descent from the col, the path turns abruptly west and contours several ridges until, quite suddenly, the imposing Franz Senn Hut stands before you. It is built of wood and is steeped in tradition, as well as being most comfortable and welcoming.

Two worthwhile objectives from the Franz Senn Hut are the Ruderhofspitz, 11,398 feet/3,474m (8 hours), and the Lusenser Fernerkogel, 10,820 feet/3,298m (9 hours).

→

The Lusenser Fernerkogel provides extensive views over the Austrain Tyrol. An ascent of this peak (10,820 feet/3,298m) is possible from Franz Senn Hut.

•DAY 7 **8 miles (13km)**

Franz Senn Hut to Potsdamer Hut

This is a delightful day's walk. The highest peaks of the Stubai are now being left behind, and the lower altitudes mean that the flowers and their accompanying butterflies are the best of the entire trek. It is more a nature ramble than a tough mountain crossing. The path winds round the hillside, in and out of ravines, with streams and cascades everywhere. The highest point of the day is the crossing of the Wildkopfscharte at 8,481 feet (2,585m).

Approaching the Potsdamer hut, the path runs through fields that are thick with juniper and azalea. It makes a fantastic sight in spring. The hut itself is of an old, wooden chalet style, and is unlikely to be busy. It is warm and comfortable with a paneled *Gastube* (bar) and a huge, tiled stove.

VARIATION: 7½ miles (12km), Franz Senn Hut to Starkenburger Hut; 3¾ miles (6km), Starkenburger Hut to Neustift or Fulpmes. A worthwhile alternative finish to the trek is to take the traverse path from the Franz Senn Hut to the Starkenburger hut and then descend the next day to Neustift or Fulpmes. Both of these villages are most attractive and have excellent services and communications.

•DAY 8 **6¼ miles (10km)**

Potsdamer Hut to Sellrain

A jeep track leads gradually down through the sweet-smelling pines to the village of Sellrain, from where a regular bus service will speed you back to Innsbruck. The contrast to the harsh mountain environment is abrupt, with charming villages of chalets, each displaying window boxes bright with geraniums and piles of logs stacked under the eaves ready for the winter.

You will be leaner, fitter and browner, and keen to explore the historic old city of Innsbruck after your successful completion of the Stubai Horseshoe.

↓ factfile

OVERVIEW

The Stubai is one of the smallest and most varied of the mountain regions that make up the Austrian Tyrol. It is easily accessible from Innsbruck. For the mountain walker, the great advantage of the Stubai is the network of paths that link a series of huts providing overnight accommodation and excellent meals. Once you have gained height on the first day of your trek, you can maintain this height for the rest of the vacation. The paths contour steep hillsides and boulder slopes, and they cross cols where new peaks and glaciers burst into view, yet you are never far from a reassuring marker cairn or red-blazed boulder. Steep and rocky sections that might prove difficult are safeguarded by fixed cables. The route recommended here involves a minimum of 8 days, with accommodation at seven different huts. More ambitious walkers might like to extend the trek by spending 2 nights at certain huts and climbing a peak on the extra day. The basic trek is about 40 miles (65km).

alpine chough

Start: Obertal in the Gschnitztal.

Finish: Sellrain.

Difficulty & Altitude: A hut to hut trek through the Stubai makes an excellent introduction to the Alps, and will be enjoyed by any fit walker who thrives on mountain scenery. However, although the paths are well marked and the weather is usually fine, sudden storms of hail and snow cannot be ruled out, even in mid-summer. A waterproof anorak and overtrousers are essential, as well as a fleece jacket, gloves and balaclava. Depending on the year, you may have to cross a few snowfields and small glaciers, so you should carry an ice axe. If you intend climbing some mountains, you must carry rope, crampons and harnesses too. The highest col crossed on the trek is at 9,450 feet (2,880m), thus altitude should not be a problem, although it will be a few days

before you are properly acclimatised. For climbers, the highest peak in the Stubai is the Zuckerhutl at 11,500 feet (3,450m). It can be climbed in a long day from the Dresdner Hut. Mountain rescue can be coordinated from the huts, all of which are on the telephone. Make sure your insurance covers mountain rescue.

ACCESS

Airports: Innsbruck airport is fairly small, but Munich is well connected to Innsbruck by bus and train.

Transport: From Innsbruck, Bahnhof trains run frequently to Steinach, from where the post bus will take you on to Obertal. Alternatively, a bus from outside the Bahnhof in Innsbruck will take you directly to Obertal. For your return to Innsbruck, take the post bus from Sellrain.

Passport & Visas: Passports only are required.

Permits & Restrictions: No permits or restrictions. Walking is Austria's national sport, and all nationalities are welcome.

LOCAL INFORMATION

Maps: Compass Wanderkarte 1:50,000 Sheet 83 Stubaier Alpen; Alpenvereinstarte 1:25,000 Sheet 31/1 Hochstubai.

Guidebooks: *Stubai Alps*, Eric Roberts (West Col); *Hut to Hut in the Stubai Alps*, Allan Hartley (Cicerone); *Walking Austria's Alps Hut to Hut*, Jonathan Hurdle (Cordee).

Background Reading: *Over Tyrolese Hills*, F.S. Smythe (Hodder).

Accommodation & Supplies: Throughout the trek, accommodation is in mountain huts. You can either bring sheet sleeping bags and use cheap, communal Matratzenlager accommodation with blankets marked *Fusse Ende* (foot end) and *Kopf Ende* (head end), or pay a little more for a small room of two or four bunks with duvets. The hut menus are varied and delicious, and you can purchase bread, cheese, salami and chocolate for lunch. Pre-booking of huts is not usually necessary and no one is ever turned away.

Currency & Language: Austrian schilling. German is spoken; most hut wardens also speak adequate English.

Photography: No restrictions.

Area Information: Full information about huts, costs, transport and telephone numbers in the Stubai can be obtained from the Austrian Alpine Club (A.A.C.), 2 Church Street, Welwyn Garden City, Hertfordshire AL8 6PQ, UK (tel.: 00 44 1707 324 835). Membership of the A.A.C. gives you reduced prices for overnight accommodation in the huts and preferential booking.

TIMING AND SEASONALITY

Best Months to Visit: June to September.

Climate: By June the paths should be free of winter snow and the alpine flowers will be at their best. Mornings can start with a slight frost, but when the sun gets up it is usually hot. For this reason prompt starts are advisable and you should plan to arrive at the next hut soon after lunch. Blizzards and snow storms, however, are not unknown at any time of the year.

HEALTH & SAFETY

Vaccinations: None required.

General Health Risks: No special precautions.

Special Considerations: The midday sun, particularly when reflected off a snowfield, can severely burn the skin. Make sure you apply plenty of sunscreen.

Politics & Religion: You will find tolerance to all religious and political beliefs.

Crime Risk: Low in the mountains, although valuables should not be left unattended in the huts.

Food & Drink: Beer, wine and soft drinks are available in the huts. Coffee and tea are quite expensive, but it is usual to bring your own coffee powder and tea bags and buy hot water known as "tee-wasser". For evening meals some huts have a set time, others operate a self-service system. Breakfast is served on demand between 5:00 and 9:00 A.M. Water in high mountain streams is usually safe to drink, but be careful of silt-laden glacier torrents, which can cause stomach problems. With the altitude and hot, dry air, it is important to drink a lot in the mountains.

HIGHLIGHTS

Local Flavour: Having arrived at the next hut by early afternoon, it is bliss to while away the rest of the day on the balcony in the sun, with a litre of beer or a carafe of wine at your elbow. The cares of the world slip from your shoulders. Later, after supper, spontaneous singing, accompanied by an accordion, often breaks out in the warm *Gastube* (bar). Evening mists gather in the valleys far below and you can watch the last rays of the sun lighting up the hanging glaciers of the higher peaks.

Wildlife & Flora: Below the snowfields and screes, grassy alps are still scythed by hand and the hay is stooked or hung out to dry. The flowers are wonderful: gentians, primula, mountain saxifrage, campanula, cinquefoil, lousewort, asters and mountain avens. Wildlife can include floppy-eared sheep, marmots that screech out warnings, chamois and choughs.

Walkers can spend an extra night at Dresdner Hut and make a straightforward ascent of the Schaufel Spitze, seen here rising behind the hut.

temperature and precipitation

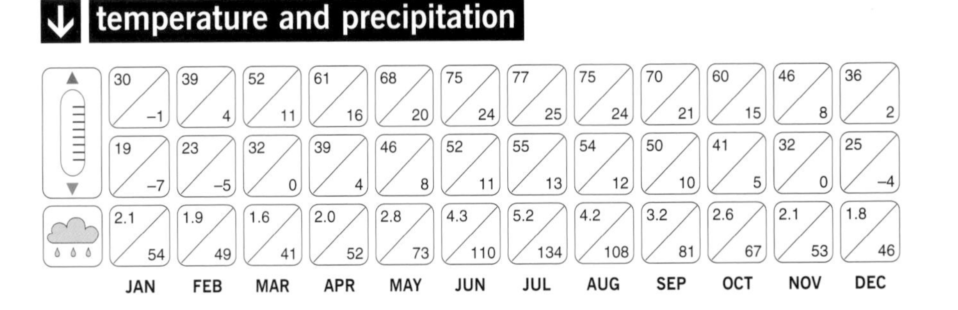

	JAN	FEB	MAR	APR	MAY	JUN	JUL	AUG	SEP	OCT	NOV	DEC
Max temp (°F / °C)	30 / –1	39 / 4	52 / 11	61 / 16	68 / 20	75 / 24	77 / 25	75 / 24	70 / 21	60 / 15	46 / 8	36 / 2
Min temp (°F / °C)	19 / –7	23 / –5	32 / 0	39 / 4	46 / 8	52 / 11	55 / 13	54 / 12	50 / 10	41 / 5	32 / 0	25 / –4
Precipitation (in / mm)	2.1 / 54	1.9 / 49	1.6 / 41	2.0 / 52	2.8 / 73	4.3 / 110	5.2 / 134	4.2 / 108	3.2 / 81	2.6 / 67	2.1 / 53	1.8 / 46

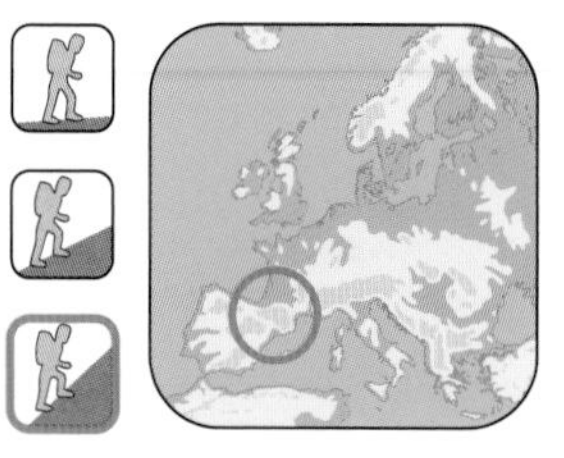

the PYRENEAN HIGH ROUTE

FRANCE/SPAIN

Kev Reynolds

THE ROUTE FOLLOWED BY THIS TREK FORMS ONE OF THE MOST SCENIC SECTIONS OF THE CLASSIC HAUTE RANDONNÉE PYRÉNÉENNE (THE PYRENEAN HIGH ROUTE), A 45-DAY TREK LEADING FROM THE ATLANTIC COAST TO THE MEDITERRANEAN.

As its name suggests, the route remains as high as possible; hugging the international frontier and crossing from one side to the other, offering the walker a constantly changing experience. Although it is mostly on good paths, this is a demanding mountain trek; particular attention must be paid to the weather, and its potential seriousness should not be underestimated.

The two sides of the mountain range are very different. The French slopes are lush and green with peaks rising steeply from the plains, whereas the Spanish side tends to offer stark terrain, the mountains bleached and bare, and with one sierra after another rolling away to blue horizons. The route begins and ends in limestone country, but with some savage granite mountains in between. Little habitation is seen on this route; there is a tremendous sense of isolation, especially south of the frontier, and the scenery is of unchallenged grandeur.

•DAY 1 5½ miles (9km)

Lescun to Cabanes d'Ansabère

On the assumption that you will arrive in Lescun after a tiring journey, this first stage is kept fairly short. The village of Lescun gazes south across a neat pastoral landscape to the hint of jagged aiguilles bristling against the frontier with Spain. The route, waymarked with red and white stripes, leads between undulating meadows and along a forestry track to a bridge over the Gave d'Ansabère (*gave* is the local name for a mountain stream, or river). The way continues in and out of woods, steadily gaining height to arrive in a mountain bowl at the foot of the Cirque de Lescun. Directly ahead, across this bowl through which streams meander, soar the finger-like Aiguilles d'Ansabère, pale limestone peaks bursting from a chaos of scree – a scene guaranteed to stop you in your tracks.

↑

Lescun, a pastoral village on the edge of the Basque country, marks the start of the Pyrenean High Route. Across the meadows are the mountains that form the frontier between France and Spain.

•DAY 2 **8½ miles (13.5km)**

Cabanes d'Ansabère to Refuge d'Arlet

On a hot day this can be a tough stage, for there's hardly any shade and only one place to refill water bottles (at Ibon de Acherito). The path climbs to a small tarn, before mounting the upper slopes to a grassy saddle on the frontier ridge at about 6,650 feet (2,027m). Spain appears as a series of gently rolling hills that fade into a haze of blue as you descend a pathless slope to the tarn of Ibon de Acherito. The route then makes a traverse before climbing through a shallow combe to open hillsides seething with insects. French slopes are regained by way of Col de Pau or Col de Burcq; good paths follow until you cross Col de Saoubathou and the way cuts across high pastures to reach Refuge d'Arlet, which overlooks another tarn.

•DAY 3 **8 miles (13km)**

Refuge d'Arlet to Col du Somport

Between the Arlet refuge and Col du Somport, the High Route remains within French territory. At first there are strange outcrops of pudding stone beside the trail, and one or two shepherds' cabanes among the high moors and mountains. The mournful bleating of sheep drifts on a still summer's day, and lammergeier send cross-like shadows rippling over the pastures. The trail reaches a ridge spur, drops to a stream, then rises steeply to cross Col Plâtrière, after which there's a long descent into and through forest on the way to the Vallée d'Aspe, north of Col du Somport. Just below the col there's a *gîte d'étape* (a walkers' hostel).

•DAY 4 **10 miles (16km)**

Col du Somport to Refuge d'Ayous

A long, slanting climb takes the route up the east flank of the Vallée d'Aspe, passing more shepherds' cabanes and the unmanned Refuge de Larry, before turning abruptly to the east, where a series of zigzags climbs to Col d'Ayous. Emerging at this craggy pass, you have a dramatic view of Pic du Midi d'Ossau, the very symbol of the Pyrenees. The lake at the Ayous refuge below offers a perfect mirror-image of this sculpted mountain, and the beauty of the sunrise from the hut defies description.

•DAY 5 **5 miles (8km)**

Refuge d'Ayous to Refuge de Pombie

This is only a short stage, but it's a strenuous one. However, with Pic du Midi in view almost every step of the way, there's no shortage of visual drama as each facet of the mountain is different. There's also a series of mountain tarns and a diamond-studded tiara of peaks, which add light, colour, and reflected glory, helping to make this one of the most memorable of all stages of the trek. From Col de Peyreget you gaze into Pic du Midi's inner sanctum, a wilderness of bronze crags and black-shadowed gullies rising to summits that appear like castle battlements. Then plunge steeply down to the hut from where you look across the little Lac de Pombie to the huge South Face. For those with sufficient time and energy, the ascent of the 9,465-foot (2,885-m) Pic du Midi is worth tackling for the immense summit panorama; the standard route is a straightforward scramble of 3 to 3½ hours – but beware of falling stones.

→

From near the head of the Ordesa Canyon, trekkers on the Faja de Pelay have a great view of Monte Perdido, the third-highest mountain in the Pyrenees. From this angle the mountain's great glacial tiers and dramatic face are hidden, but its mere size means it cannot fail to impress.

• DAY 6 **4½ miles (7.5km)**

Refuge de Pombie to Refuge d'Arrémoulit

East of Pic du Midi, across the depths of a valley draining north from Col du Pourtalet, the big granite Balaitous massif boasts the first of the 9,800-foot (3,000-m) summits. You must descend to the bed of the valley, and then make a steady ascent on the eastern side through a narrow cleave, which leads to an exciting and airy traverse across the face of a rocky peak. The so-called Passage d'Orteig is a narrow man-made ledge high above a lake, but for trekkers fearing an attack of vertigo there is an easier alternative. Both paths arrive at Refuge d'Arrémoulit.

• DAY 7 **3 miles (5km)**

Refuge d'Arrémoulit to Refuge de Larribet

Don't be fooled by the modest distance of this stage. It's a tough route crossing two high passes, rough and rocky underfoot, and should be avoided altogether if there's any threat of storm. The route leads from the hut door over a sparkling granite boulderscape and up screes in a little hanging valley to gain Col du Palas, which forms the Franco–Spanish border. Across the col a traverse is made of scree and rocks at the head of a horseshoe to locate another col in the opposite wall. Port du Lavedan is often part-choked with snow and demands care descending the north-eastern side. Paint flashes direct the route among boulders, scree blocks, and around black tarns in a haunted landscape, before coming to the Larribet hut north of Balaitous.

• DAY 8 **9 miles (14.5km)**

Refuge de Larribet to Refuge Wallon

There's a world of difference between the raw granite of the Balaitous and the seductive pastures of the Vallée du Marcadau, and this stage makes the transition from one to the other in crossing Col de Cambales. A descent is made northwards from Larribet, then the route arcs round the footstool of the Balaitous to climb towards Port de la Peyre-St-Martin, a pass that offers a way back into Spain. Shortly before this pass, however, the High Route breaks off to the east, enters a savage region of rocks and precipices, and eventually climbs to the Col de Cambales, gateway to the Marcadau. To the south rises Pic de Cambales, a fabulous viewpoint on the frontier crest, but our way ignores this and weaves a journey between dozens of small tarns and down to a pastureland basin and the Wallon refuge.

• DAY 9 **7 miles (11.5km)**

Refuge Wallon to Refuge de Bayssellance

A direct view through the Arratille glen shows the fish-tail summit of the Vignemale above converging ridge systems. The route to the Bayssellance hut rises through this glen, crosses Col de Arratille, then skirts the headwaters of the

The Marcadau Valley forms an oasis of tranquility between the wild massifs of the Balaitous to the west and Vignemale to the east.

walk profile

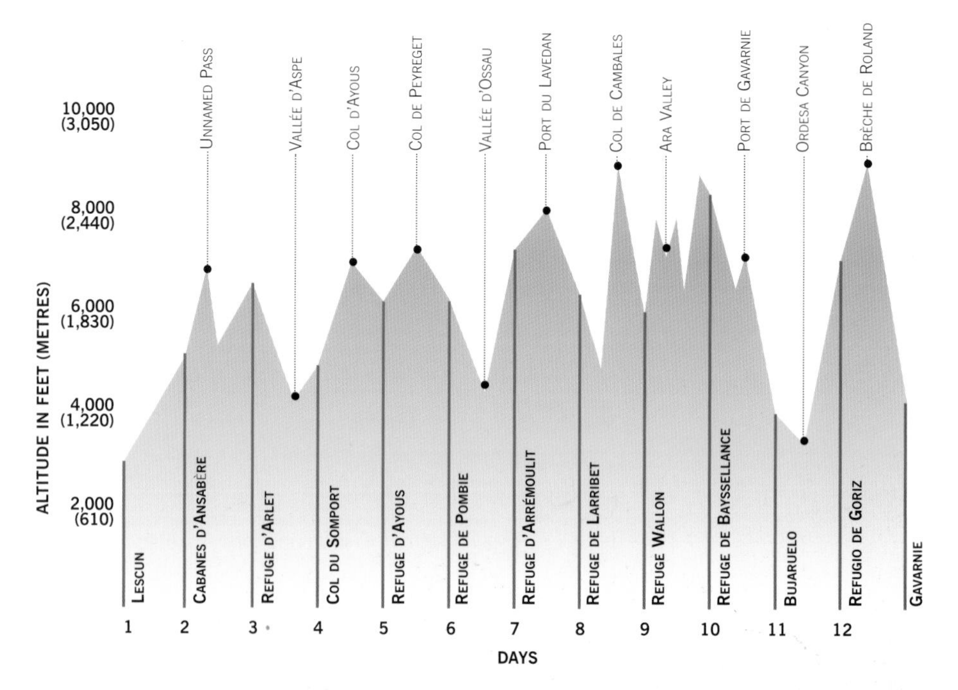

Spanish Ara Valley, and returns to France by way of Col des Mulets. A descent is then made in full view of the tremendous North Face of the Vignemale, the finest outburst of rock in all the Pyrenees. By ignoring this and descending the length of the Ara Valley instead, you could shorten the trek by a day. But it's better to enjoy the Vignemale, crossing the little flood plain at its feet and climbing up and over Hourquette d'Ossoue on the eastern side. Refuge de Bayssellance lies just below the pass, a beehive of a building and the highest manned hut on this trek.

•DAY 10 11 miles (17.5km)

Refuge de Bayssellance to Bujaruelo

Descend the trail from the Bayssellance hut and ahead of you is the famed Cirque de Gavarnie. The path leads all the way to Gavarnie, where the trek is due to end, but our route will divert to the south and make a loop through the Ordesa Canyon. Shortly after leaving Bayssellance, the way passes two caves blasted in the mountain below the Ossoue Glacier by a late nineteenth-century eccentric. Then, at a small dammed lake, the High Route breaks away from the main →

→ *Looking west from the summit of Pic d'Ossau, the Lacs d'Ayous are clearly visible.*

track and climbs to a spur leading to the Port de Gavarnie and an ancient descent route on the Spanish flank. At the base of this you come to Sant Nicolas de Bujaruelo, an old hospice turned private refuge.

•DAY 11 **11 miles (18.5km)**

Bujaruelo to Refugio de Goriz

The spiritual heart of the Parque Nacional de Ordesa is the Ordesa Canyon, one of the awe-inspiring sites of the Pyrenees. Huge, multi-coloured cliffs erupt from deciduous forests, facing one another across a chasm through which waterfalls cascade in rainbows of spray. At its head stands Monte Perdido, the "lost mountain" and the third-highest in the Pyrenees, on whose lap sits the Goriz hut. One can either trek to it through the canyon beside the waterfalls, or climb a monstrously steep trail up the south wall to gain a high balcony path, the Faja de Pelay – a tough option on a long day's walk. The two trails unite at the head of the canyon, and from there wind up to the hut.

•DAY 12 **8½ miles (13.5km)**

Refugio de Goriz to Gavarnie

Whereas Ordesa's canyon is rich in vegetation, the landscape above Refugio de Goriz becomes increasingly barren. This is thirsty limestone country, and as it approaches the famed Brèche de Roland the trail works its way into a scene of utter desolation, a graveyard of dead mountains. Yet the moment you enter the huge cleft of the Brèche and look to the north, so a new, fresh and vibrant world is revealed. You are standing on the threshold of France, in the rim of the Cirque de Gavarnie itself. A glacial napkin is draped across the face of the cirque just below; long waterfall ribbons pour silently into unseen depths, and distant foothills are green and enticing. The descent to Gavarnie is made via a brief, crevasse-free glacier to the Refuge de la Brèche, then the way swings to the right and continues down a very steep stairway of rock to the base of the cirque. Here, there is a hotel, with cool drinks and a view to gasp at. Finally the trek ends along a track busy with a procession of ponies carrying tourists who may barely dream of the scenes through which you have trekked over the past 12 days.

↓ factfile

OVERVIEW

A west to east traverse of a section of the High Pyrenees, crossing and recrossing the borders of France and Spain, and visiting two national parks on the way. The route is some 91 miles (146km) long and takes 12 days to complete.

Start: Lescun, on the western flank of the Vallée d'Aspe.

Finish: Gavarnie, at the headwaters of the Gave de Pau.

Difficulty & Altitude: A demanding trek mostly on good mountain paths, although some sections are virtually trail-free. In poor conditions the route takes on an additional edge of seriousness; basic hill-walking skills are essential. There are several passes in excess of 8,200 feet (2,500m) to cross, with the maximum altitude reached being 9,209 feet (2,807m) at the Brèche de Roland. Mountain rescue on the French side of the frontier is well-organised and efficient; it is less so in Spain. The service is not free, and trekkers are advised to have adequate mountain rescue insurance.

ACCESS

Airports: The nearest international airport is Tarbes–Lourdes (30 miles/48.5km to the north), with onward rail and bus routes into the mountains.

Transport: To get to the start of the trek take a train to Oloron-Sainte-Marie and bus from there to Lescun-Cette-Eygun (the Canfranc bus), followed by a 3½-mile (6-km) walk to Lescun. Gavarnie, the journey's end, has a bus service to Lourdes, where there are mainline trains, and a connecting service to Tarbes–Lourdes airport.

Passport & Visas: Passports are needed for France and Spain. Citizens of Australia and New Zealand require tourist visas for both France and Spain; UK and US visitors do not require visas.

Permits & Access: No permits needed. There is free access to all areas, but note that wild camping is not allowed within 1 hour's walking distance of a road.

LOCAL INFORMATION

Maps: The French TOP 25 series of maps at a scale of 1:25,000 give good coverage, with sufficient overlap on the Spanish side of the mountains for most of the route to be shown: sheets 1547 OT, 1647 OT and 1748 OT. For the Ordesa section of the

griffon vulture

trek, Editorial Alpina Sheet Ordesa at 1:40,000 is needed.
Guidebooks: *Pyrenees High Level Route*, Georges Véron (Gastons/West Col); *Walks & Climbs in the Pyrenees*, Kev Reynolds (Cicerone Press).
Accommodation & Supplies: Mountain huts (refuge or refugio), or *gîtes d'étape* occur throughout the trek. Camping is possible, but note the restrictions mentioned above. The only official campsites on this route are in Spain's Ara Valley. After leaving Lescun there are no opportunities for restocking with supplies.
Currency & Language: Franc (France), peseta (Spain). French and Spanish.
Photography: No restrictions.
Area Information: Randonnées Pyrénéennes, 742-29 rue Marcel Lamarque, 65007 Tarbes Cedex, France (tel.: 00 33 5 62 93 66 03); Parc National des Pyrénées, 59 rue de Pau, 65000 Tarbes, France (tel.: 00 33 5 62 93 30 60); Parque Nacional de Ordesa, Huesca, Spain (tel.: 00 34 974 22 04 62).

TIMING & SEASONALITY

Best Months to Visit: Late June to September inclusive. Any earlier, and the route may be too dangerous because of avalanche. After September most huts will be unmanned. October would be possible if you are prepared to camp and carry sufficient food for the complete trek.
Climate: The summer months of June to September in general enjoy reasonably settled weather conditions, although, because of the alignment of the Pyrenean range, the northern slopes are affected by moist airstreams from the Atlantic. Expect some rain and occasional electrical storms. Daytime temperatures can hover around 85°F (30°C) on the Spanish slopes, but by late September a light ground frost overnight is not uncommon.

HEALTH & SAFETY

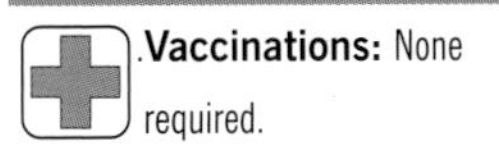

Vaccinations: None required.
General Health Risks: No special precautions necessary. Citizens of EC countries are entitled to medical attention on the same terms as French and Spanish citizens if they carry form E111. Non-EC trekkers should have adequate travel insurance.
Special Considerations: As with all mountainous regions, beware stonefall, slippery and exposed paths, old snowfields, patches of ice, stream crossings, etc. Also beware exposed ridges when a storm approaches. The venomous asp viper is sometimes encountered on the Spanish slopes; if you get bitten by one seek urgent medical attention. In summer flies can be a nuisance, especially the large horsefly.
Politics & Religion: Both France and Spain are politically stable, and are strongholds of the Roman Catholic faith.
Crime Risks: Low.
Food & Drink: Food on offer in mountain huts has a high nutritional value, but the choice of menu is restricted. Streamwater in the mountains should be safe for drinking, except below habitation, grazing animals or other obvious pollution sources.

HIGHLIGHTS

Scenic: Below the rocky peaks and ridges, the Pyrenees offer pastoral landscapes of great beauty and diversity. Tiny lakes lie trapped in the most idyllic places; trout fill mountain streams; shepherds graze their flocks in remote valleys and spend their summers in simple stone hovels, or cabanes; and you may meet the occasional muleteer leading his animal laden with supplies toward a distant hut.
Wildlife & Flora: Chamois (known here as the izard) and marmots are the most commonly seen mammals in the Pyrenees, and red squirrels and red or roe deer may sometimes be spotted in the forest. The marmot, whose distinctive, shrill whistle often signals his presence, inhabits most of the valleys visited on this trek. As for birdlife, the lammergeier, or bearded vulture, with its 10-foot (3-m) wingspan, may be seen soaring overhead, along with griffon vultures, golden eagles and the acrobatic kite. In this "Flower Garden of Europe" – there are 142 endemic species in the Pyrenees – the mountain flora will be a highlight of any visit, but particularly in June, when the hillsides are carpeted with gentians, iris, narcissi and a variety of orchids.

pheasant's eye narcissus

The Grande Cascade, said to be the longest waterfall in Europe, sprays forth from the Cirque de Gavarnie.

temperature and precipitation

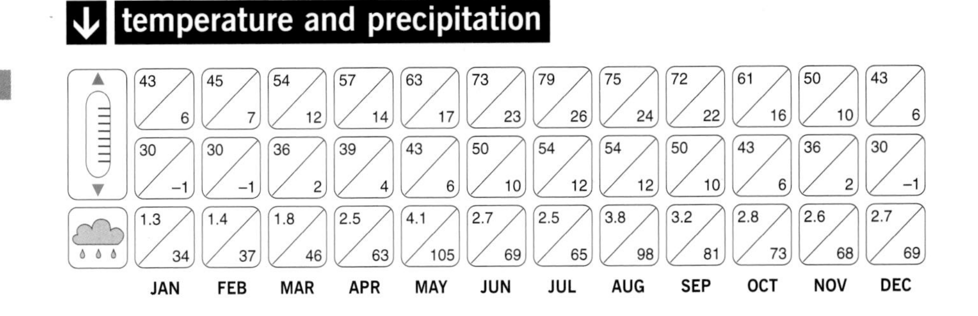

	JAN	FEB	MAR	APR	MAY	JUN	JUL	AUG	SEP	OCT	NOV	DEC
Max temp (°F / °C)	43 / 6	45 / 7	54 / 12	57 / 14	63 / 17	73 / 23	79 / 26	75 / 24	72 / 22	61 / 16	50 / 10	43 / 6
Min temp (°F / °C)	30 / −1	30 / −1	36 / 2	39 / 4	43 / 6	50 / 10	54 / 12	54 / 12	50 / 10	43 / 6	36 / 2	30 / −1
Precipitation (in / mm)	1.3 / 34	1.4 / 37	1.8 / 46	2.5 / 63	4.1 / 105	2.7 / 69	2.5 / 65	3.8 / 98	3.2 / 81	2.8 / 73	2.6 / 68	2.7 / 69

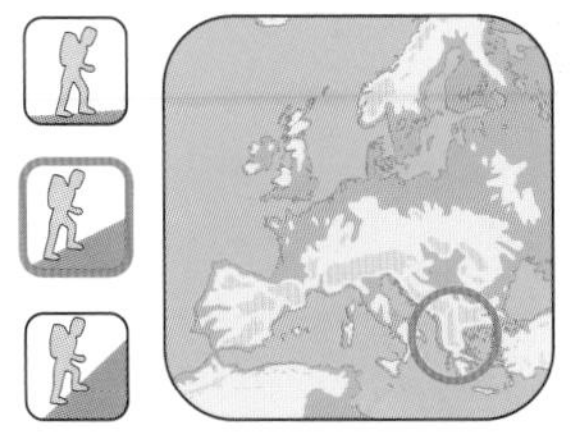

the PINDOS HORSESHOE

NORTHERN GREECE

Judy Armstrong

THE PINDOS MOUNTAINS FORM THE CENTRAL SPINE OF GREECE, STRETCHING FROM MOUNT GRAMMOS, ON THE ALBANIAN BORDER SOUTH TO THE GULF OF CORINTH, AND ARE DOTTED WITH DENSE WOODLAND, STONE VILLAGES, DEEP GORGES AND JAGGED LIMESTONE TOWERS.

Within the vast tracts of the Pindos – the best-known area for trekking is in the north – overnight accommodation has been provided for trekkers and trails are waymarked; it is also home to the Vikos National Park.

This six-day trek includes a superb walk down the deepest gorge in the world, and a foray through limestone badlands on an alpine plateau that is snow-covered for at least four months of the year. Much of the enjoyment is in the detail: the rainbow of wildflowers in May and June, the clouds of butterflies, the ponderous tortoises and the yellow-bellied frogs basking in springs. Much, too, can be made of the incomparable scenery: towering orange and grey limestone cliffs, gaping caves, lakes and rivers so pure that instinct drives you to dip your head and drink directly from the flow.

Daily distances are short, so time can be spent absorbing the views, dozing in the shade, photographing details. This also leaves scope for excursions, best tackled once the rucksack has been dumped at the pension, thirst has been slaked, and fat, marinated olives consumed.

itinerary

• DAY 1 7½ miles (12km)

Vitsa/Monodendri to Vikos

You will find it hard to leave the stone village of Vitsa, not just because of its tranquility, but because no one can tell you the best way to the adjacent village of Monodendri without slogging along the tarmac road. It's a great introduction to the North Pindos: the first lesson is that getting out of the villages is the key to the day's trekking. In this case just follow your nose uphill and, about half an hour later, Monodendri will appear as a stone huddle on the horizon. Don't leave this much-photographed village without visiting the uninhabited monastery of Ayias Paraskevis perched on the edge of the Vikos Gorge.

The descent to the gorge passes a Guinness Book of Records sign that gives vital statistics for this "Deepest Gorge in the World". The figures are the wrong way around: for the record, the gorge's maximum dimensions are 5,410 feet (1,650m) deep and 3,215 feet (980m) wide.

The track down to and along the gorge has traditionally been treacherous; in 1998 serious work was carried out to improve it, and it is now virtually impossible to get lost. It dives from sun-scorched limestone walls to cool woodland, to the riverbed littered with car-sized boulders. All around, cliffs soar skywards and the view changes constantly as the gorge sweeps through bends.

The climb to Vikos village (called Vitsiko on some maps) follows an ancient *kalderimi*, or cobbled track, which switches up the gorge side. The village squats on a breezy saddle, with panoramic views in all directions.

• DAY 2 3 miles (5km)

Vikos to Mikro Papingo

A short day, but with plenty of climbing and afternoon ambles from Papingo. The *kalderimi* twists downhill, past a bone-white chapel on a blue riverside bend, to a gorge. You're aiming for the Voidhomatis Spring – a torrent of icy water that has spent a week filtering 4,250 feet (1,300m)

↑

From a vantage point on the hillside above Voidhomatis Spring, the bottom of the Vikos Gorge is just a tumble of red rocks, white boulders and surging blue water. From here the walk heads right, towards the limestone towers and gaping caves of Mikro Papingo.

←

The footpath through the Vikos Gorge follows the waymarked national 03 trail. Woodland tracks, bare river boulders, and grassy riverbanks make for pleasant walking through the deepest gorge in the world.

down limestone cliffs. This is often smothered in butterflies and is an idyllic place to linger.

The climb to Mikro Papingo is steady, with far-reaching views, scree slopes, sun-dappled woodland, and glimpses of caves. Mikro Papingo and its sister, Megalo Papingo, are protected traditional communities. Wandering down the narrow, cobbled lanes, dodging goats and greeting women clad in black, is like stepping back in time. From Mikro Papingo walk to Megalo via the old Zagorian arched bridge, an architectural delight in a limestone gorge. From here the Pirgi, or Papingo Towers, are neck-craningly impressive, looming over the villages like benign giants. Follow this gorge upstream to find a series of natural swimming holes – caramel-coloured basins scooped out of the limestone.

•DAY 3 **5½ miles (9km)**

Mikro Papingo to Astraka Refuge

The climb from Mikro Papingo, up a sandy mule track, takes you above the treeline, with rural Greece spread like a carpet at your feet. The Astraka cliffs dominate the view upwards: gray, serrated hulks striped with snow gullies well into June. Astraka Refuge is a solid, white building on a windy col at 6,397 feet (1,950m). In the summer of 1999 an additional hut was being built, designed to accommodate winter climbers and skiers, and summer trekker overflow. The main hut is basic, but the location is magnificent and ripe for exploration.

Greece seethes with myths and legend, and there's one on the doorstep of Astraka Refuge. From the terrace a path trickles down a boulder bank, skirts a snowmelt pond and summer shepherds' huts, crosses an alpine meadow, and climbs to Dhrakolimni, or Dragon Lake. It is bottomless, the whisper goes, and its guardian dragon lives in the depths.

•DAY 4 **6½ miles (10km)**

Astraka Refuge to Gamila 1 and Astraka, and back

This is a day for peak-bagging: Gamila 1 at 8,192 feet (2,497m) and, if you have the energy, Astraka at 7,992 feet (2,436m). There are several approaches to Gamila, the shortest following a slim trail across a limestone moonscape, at the foot of the Astraka cliffs. A lone tree is the signal to turn left, into a shallow valley that leads up to Gamila's ankles. A steady hike up the broad flank brings you to the flat, rocky summit. The point is not the walk, but the views across the valley of the River Aóos to Mount Smolikas, the second-highest mountain in Greece at 8,651 feet (2,637m).

The climb up Astraka is more hands-on, backtracking from the refuge toward Papingo, then weaving left round the cliff edges to follow barrel-like cairns to the summit. A panorama shows mountains bleeding into valleys, into mountains, into valleys. Both the scale and the wilderness are huge and, up on Astraka, it's easy to see why more than three-quarters of Greece is classified as montane, non-arable, or otherwise uninhabitable.

→

•DAY 5 **7½ miles (12km)**

Astraka Refuge to Tsepelovo

The route dives off the col towards Gamila, then rolls up and down grassy hillocks, through jagged limestone pavements, past torrent-carved streams, over beds of snow-damp crocus. Listening out for hysterical canine welcoming committees, the trekker descends past little lakes, over alpine meadows, and down limestone rock bands to peer into the upper reaches of the Megas Lakkos Gorge.

The path drops into the infant gorge and meanders up the left bank, under steep cliffs and past banks of wildflowers. Well waymarked, the path climbs out of the gorge when the walls narrow in preparation for joining the mighty Vikos. It wanders over lumpy rises, rocky cols, through streams, and finally down a grassy meadow to the village of Tsepelovo. This is a stone enclave on the edge of the mountains, snowbound in winter and a cool haven in summer. Men gather in the main square to posture, rattle worry beads, and play backgammon. The women stay indoors.

•DAY 6 **6¼ miles (10km)**

Tsepelovo to Vradheto, and back

A decaying *kalderimi* – the old route to Vradheto, a village that until recently was accessible only by foot – leads out of the village and up the right side of a deep-slit ravine.

The climb is steep, through limestone bands and across grassy bowls, until the new road is reached. Soon, however, a track filters off the tarmac to traverse blonde-tussock hillsides. It leads eventually to the Beloi viewpoint over the pancake rock stacks of the Vikos Gorge. With Monodendri behind your left shoulder, the view is along the length of the gorge, with Vikos village just visible as a clutch of grey buildings. The scale is terrific.

Vradheto itself, just half an hour away, is a very small village with an active taverna. Its claim to fame is the *kalderimi*, or Vradheto Steps, which dives through a rock palisade in an intricate series of twists and turns, to a Zagorian bridge and the modern road.

Modern…it's all relative, isn't it? The Pindos Mountains mixes modern with ancient in such a way that it becomes timeless – a paradise found.

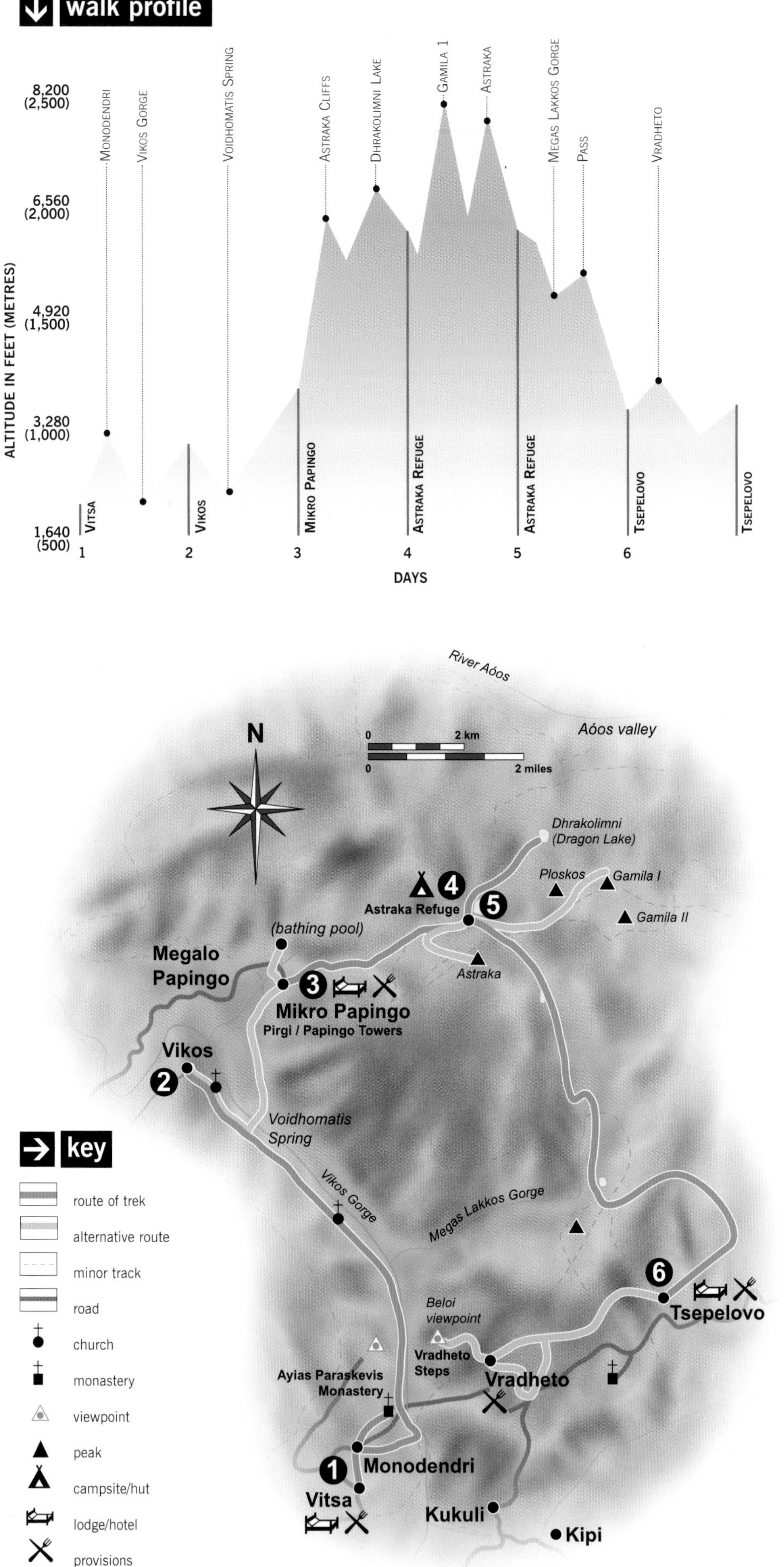

factfile

OVERVIEW

An almost circular walk through a picturesque and remote area of the Pindos Mountains in Northern Greece. This route takes in the deepest gorge in the world, villages with protected traditional community status, and high alpine walking that includes an optional ascent of Gamila 1 at 8,192 feet (2,497m). The route is approximately 36 miles (58km) long and takes a leisurely 6 days.

Start: Vitsa or Monodendri (adjacent villages), North Pindos.

Finish: Tsepelovo, North Pindos.

Difficulty & Altitude: Paths are generally well marked, with red paint blobs on rocks and trees marking the 03 national trail for most of the route, and blue and yellow markings marking the final leg. Some difficulties may be encountered on the alpine section around Gamila in May and early June. There are considerable ascents and descents, and the highest altitude on the route is Gamila 1, at 8,192 feet (2,497m); Astraka Col, where the refuge is located, is at 6,397 feet (1,950m).

ACCESS

Airports: International flights to Preveza airport (2½ hours from Vitsa). Alternatively, fly to Corfu and transfer to Igoumenitsa by ferry. There is a local airport at Ioannina.

Transport: Ioannina, the area capital, is well serviced by buses, with regular services onwards to the Zagoria villages.

Passport & Visas: A valid passport; visas are not required for stays of up to 3 months.

Permits & Access: None.

LOCAL INFORMATION

Maps: Map availability is a problem, although the trek described follows waymarked trails. Military maps are not available either. Your best bet is to purchase maps from the Hellenic Alpine Club (from the Korfes magazines), showing contour intervals at 200m (656 feet).

Guidebooks: *Trekking in Greece* (Lonely Planet) – an excellent companion, but out of print; *Greece on Foot*, Marc Dubin (Cordee) – covers this route but some track descriptions will be out of date; *Mountains of Greece*, Tim Salmon (Cicerone) – gives an overview of the Pindos, focusing on a month-long trek from Delphi to the Albanian border.

fig tree

Background Reading: *Greek Phrasebook* (Lonely Planet); *The Olive Grove: Travels in Greece*, Katherine Kizilos (Lonely Planet); *Roumeli: Travels in Northern Greece*, Patrick Leigh Fermor (Penguin).

Accommodation & Supplies: Accommodation is available in small, family-run hotels and guesthouses, with 2 nights in a mountain hut on Astraka Col. Morning and evening meals are available, but day food can be difficult to find. Wild camping is not really an option – it is illegal and impractical.

Currency & Language: Drachma. Greek (very few people in this area speak English).

Photography: No restrictions.

Area Information: Australia: 51 Pitt Street, Sydney, NSW 2000 (tel.: 00 61 2 241 1663). UK: National Tourist Office of Greece, 4 Conduit Street, London W1R 0DJ (tel.: 00 44 171 734 5997). USA: Olympic Tower, 645 Fifth Avenue, New York, NY 10022 (tel.: 001 212 421 5777); 611 West Sixth Street, Suite 2198, Los Angeles, CA 92668 (tel.: 001 213 626 6696). Greece: Ioannina EOT (tourist office), Napoleonda Zerva 2, Ioannina; Greek Alpine Club, EOS, Despotatou Ipirou 2, Ioannina.

Politics & Religion: Despite being in a militarily sensitive area, there are no obvious political undercurrents. The Greek Orthodox Church holds sway in Zagoria.

Crime Risks: Low.

Food & Drink: Breakfasts tend to be fairly basic; evening meals are more satisfying – often there's no choice, but the food is usually excellent. Greek-brewed beer and good Greek wine is available everywhere. Tsipuro is a popular spirit, and ouzo is the usual aperitif. Water is safe to drink from marked springs and from river sources; tap water is also safe to drink.

Wildlife & Flora: Beware flocks of grazing sheep: they are guarded by leggy dogs that are not cuddly collies. The wildflowers are breathtaking in May and June. Maple, beech, fig and oak make up woodland; huge spreading plane trees take pride of place in villages. Wildlife is abundant but on a small scale: tortoises, frogs, newts and exotic insects. More elusive are the eagles, bears, wolves and lynx.

TIMING & SEASONALITY

Best Months to Visit: May, June, September, October.

Climate: Temperate in May and June with sporadic thunderstorms; scorching hot in July and August; temperate in September; damp in October; snow-covered from November to April.

HEALTH & SAFETY

Vaccinations: None required; tetanus and rabies could be considered.

General Health Risks: No special precautions necessary.

Special Considerations: Take plenty of insect repellent: mosquitoes are voracious in woodland, and a variety of biting insects appear at night. Sunblock should be worn at all times, even on cloudy days. Antihistamine is useful against nettle stings and bites.

HIGHLIGHTS

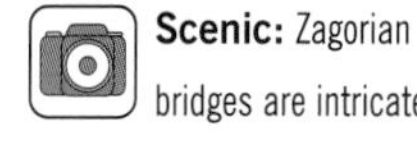

Scenic: Zagorian bridges are intricate stone structures, often spanning boulder-strewn canyons that are dry in summer but become torrents after storms. The ancient cobbled tracks that cover much of this route are called *kalderimi* – once the domain of mules and their drivers, in many places they are now considered works of art. Villages are built almost entirely from local stone: a soft white rock from quarries and a harder black stone from rivers.

The Vradheto Steps, an ancient kalderimi *or cobbled mule track, were once the only way to reach the remote village of Vradheto.*

temperature and precipitation

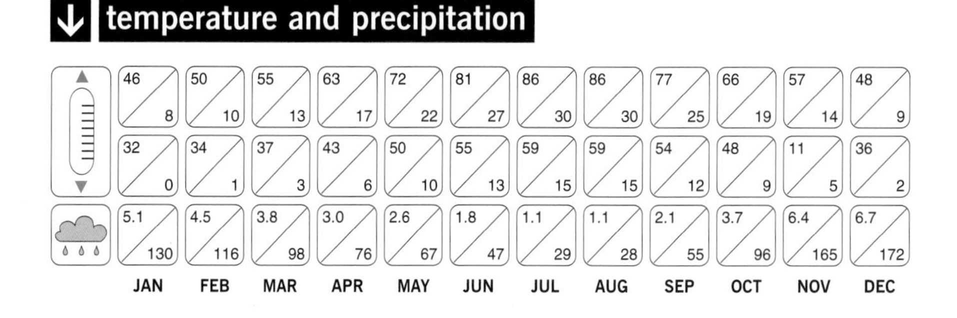

	JAN	FEB	MAR	APR	MAY	JUN	JUL	AUG	SEP	OCT	NOV	DEC
Max temp (°F / °C)	46 / 8	50 / 10	55 / 13	63 / 17	72 / 22	81 / 27	86 / 30	86 / 30	77 / 25	66 / 19	57 / 14	48 / 9
Min temp (°F / °C)	32 / 0	34 / 1	37 / 3	43 / 6	50 / 10	55 / 13	59 / 15	59 / 15	54 / 12	48 / 9	11 / 5	36 / 2
Precipitation (in / mm)	5.1 / 130	4.5 / 116	3.8 / 98	3.0 / 76	2.6 / 67	1.8 / 47	1.1 / 29	1.1 / 28	2.1 / 55	3.7 / 96	6.4 / 165	6.7 / 172

Asia

Asia constitutes the world's largest land mass, with many of its countries meeting amid a vast range of mountains, the high Himalaya. Here lies some of the most spectacular scenery on this planet; it is no wonder that four of these five treks take place in and around this impressive range of mountains. The fifth trek rises from the exotic jungle of Malaysia in the form of Mount Kinabalu.

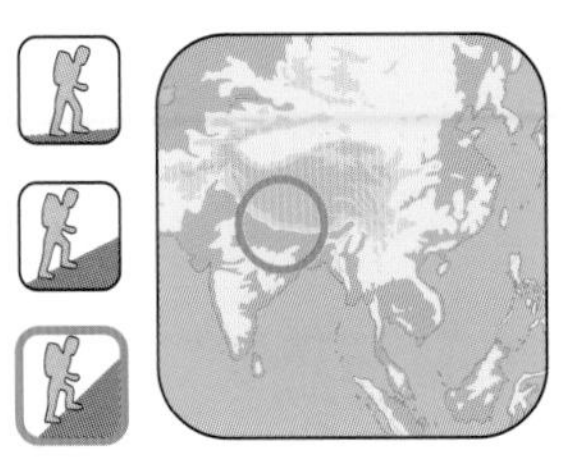

the MANASLU CIRCUIT

NEPAL

Kev Reynolds

CIRCLING THE WORLD'S EIGHTH-HIGHEST MOUNTAIN, THE MANASLU TREK IS WITHOUT QUESTION ONE OF THE MOST VARIED AND SCENICALLY EXCITING IN ALL NEPAL.

From the lush foothills, with their innumerable terraces that lead the eye to the vertical Arctic wall of the distant Himalaya, the trek enters the valley of the Buri Gandaki, the river whose course the route will follow for eight days.

For several days the trail is forced through a series of spectacular gorges, the river choked in places by house-sized boulders, and waterfalls cascade from soaring cliffs. But when you emerge from these gorges, you are in a very different world. The way now leads through an avenue of Himalayan peaks; glaciers hang inside glens, and tumble from 23,000–26,000-foot (7,000–8,000-m) mountains. Above Samdo, the highest village – inhabited by Tibetans who fled over the mountains in the wake of the Chinese invasion – the trek curves round the back of Manaslu to cross the Larkya La before descending to the warm foothill country again.

The route was only opened to trekkers from the West in 1991, with a strict limit of 400 permits a year. This limit appears to have been forgotten, but the trek remains much less-travelled than the neighbouring Annapurna Circuit. It is very much a route to treasure.

itinerary

•DAY 1 2½ miles (4km)

Gorkha to Kalikastan

Since the journey by road from Kathmandu to Gorkha will take at least half a day, and time is then needed to arrange porter loads, this first stage is a very short one. However, a steep ridge spur has to be crossed, and on the crest stands the palace of Prithvi Narayan Shah, who unified the country in 1769. The palace is now a Hindu temple, from which a stunning view north shows the Manaslu Himal through a framework of trees. The camp is sited just below the ridge.

•DAY 2 6 miles (9.5km)

Kalikastan to Koyapani

The trail winds among terraces of rice and millet, visits several villages, and enjoys fine views of distant snow peaks. It's an up and down route, which follows a vegetated ridge, dips into wooded basins, and crosses a spur or two before finding a camp on grassy terraces below Koyapani.

•DAY 3 5 miles (8km)

Koyapani to Santi Bazar

The path descends beside a stream, which grows into the Mukti Khola river as the way progresses. Where the river is too wide to negotiate by stepping stones or wading, there is a suspension bridge. The trail leads through the busy little township of Arughat Bazar, where permits must be produced at the police checkpoint, and continues to the teahouses of Santi Bazar, on the banks of the Buri Gandaki.

•DAYS 4–7 26½ miles (42.5km)

Santi Bazar to Deng

On Day 4 the trek encounters the first of the Buri Gandaki's gorges, and for the next few days it becomes a true helter-skelter of a route; sometimes along what appear to be ledges hacked from the steep, rock walls, sometimes in the very bed of the gorge itself; climbing out again on a rough trail that goes in and out of the sun. The way crosses and

↑ *This Buddhist monastery, Sama Gompa, stands on the edge of a yak pasture near Samagaon. The highest gompa in Buri Gandaki's valley, it faces Manaslu as if to receive the mountain's benediction. Seventy monks and nuns live in a collection of small cell-like buildings nearby.*

← *Beyond Samdo, the Buri Gandaki is crossed for the very last time. When it was first met near Arughat Bazar, it was a broad, rushing river, but now it's a modest stream, easily spanned by a wooden bridge. In the background are the mountains that form the border between Nepal and Tibet.*

recrosses the river on suspension bridges and, reaching Jagat, comes to the first sign of Buddhist culture. At the end of Day 7 you arrive in Deng, with views down the valley to Ganesh IV, and upstream to the ice crest of Shringi Himal.

•DAYS 8–9 **15½ miles (25km)**

Deng to Syala

Waterfalls spray from high cliffs in the last of the gorges, and tiny watermills dot the way towards Ghap. Here you pass through a *kani* (an entrance gateway) with decorated inner walls and ceiling, and come to a *mani* wall (made with stone slabs with prayers carved upon them) engraved with pictures of the Buddha. Namru has a good campsite in a woodland glade below the village, and next day you pass through villages with magical names: Ligaon (with a *gompa*, Buddhist monastery), Sho (with a large prayer wheel), and Lho which has a *gompa*, impressive *kani*, and a stunning view of Manaslu. The yak pasture at Syala rewards you with one of the most magnificent panoramas of the whole trek – sunset and sunrise views are especially fine, when the peaks are set alight with colour.

•DAY 10 **5½ miles (9km)**

Syala to Samdo

Now the way goes through a broader, gentler and more stony moat below glaciers, icefalls and soaring mountains. Samagaon and its calm-inducing *gompa* lies at the very foot of Manaslu. Just beyond, yaks graze the scant pastures, then the trail rises to Samdo, built on a levelled spur of moraine at 12,631 feet (3,850m), backed by a steep hill from whose upper slopes a view west shows the wedge-shaped Larkya Peak and the saddle of the Larkya La – the crux of the route. Down the valley stand Manaslu, Ngadi Chuli, and Himalchuli, and, if you've pitched your tent in a chosen spot, you'll be able to gaze at the stars that sit upon them. Two nights should be spent at Samdo to acclimatise.

•DAYS 11–12 **10½ miles (17km)**

Samdo to Bimtang

After Samdo, the way to the Larkya La goes a short distance up the valley, then leaves the Buri Gandaki to head west along the north flank of the valley, crossing an exposed landslip and climbing to a shallow glen at 14,600 feet →

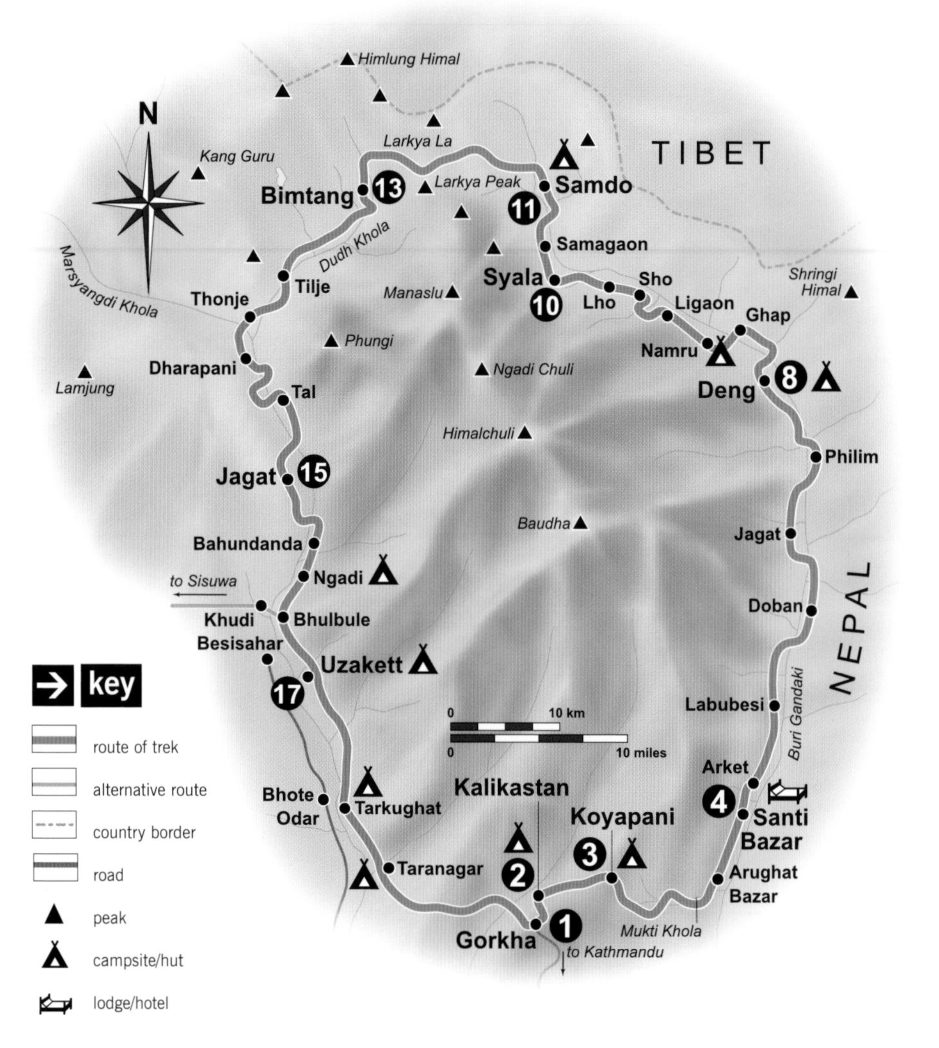

(4,450m), where you can camp. This crossing begins before dawn so that the pass is reached before the bitter winds of late morning sweep across it. The climb to the Larkya La is generous, but the descent is not. The trail is a steep one, demanding care – especially in wintry conditions. But 4,898 feet (1,493m) below lies an open, flat meadow banked by a long moraine wall and with incredible views both up and down the valley. This is Bimtang.

•DAYS 13–14 **17½ miles (28km)**

Bimtang to Jagat

A good day's trekking from Bimtang takes you out of Manaslu's clutches and leads to Thonje, at the confluence of the Dudh Khola and Marsyangdi Khola. Here, the trek joins the Annapurna Circuit and you have a choice. Either turn right to trek to Jomosom or bear left and descend through the Marsyangdi's valley to continue with the Manaslu Circuit, passing, on Day 14, through Dharapani and Tal on the way to Jagat, high on the river's right bank.

•DAYS 15–16 **13 miles (21km)**

Jagat to Uzakett

Terraced fields and thatched houses are met once more as the route continues downstream alongside the Marsyangdi, the temperature rising almost by the hour as height is lost. Shortly after leaving the village of Ngadi on the morning of Day 16, the trail forks in the bazaar village of Bhulbule. The route to Gorkha remains on the left bank, but an alternative option is to cross the river to Khudi and trek cross-country for 2½ days to Sisuwa (Begnas Tal). The Manaslu Circuit, however, stays in the Marsyangdi's valley.

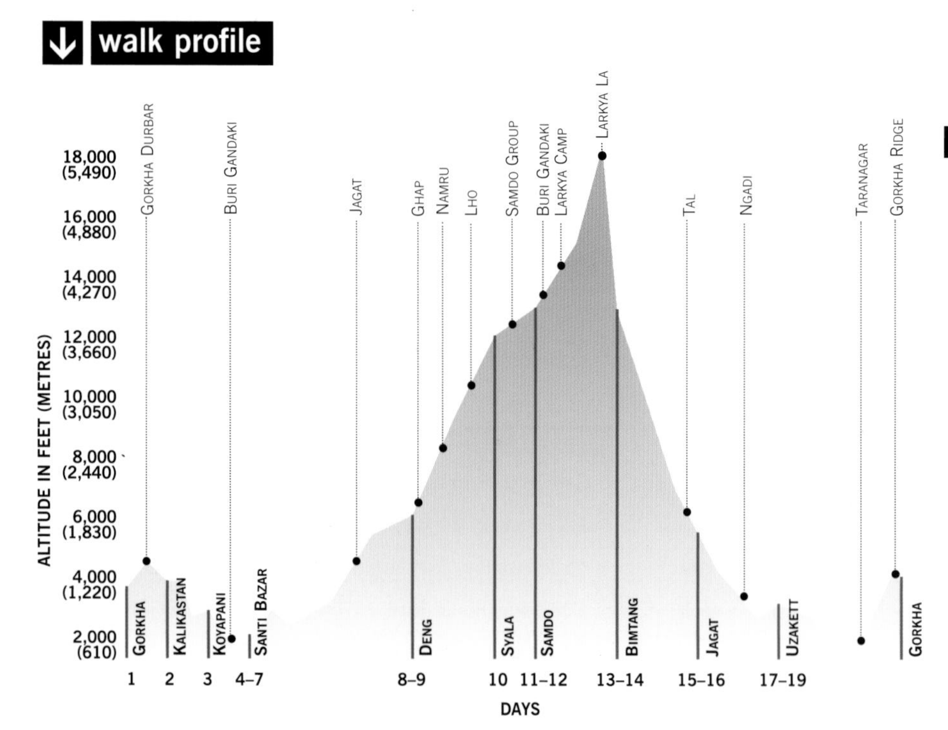

•DAYS 17–19 **20½ miles (33km)**

Uzakett to Gorkha

Within minutes of leaving Uzakett, the trail climbs a spur to enable you to gaze across a patchwork of rice terraces to the snowpeaks of the Annapurna and Lamjung Himals, then down to banana groves and ochre-walled houses. For 2 days the trek works its way through a land of rice and millet, crossing tributary streams and edging along narrow causeways in a series of landscapes. Only rarely do snow peaks show themselves, but from a riverside camp at Taranagar at the end of Day 18, the alpenglow blazes pink on distant peaks. On the last morning, a 3-hour walk brings the trek full circle to Gorkha and the road to Kathmandu.

factfile

OVERVIEW

A 19-day, 102-mile (164-km), counter-clockwise circuit of the Manaslu Himal in central Nepal. The route heads north through the deep valley of the Buri Gandaki, crosses the glaciated Larkya La, and descends to the foothills via the Dudh Khola and Marsyangdi valleys.

Start/Finish: Gorkha, midway between Kathmandu and Pokhara.

Difficulty & Altitude: This is a strenuous trek which demands much height gain and loss on most days. Descent from the high pass is often on snow or ice for some way, and the trail is extremely steep in places. The trek begins in the foothills at around 3,700 feet (1,128m), and reaches its highest point on the Larkya La at 17,103 feet (5,213m). The ability to acclimatise to the altitude and cold is an important consideration.

ACCESS

Airports: Kathmandu (international airport).

Transport: As this trek is available only to organised groups, your transport to the start will be arranged by your trekking company.

Passport & Visas: All visitors to Nepal, other than Indian nationals, need a passport and visa.

Permits & Restrictions: Your trekking agent will arrange all the necessary permits. A government officer travels with you to ensure restricted areas are not entered.

LOCAL INFORMATION

Maps: There are no strictly accurate maps available for this trek. The following, however, will suffice, as you will have professional sherpa guides for most of the route. Latest Trekking Map Kathmandu to Manaslu Ganesh Himal (Mandala Trekking Maps, Kathmandu); Nepa Maps ("For Extreme and Soft Trekking") The High Route Around Manaslu. Namaste Trekking Maps 1:125,000 sheet Manaslu (Pilgrims Book House, Kathmandu).

Guidebooks: *Manaslu – A Trekker's Guide*, Kev Reynolds (Book Faith India) – the only guide devoted to this route; *Trekking in the Nepal Himalaya*, Stan Armington (Lonely Planet) – includes a brief route outline.

Accommodation & Supplies: There are a few simple lodges in the Buri Gandaki Valley, and trekkers' lodges in the Marsyangdi Valley, but it is essential to be self-sufficient in regard to camping and food. Porters will carry the loads.

Currency & Language: Rupee (Rps). There is a "standard" Nepali language in common use throughout the country, and many dialects. Your trekking crew will understand some English.

Photography: No restrictions.

Area Information: None, other than that found in guidebooks.

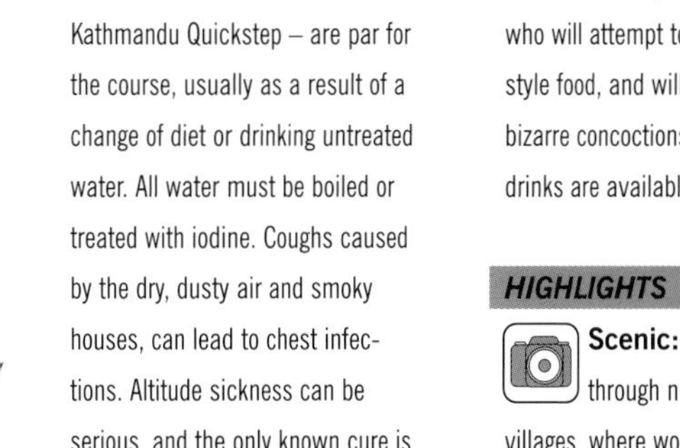

black-faced langur monkey

TIMING & SEASONALITY

Best Months to Visit: Trekking in Nepal falls within two seasons – pre- and post-monsoon. Pre-monsoon, March and April, are fine, but in the low foothills the heat can be oppressive in the middle of the day, when haze obscures the views. October to mid-December (post-monsoon) offer the best views and trekking conditions. Heavy snow can occur throughout the year.

Climate: The monsoon – June to September – translates into snow on high ground. Temperatures gradually build from March until the monsoon. From October to mid-December, the foothills enjoy pleasant daytime temperatures, slightly cooler at night. In the upper valleys, daytime temperatures in the post-monsoon months are cool, becoming much colder at night. At the highest camps expect night-time temperatures below freezing.

HEALTH & SAFETY

Vaccinations: Trekkers ought to be vaccinated against tuberculosis, tetanus, meningitis, polio and hepatitis A., and take anti-malarial tablets.

General Health Risks: Upset stomachs and diarrhoea – the Kathmandu Quickstep – are par for the course, usually as a result of a change of diet or drinking untreated water. All water must be boiled or treated with iodine. Coughs caused by the dry, dusty air and smoky houses, can lead to chest infections. Altitude sickness can be serious, and the only known cure is descent to lower altitudes. Gain height slowly, allow for rest days, and drink plenty of liquids.

Special Considerations: In the foothills take precautions against sunstroke, and in the high country be prepared for sub-zero temperatures. Sections of trail are exposed, there are rivers to cross by wading, and there will no doubt be landslide areas that need to be traversed; so remain alert at all stages of the trek.

Politics & Religion: Nepal is a democracy, and there are no real concerns. The majority of Nepalis are Hindus, but for much of the trek you will be in Buddhist areas.

Crime Risks: Low.

Food & Drink: On trek you will be catered for by a sherpa crew who will attempt to serve western-style food, and will produce some bizarre concoctions. Tea and soft drinks are available in the villages.

Trekking in Nepal depends on the employment of porters who carry tents, personal belongings and equipment in large bamboo dokos.

HIGHLIGHTS

Scenic: The trek passes through numerous villages, where women work the fields, men plough with buffalo or carry loads to a distant market, and children pad barefooted to school. In the higher country, houses are built of stone with shingles on their roofs weighted down with rocks. Prayer flags adorn the houses, and some villages have their own *gompa* (Buddhist monastery).

Wildlife & Flora: Black-faced langur monkeys are frequently seen in foothill forests, and the bharal (blue sheep) may be seen in small herds among the stark terrain near the Larkya La. Birdlife abounds in the foothills – brightly colored finches, warblers, and the conspicuous river chat or white-capped redstart. Both the lammergeier and golden eagle circle the high valleys. Pre-monsoon, rhododendrons brighten the forest; in autumn, frangipani and bougainvillaea are spectacular.

temperature and precipitation

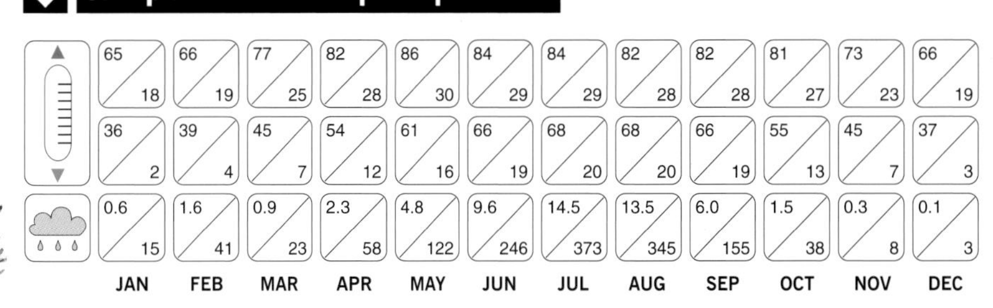

	JAN	FEB	MAR	APR	MAY	JUN	JUL	AUG	SEP	OCT	NOV	DEC
Max temp	65 / 18	66 / 19	77 / 25	82 / 28	86 / 30	84 / 29	84 / 29	82 / 28	82 / 28	81 / 27	73 / 23	66 / 19
Min temp	36 / 2	39 / 4	45 / 7	54 / 12	61 / 16	66 / 19	68 / 20	68 / 20	66 / 19	55 / 13	45 / 7	37 / 3
Precipitation	0.6 / 15	1.6 / 41	0.9 / 23	2.3 / 58	4.8 / 122	9.6 / 246	14.5 / 373	13.5 / 345	6.0 / 155	1.5 / 38	0.3 / 8	0.1 / 3

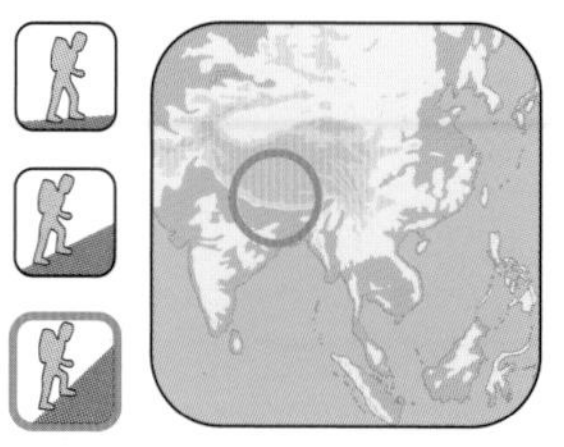

the ULTIMATE EVEREST TREK

NEPAL

John Manning

THE REWARDS OF TREKKING AMONG THE HIGHEST MOUNTAINS ON EARTH HARDLY NEED SPELLING OUT. TO DO SO AMONG THE WELCOMING, GENUINELY FRIENDLY SHERPA VILLAGES OF THE KHUMBU REGION IS DOUBLY REWARDING. NO WONDER THEN THAT THIS IS ONE OF THE MOST POPULAR TREKKING DESTINATIONS ON THE PLANET.

On a busy day there are perhaps almost as many people on the Kala Patta ridge, admiring the classic view of Everest, as there are on Scafell Pike in the English Lake District. Few, however, combine the classic Everest trek with the equally rich excursion to the frozen lakes of Gokyo, linking the quieter Dudh Kosi Valley with that of the Khumbu glacier by crossing the frozen Cho La pass.

Western explorers who came to the region in the early 1900s, looking for ways to scale Sagarmartha, as Nepalis call the highest mountain on Earth, found tranquil valleys populated by friendly, peaceful Sherpa people whose lifestyles belonged in the Middle Ages. Today, the spread of satellite televisions and telephones is moving things on a little, but the tranquility and friendly welcomes are undiminished; and the mountains will be there for ever.

itinerary

•DAY 1 **5 miles (8km)**

Lukla to Phakding

Half the day will be spent on the exhilarating flight from Kathmandu to Lukla. Clear skies will permit views of such Himalayan giants as Ama Dablam, Everest, and Nuptse.

Phakding, by the Dudh Kosi River, is roughly 650 feet (200m) lower than Lukla; that, and the exercise involved walking there, will help acclimatisation. If Phakding is busy, try Benkar (8,875 feet/2,705m), an hour beyond. Both have accommodation.

•DAY 2 **6 miles (10.5km)**

Phakding to Namche

Beyond Phakding the trail crosses to the right bank of the Dudh Kosi, through the forested valley. At Mondzo (9,302 feet/2,835m) the trail enters the Sagarmartha National Park, where an entrance fee is payable. The rock walls and huge boulders in this area have been ornately carved with Buddhist prayers. Beyond Jorsale (9,100 feet/2,770m) the trail leads up to Namche. You might spot Impeyan pheasant, Nepal's national bird, among the shrubbery.

•DAY 3 **Rest day**

Namche

Namche is a good spot to rest and acclimatise for a day. There are plenty of shops to explore, bars, and a famous Saturday market, which draws traders from Tibet. A good leg stretch would be a walk to the nearby Sherpa villages of Kunde (12,602 feet/3,841m), where there is a small hospital built by Sir Edmund Hillary's Himalayan Trust, and Khumjung (12,400 feet/3,780m), home to the Mount Everest bakery, where you can buy croissants and Danish pastries. Your walk could also take in the Everest View Hotel (12,700 feet/3,870m), built to accommodate wealthy tourists who fly direct to nearby Syangboche and promptly suffer the severe effects of altitude.

↑ *The route from Lukla airstrip to the village of Phakding is alive with vegetation. Sadly, this environment has often been affected by the demand for firewood, to the extent that the Nepalese government has introduced laws banning the harvesting of wood for fuel in Sagarmartha National Park.*

← *Everest, the highest point on the planet – a lure that draws hundreds of trekkers to the Khumbu region every year.*

•DAY 4 **4 miles (6.5km)**

Namche to Phortse Tenga

The easily followed trail heads north-east, high on the valley side above the Dudh Kosi. Watch out for herds of wild goat on the slopes. The walk offers views of Ama Dablam (22,300 feet/6,797m), Everest, Lhotse (27,890 feet/8,501m), and Nuptse, and should take about 5 hours.

•DAY 5 **2½ miles (4km)**

Phortse Tenga to Dole

Though only a couple of miles' walking is involved, it's important to take it easily to aid acclimatisation. You're going from 11,950 feet (3,643m) to 13,400 feet (4,084m). You will also be leaving well-vegetated slopes behind for open mountain scenery.

•DAY 6 **3 miles (5km)**

Dole to Machhermo

Another short day, with little trouble trail-finding but another ascent of more than 1,000 feet (300m), at altitude. Machhermo was the scene of a much-reported yeti attack on grazing yaks some years ago. Sherpas in the area are adamant that the creature exists.

•DAY 7 **Rest day**

Machhermo

There's plenty to occupy those who wisely choose to spend an acclimatisation day here. Explore the ridge behind the lodges or head west up the mountain-rimmed tributary valley. Solitary musk deer might be approached to within a few feet; Bhuddists don't hunt, so the deer have no reason to fear humans.

→

•DAY 8 **4½ miles (7km)**

Machhermo to Gokyo

Views of Cho Oyu draw you up the valley past Pangka (14,925 feet/4,549m), the site of a fatal avalanche in 1995. The first of the Gokyo lakes, Longpongo, is reached after about 2½ hours. Gokyo village (15,719 feet/4,791m) sits on the side of the third lake.

•DAY 9 **Rest day**

Gokyo

Another rest and acclimatisation day is in order. For the active, the ridge up to Gokyo Ri (17,520 feet/5,340m), festooned in prayer flags, provides one of the classic views across to Everest and a mountain panorama encompassing, among others, Lhotse, Makalu, Thamserku and Kangtega, Ama Dablam, Cho Oyu, and Nepal's longest glacier, Ngozumpa, which flows down behind the village from Cho Oyu.

The village has a secondhand bookstore and several other small stores, but the lake – subject to a constant freeze/thaw action throughout the day – also provides entertainment; trapped air beneath the creaking ice makes eerie sounds reminiscent of whale song.

Foreign aid has helped relieve some of the poor conditions in the Sherpa villages of Khumjung (in the foreground) and Kunde (in the distance) – Kunde has a health post, and Khumjung has a bakery.

A sherpa drives his three yaks across a metal bridge suspended high above the Bhote Khosi, near Namche. Yak are a vital commodity in the region; they are widely used as beasts of burden, and their dung is burned as fuel.

•DAYS 10–11 25 miles (40km)

Gokyo to Cho La pass, and on to Lobuche

This section of the trail can be difficult and should be attempted only by self-sufficient, experienced groups, and then only in good weather. Lodges exist but are not always open, even at peak season. The alternative is to head back down the Dudh Kosi Valley to Phortse (12,598 feet/3,840m), then up the Imja Khola Valley, branching north up the Lobuche Khola Valley, and joining this route again at Lobuche (16,175 feet/4,930m). The detour need take no longer than the Cho La crossing.

The trail across the grey, moraine-strewn Ngozumpa Glacier is usually obvious in fine weather. Before reaching the Cho La, the trail climbs through pastures at Thagna, crosses a ridge, and fights its way across an extensive boulder field before it's time to camp among the boulders, where flat sites are at a premium. Take care on the rocky scramble up to the Cho La as dislodged rocks can be dangerous. Frostbite has also affected trekkers on the pass.

The second day's journey over the pass to Lobuche is long and tiring, climbing steeply over the glaciated Cho La (17,783 feet/5,420m), then dropping, high above the frozen Cholatsho Lake and through seasonal grazing areas, to the Lobuche Khola Valley; Lobuche is still a good walk north of here. Self-sufficient groups might take another day to complete the trip to Lobuche.

•DAYS 12–14 8 miles (13km)

Lobuche to Kala Patta or Everest Base Camp, and back

The highlight of the trek, the view from Kala Patta, is easily reached from Lobuche, though it's a long day's walk parallel to the hidden, shattered Khumbu glacier there and back. The trail continues up from Lobuche to cross the Changri glacier to Gorak Shep (17,002 feet/5,182m), where you'll find a small teahouse and might spot Tibetan snow cocks – large, grey grouse-like birds, quite accustomed to man.

At this point you have a choice: Kala Patta, which offers the closest view trekkers can have of Everest from the Nepal side; or the longer but flatter walk to Everest Base Camp. The latter is scenically unrewarding, the trail can be easily lost, and there are no facilities there whatsoever. Attempting both in one day would be too demanding for all but the fittest. Base Campers would be better scaling Kala Patta, spending the night at Gorak Shep, and heading up to Base Camp in the morning, returning to Lobuche afterwards.

The trail up the bare Kala Patta ridge, a lowly spur of Pumori (23,443 feet/7,145m), is visible from the Gorak Shep teahouse. It looks easy, but allow yourself plenty of time to cope with the altitude. The view across to Everest and Nuptse from the cairned high-point is unequalled. Having left an offering to the mountain gods among the prayer flags flying from the summit ridge cairns, you can retrace your steps to Lobuche.

walk profile

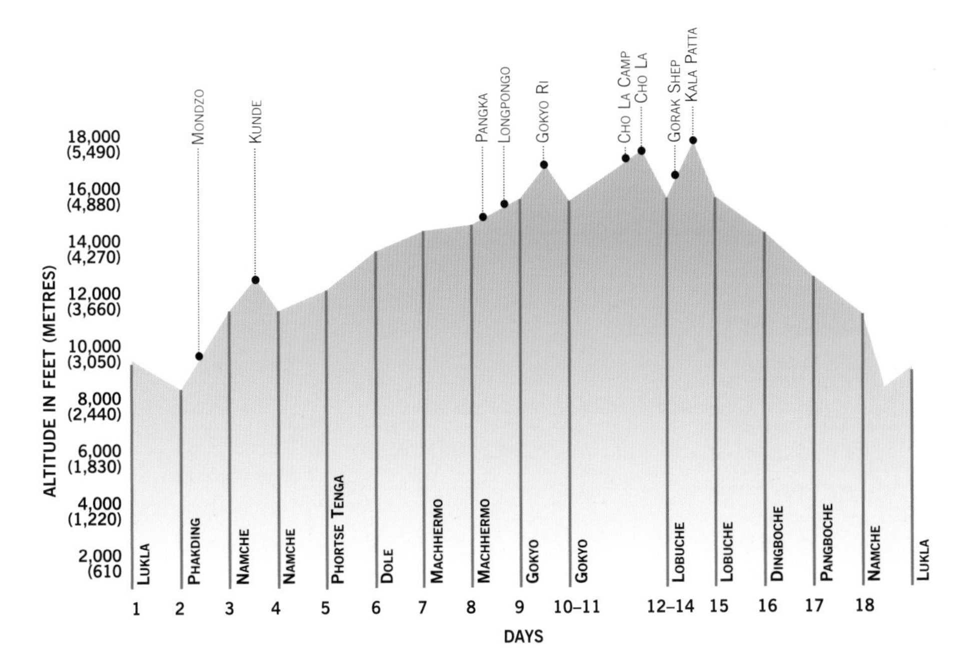

→

Trekkers approach the thumb-like profile of Pumori, which rises dramatically behind the ridge of Kala Patta.

•DAY 15 **6 miles (9.5km)**

Lobuche to Dingboche

The walk down the valley to Dingboche (or Pheriche) is relaxing – you're heading downhill at last, through alpine-like meadows and across frozen streams. Following the main trail, the mountains – such as Tawoche (21,462 feet/6,542m) on the west side of the valley – seem closer and more distinct than the giant ranges to the north were. The friendly villages offer accommodation, food and camping areas; there is a health post at Pheriche.

•DAY 16 **3 miles (5km)**

Dingboche to Pangboche

We now head south-east along the Imja Khola Valley, with the vegetation returning as we lose altitude. The woods are home to deer and pheasant, and there are several Bhuddist monasteries in the area.

If you need to make up time, this would be the stretch on which to do so; acclimatised and fit, you could reach Pangboche from Lobuche within a day.

•DAY 17 **15 miles (24km)**

Pangboche to Namche

Plenty of ascending and descending today, along clear trails in dense forest. Take time, however, to visit Tyangboche monastery, one of the region's most important. Founded in 1916, the building has been destroyed twice: by an earthquake in 1934 and by fire in 1989. It was rebuilt with local and foreign money, including funds from the Himalayan Trust. The large friezes in the main temple relate Sherpa and Bhuddist legends. Trekkers are welcomed at certain times and it is customary to leave an offering (usually a few notes of currency).

•DAY 18 **12 miles (19.5km)**

Namche to Lukla

The return to Lukla from Namche, nearly all downhill, should be comfortably achieved in one day, ready for the flight back to Kathmandu the following morning.

↓ factfile

OVERVIEW

A physically demanding, 95-mile (152-km) trek through the Sagamartha National Park in the Khumbu region of Nepal to the foot of Everest, at 29,028 feet (8,848m) the highest mountain on Earth. This is one of the most spectacular routes of the region and involves the crossing of the Cho La, a high, glaciated pass that is physically demanding, and after poor weather might well be impossible; even after moderate snow it becomes a technical route.

Start/Finish: Lukla air strip (9,350 feet/2,850m).

Difficulty & Altitude: Most sections of this high-level route are on well-trodden paths. However, trekking at altitude is extremely strenuous, and most people limit themselves to a few hours a day. Allow at least 18 days for the trek, including several rest/acclimatisation days. It is important to take the walk at a steady pace, allow for adequate rest, and drink plenty of boiled and/or treated water. Keeping well hydrated is a key factor in proper acclimatisation. Trails are generally easily followed in good weather and maps are adequate for general route-finding in such conditions. This trek should not be attempted in poor weather. The Cho La section should not be underestimated; self-sufficient groups will have little problem in fine weather, but ropes and ice-axes are recommended in all conditions. Individuals, those with little experience, or those having difficulty with acclimatisation should take the suggested alternative.

yak

ACCESS

Airports: Lukla airstrip, reached by air from Tribhuvan International Airport, Kathmandu, is the most popular starting place for this route. Inclement weather frequently causes delays.

Transport: Lukla, the start of the trek, can be reached on foot from Jiri (6,250 feet/1,905m), to which there is a bus service from Kathmandu. This will add several days to the trek but will assist with acclimatisation.

Passport & Visas: Full passport and entry visas for all visitors.

Permits & Restrictions: Until recently, trekking permits have been required in all parts of Nepal. These are available from the Central Immigration Office,

Tridevi Marg, between Kantipath and Kathmandu in Thamel. Please check in case they are reinstated. If necessary, trekking agencies can assist you in getting the right permit and save you from the queues.

LOCAL INFORMATION

Maps: Mandala Maps 1:75,000 Khumbu, Mount Everest – features a 1:50,000 trekking map of the route on the back. Place names, contours and spot heights are often inaccurate.

Guidebooks: *Everest, A Trekkers Guide*, Kev Reynolds (Cicerone); *Trekking in Nepal – a Traveler's Guide*, Stephen Bezruchka (Cordee, UK/The Mountaineers, USA); *Trekking in the Nepal Himalaya*, Stan Armington (Lonely Planet); *Trekking in the Everest Region*, Jamie McGuiness (Trailblazer).

Background Reading: *View From the Summit*, Sir Edmund Hillary (Doubleday) – the autobiography of the first man to stand on the summit of Everest, with a wonderful account of that expedition as well as a wealth of information on the work of the Himalayan Trust, a charity established by Hillary to better the lot of the Sherpa people of the Khumbu region.

Accommodation & Supplies: As this is one of the most popular treks in the Himalaya, there is a good infrastructure of lodges, teahouses, and stores along much of the way. There is, however, a good argument for attempting this route with a commercially organised trekking group, as there is very little in the way of accommodation or teashops during the 2-day (at least) crossing of the Cho La pass, when self-sufficiency is necessary.

Currency & Language: Nepalese rupee; can only be bought in Kathmandu. Receipts for foreign currency should be retained if you wish to trade your unspent rupees at the end of the trip. US dollars (US$) are also very useful to have. Nepalese and various dialects are spoken throughout Nepal; the Sherpas of the Khumbu region have their own language. English is widely spoken.

Photography: The government limits the amount of film visitors can bring into the country.

Area Information: Ministry of Tourism, Tripureshnawar, Kathmandu, Nepal (tel.: 00 977 1211 286).

TIMING & SEASONALITY

Best Months to Visit: Autumn and spring (October to May). Outside the main trekking season poor weather can prevent any views of the mountain.

Climate: Outside the months of October to May temperatures can be too low. Between the end of November and March deep snow can fall, closing parts of the route.

HEALTH & SAFETY

Vaccinations: None required, but the following should be considered: tuberculosis, typhoid, hepatitis, meningitis and, of course, tetanus.

General Health Risks: Rabies is present in Nepal and a good precaution is to carry a stick to fend off dogs, and not to approach animals. You should consult your own doctor at least 3 months before you travel about any currently recommended precautions, including those for malaria if you intend to visit other parts of Nepal.

Special Considerations: Diamox can help lessen the ill-effects of acclimatisation. It is not, however, a cure-all, and the best cure for poor acclimatisation is to descend as soon as possible. You should make yourself aware of potentially fatal high-altitude disorders before setting off. Anyone with heart or lung problems should consult their doctor before trekking at altitude. Hands should be kept very clean as the most common form of infection is from trekkers or trek staff not washing their hands adequately after visiting the toilet, leading to others being infected and themselves being reinfected. Suitable medicines, such as Immodium, should be carried in case of diarrhea.

Politics & Religion: Nepal is a Hindu country but the Khumbu region is predominantly Bhuddist; the two faiths exist happily side by side in many parts of Nepal. Please note that Nepalis will not step over outstretched legs, and it is considered rude to point the soles of your feet at someone. It would be good manners, especially for women, to remain covered up in villages and especially temples. If you want to visit a *gompa* (Buddhist monastery) respect any visiting times and don't stray beyond the area tourists and trekkers are allowed to visit. Finally, try to learn a little of the language. Most Sherpa guides in the region speak passable English but interaction with porters and local families will be all the richer for knowing a few key words.

Crime risks: Low.

Food & Drink: All water should be boiled or purified. All food should be properly cooked; fruit and vegetables should be peeled and boiled; salads should be washed only in treated water.

musk deer

HIGHLIGHTS

Scenic: The scenery on this planet doesn't get much more spectacular than the Khumbu region. Mountains, seraced, glaciated, fluted, many over 26,000 feet (8,000m), dominate throughout the trek, the clear mountain air granting panoramas of breathtaking peaks in every direction. The valleys are initially lush, green forests of blue pine and rhododendron, and later become glacier-carved gorges through which mineral-rich, grey waters rush down from the ice flows over contorted boulders. Moraine walls lead the eye to sheer walls of rock and ice, and streams of fluttering prayer flags adorn the steep, grassy ridges that lead to higher peaks. At the valley heads wide, shattered and rock-strewn glaciers transfix the viewer.

Wildlife: Yaks, mostly domesticated, seem to symbolise the wildlife of the Khumbu region, but there's plenty more: musk deer and wild goats can be seen in the higher region, and in the forests small mammals can be glimpsed. The birdlife is rich, with several species of pheasant in wooded areas, and snow cocks and choughs easily spotted higher in the valleys. This is also said to be yeti-country (otherwise known as the Abominable Snowman).

←

Colourful rugs are displayed from the balcony of a typical two-story Sherpa home in Namche.

temperature and precipitation

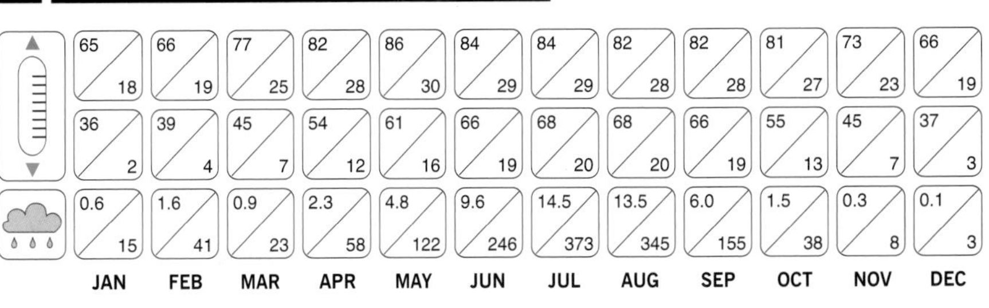

	JAN	FEB	MAR	APR	MAY	JUN	JUL	AUG	SEP	OCT	NOV	DEC
Max (°F / °C)	65 / 18	66 / 19	77 / 25	82 / 28	86 / 30	84 / 29	84 / 29	82 / 28	82 / 28	81 / 27	73 / 23	66 / 19
Min (°F / °C)	36 / 2	39 / 4	45 / 7	54 / 12	61 / 16	66 / 19	68 / 20	68 / 20	66 / 19	55 / 13	45 / 7	37 / 3
Precipitation (in / mm)	0.6 / 15	1.6 / 41	0.9 / 23	2.3 / 58	4.8 / 122	9.6 / 246	14.5 / 373	13.5 / 345	6.0 / 155	1.5 / 38	0.3 / 8	0.1 / 3

the SOURCE of THE GANGES

INDIA

Steve Callen

INDIA IS A MYSTICAL, EXCITING, CHAOTIC LAND, WHERE THE SENSES ARE STRETCHED TO THEIR LIMITS. RELIGION IS AT THE CORE OF INDIAN LIFE, AND THIS CLASSIC TREK BEGINS BY FOLLOWING IN THE FOOTSTEPS OF THOUSANDS OF PILGRIMS TO REACH THE SOURCE OF THE HOLY GANGES AT GAUMUKH, THE SYMBOLIC BIRTHPLACE OF HINDUISM, THE MAIN RELIGION OF INDIA.

The source is known as "the mouth of the cow", and refers to the ice cave at the snout of the Gangotri glacier where the Bhagirathi River, the main tributary of the Ganges, comes flowing out. Bathe in these icy cold waters and, so legend has it, all earthly sins are removed and spiritual immortality is attained, ready for the next life.

These first three days, from the bustling road head of Gangotri to the terminus of the Gangotri glacier, are straightforward, but should not be rushed if you are slow to acclimatise. Beyond Gaumukh you leave the majority of pilgrims behind and head for Nandavan and Tapovan, whose idyllic green meadows and clean glacial streams are in stark contrast to the rubble-strewn glacier you've just crossed. Here too, surrounding you in all directions, are some of the most spectacular and mighty peaks of the whole Himalaya – the Bhagirathis, Meru Peak, Shivling, and Kedarnath.

itinerary

•DAY 1 5½ miles (9km)

Gangotri to Chirbasa

Gangotri is at the road head, a bustling small town full of eating houses, trinket stalls, and hotels dwarfed by imposing cliffs on both sides. Above the noise of the buses arriving and leaving, and the bells and horns of the Hindu and Buddhist temples, is the incessant roar of the Bhagirathi River as it starts its turbulent journey to the Bay of Bengal.

Leave Gangotri via a flight of steps beside the temple and head gently uphill amid the smell of the sweet deodar, blue pine and juniper. The snow-crested peaks of Matri, Chirlas and Mandha soon appear, and suddenly all the stresses and strains of working in the West seem a million miles away. The first settlement you come to is Chirbasa, at 11,535 feet (3,516m). If you are not acclimatising well, camp the night here. There are several eating houses, and the sight of the Bhagirathi peaks should dull any pains you may have from the altitude or from trekking.

•DAY 2 3½ miles (5.5km)

Chirbasa to Bhujbasa

Past Chirbasa and through the rhododendron forest, you start to catch glimpses of Shivling (21,466 feet/6,543m). All the peaks in this area have religious connotations, but Shivling is the most impressive of all, representing the phallus of Lord Shiva – origin of the whole universe. The path now climbs above the tree line and soon becomes a high mountain desert. As you round a shoulder you catch your first glimpse of the Gangotri glacier, a huge, sweeping highway of rubble and ice cutting and sweeping its way through the mountains. The Bhagirathi peaks are starting to make their presence felt, as is Shivling, standing sentinel at the western end of the glacier. Bhujbasa (12,467 feet/3,800m) is the settlement on the valley floor and another night stop. Basic accommodation is available as well as campsites. Again, because of the altitude, take a night's rest and don't go higher until you feel well enough to.

↑

Gaumukh, source of the Ganges and one of the holiest places in the whole of India, is an isolated spot in a rock-strewn landscape, towered over by the mighty Bhagirathi range.

←

The Bhagirathi peaks draw trekkers on towards the settlement of Chirbasa.

•DAY 3 3 miles (5km)

Bhujbasa to Gaumukh

If you are trekking only as far as Gaumukh, you can leave your gear here while you make the 3-hour trek to the source. Otherwise, head slowly up the well-trodden path, perhaps taking time to wash and bathe in the crystal-fresh streams that cross the path.

As you climb over the moraine and get closer to Gaumukh, you will notice colourful prayer flags blowing in the breeze. You pass a couple of chai shops festooned with flags and religious icons and then the blue, tottering ice cliffs of Gaumukh, source of the River Ganges, stand before you. Other pilgrims will be there, some fully immersing themselves in the glacial cold waters of the Bhagirathi River. Take care not to stand too close to the cave as huge blocks of ice are continually dropping off into the river.

•DAY 4 4½ miles (7.5km)

Gaumukh to Nandavan

You now have to choose between the meadows of Nandavan or Tapovan – unless you can do both, that is. Whichever you head for, you have to cross a glacier. It can be tricky finding the route, and it's dangerous weaving your way through the glacial moraine because of the hidden crevasses, so you should have a guide for this section.

If you are heading for Nandavan, Meru Peak (21,161 feet/6,450m) starts to appear just above Gaumukh on your right. Its impressive Shark's Fin rock buttress, a seemingly impregnable feature, has seen some very strong climbers attempting to scale its walls – to date with no success.

By now your head will be dizzy, both with the altitude and with the heights of the surrounding peaks. Cross the moraine of the Raktvarn Bamak glacier and follow the left-hand bank of the Gangotri glacier until you see a cairned path heading up a steep, grassy slope. A breathless slog up here will land you at the idyllic Nandavan, base camp for the Bhagirathis. Clear glacial streams run across the meadow, in which there are numerous camping spots. The skyline is dominated by the huge bulk of Shivling and the snow dome of Kedarnath (22,770 feet/6,940m). Get your tent pitched and wait for the late afternoon colour display as the peaks change from white to orange and different hues of red and pink.

•DAY 5 7½ miles (12km)

Nandavan to Mana Parbat and back

It's certainly worth resting here for a few days and taking some day excursions. Your first excursion might be up the right bank of the Chaturangi Bamak glacier to the foot of Mana Parbat (22,291 feet/6,794m). This is a track primarily used by mountaineers on their way to the Bhagirathis. Make the effort to visit this wild, peaceful, unspoilt spot; the path is easy to follow.

→

Gangotri
Bhagirathi River
Chirbasa
Bhujbasa
Gaumukh
Sri Kailas
Yogeshwar
Sundarshan
RAKTVARN BAMAK
Mana Parbat
CHATURANGI BAMAK
Nandavan
Tapovan
GANGOTRI GLACIER
Kedar Tal
KEDAR BAMAK
BRIGUPANTH BAMAK
Mandha
Jogin
Brigupanth
Meru
Shivling
Thalay Sagar
Bhagirathi
Satopanth
Swachand
Kedarnath Dome
Kedarnath
N
0 4 km
0 4 miles

key

route of trek
alternative route
minor track
road
peak
campsite/hut
lodge/hotel
provisions

•DAY 6 5 miles (8km)

Nandavan to Tapovan

On the opposite side of the Gangotri glacier you will spot Tapovan, which can be reached on another day excursion by a faint, sinuous track that crosses the glacier – take care! There are two *ashrams* (places of meditation) at Tapovan that are worth a visit (take your own food and some offerings). On the Tapovan side of the glacier you have another chance to marvel at something man cannot corrupt or deface: the Bhagirathis, in all their rugged, towering glory.

Keep a look out for herds of ibex, which roam freely in this area. A keener eye may spot the Himalayan marmot, but you'll have little chance of seeing the elusive snow leopard, which has nearly been hunted to extinction for its fur.

Consider staying on another day for an excursion up one or other banks of the Gangotri glacier for better views of Kedarnath Dome and its impressive East Face. You'll be surprised to see that Shivling has twin, symmetrical summits.

•DAYS 7–8 16½ miles (26.5km)

Tapovan to Gangotri

The return journey is just a retracing of the way you came up. As you descend, so the air becomes thicker and warmer, and walking becomes much easier again. You leave the barren glaciers to join the high, luscious, alpine meadows. Perhaps the saddest part of the trip is leaving behind that peace and serenity, but whether or not you are Hindi, or religious in any way, this trek will touch you spiritually like no other.

A camp is made in the meadow of Nandavan, with the upward-thrusting peak of Shivling as a backdrop.

walk profile

ALTITUDE IN FEET (METRES)
22,000 (6,710)
20,000 (6,100)
18,000 (5,490)
16,000 (4,880)
14,000 (4,270)
12,000 (3,660)
CHATURANGI BAMAK
MANA PARBAT
GANGOTRI
CHIRBASA
BHUJBASA
GAUMUKH
NANDAVAN
NANDAVAN
TAPOVAN
GANGOTRI
1 2 3 4 5 6 7–8
DAYS

factfile

OVERVIEW

A trek from Gangotri to Gaumukh, the source of the holy Ganges, could take just 3 days. But the more adventurous can extend the trek to the Gangotri glacier and the meadows of Tapovan and Nandavan, and make day excursions up the Chaturangi Bamak glacier. This adds up to an 8- or 9-day trek of some 45 miles (72km).

Start/Finish: Gangotri in Uttar Pradesh.

Difficulty & Altitude: The trek to Gaumukh follows a well-worn path, but you must acclimatise properly as it attains an altitude of 13,058 feet (3,980m). Above Gaumukh, a guide is essential as the glacial crossings are loose and the path is ill-defined. Nandavan is at 14,764 feet (4,500m) and side excursions from here can take you close to 16,400 feet (5,000m); the going is hard.

ACCESS

Airports: The nearest airport is Jolly Grant Airport, 15 miles (24km) east of Dehra Dun. Flights from Delhi can be expensive and very unreliable.

Transport: The overnight Mussoorie Express, plus the daytime Shatabdi Express, runs from Delhi to Dehra Dun and Rishikesh, where buses can be caught to Uttarkashi and Gangotri. If you are travelling by bus from Delhi, allow 2 days.

Passport & Visas: Passports and tourist visas are required by all visitors.

ibex

LOCAL INFORMATION

Maps: Use the Indian Himalaya Map Series published by Leomann Maps: Sheet 7 Garhwal (Gangotri, Har-ki-Dun and Mussoorie).

Guidebooks: *Lonely Planet Walking Guide: Trekking in the Indian Himalaya* (Lonely Planet); *Garhwal and Kumaon: A Trekker's and Visitor's Guide*, K.P. Sharma (Cicerone).

Background Reading: *The Rough Guide to India* (Rough Guides).

Accommodation & Supplies: From Gangotri you must carry a tent, fuel, food, and cooking supplies. The limited accommodation as far as Gaumukh may well be taken by the many pilgrims who travel this route. The last place to purchase supplies, and the best place to organize a guide and porters, is Uttarkashi. Snacks can be purchased from teahouses as far as Gaumukh.

Currency & Language: Indian rupee (Rps). Banks in major towns and cities will accept traveller's cheques, although the State Bank of India may not accept American Express traveler's checks. In other towns you will have to use cash – US dollars ($US) and pounds sterling (£) are easily exchanged. In villages, only small denominations of the rupee are accepted. In this region Hindi is the main language, as well as Garhwali and various dialects; English is spoken by some.

Photography: Do not take photographs of anything to do with the Indian military (including road bridges), or the inside of a temple and any public bathing places close to the temple. Women may object to close-up photographs.

Area Information: In New Delhi the Government of India Tourist Office is at 88 Janpath.

TIMING & SEASONALITY

Best Months to Visit: Pre-monsoon: May and June; post-monsoon: September to November.

Climate: Be prepared for all extremes as winter snows can come early and be late to melt, and the monsoon can also come early. Autumn tends to have long clear spells, when it will be warm in the valleys and cooler higher up; at higher altitudes night-time temperatures will drop to just below freezing. Above 11,500 feet (3,450m), expect light snow at any time.

HEALTH AND SAFETY

Vaccinations: Typhoid, tetanus, polio, hepatitis and meningitis are all required.

General Health Risks: "Delhi Belly" is a common traveller ailment. Take plenty of rehydration drinks and keep to a simple, non-fatty diet. If symptoms persist, take a prescribed antibiotic. Do not trek too high too quickly; learn to recognise the symptoms of altitude sickness; if they are more than mild, descend as fast as possible and take some rest days before ascending slowly again. Ask your doctor about currently recommended malaria pills. Make sure you have good travel insurance cover, with helicopter rescue included.

Special Considerations: Take a full first-aid kit so you are self-sufficient in case of injury or illness. The road head is never more than 3 days away, and you should send for a helicopter only in extreme circumstances.

Politics & Religion: India's politics are volatile, so seek the latest advice from the appropriate overseas department in your own country. The main religion of India is Hinduism. Other religions are Islam, Christianity and Buddhism.

Crime Risks: Crime is generally low and once you are trekking there will be few problems. Keep your valuables in a money belt while travelling, and watch your bags at bus and train stations.

Food & Drink: Indian food is famous for being spicy. Vegetarians are well catered for. Be wary of buying food from street stalls where there are many flies. Never drink untreated water. Check the seal on bottled water. Don't eat ice cream or drink anything with ice.

HIGHLIGHTS

Scenic: The trek takes you past some of the most spectacular mountains in the world. Shivling and the sheer rock pillars of the Bhagirathis will take your breath away.

Wildlife & Flora: As you travel to Gangotri you pass through subtropical, temperate and alpine habitats that support a whole range of wildlife and flora. Once on the trek, keep an eye out for the herds of blue sheep and ibex. Looking skywards, you are sure to spot Himalayan griffin vultures.

Pilgrims to India's holiest site disembark at the bustling town of Gangotri.

temperature and precipitation

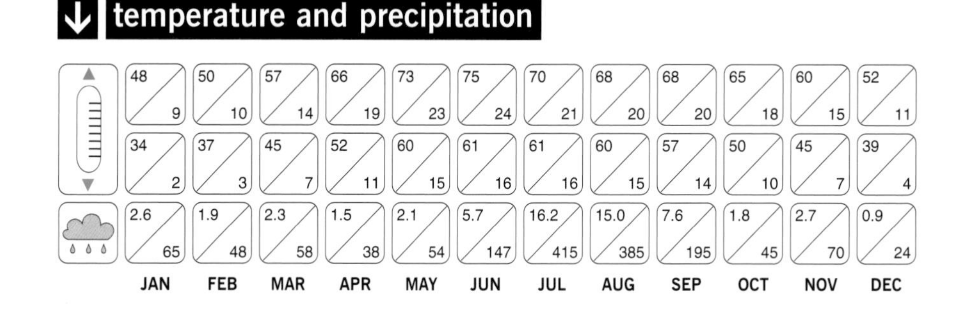

	JAN	FEB	MAR	APR	MAY	JUN	JUL	AUG	SEP	OCT	NOV	DEC
Max temp (°F / °C)	48 / 9	50 / 10	57 / 14	66 / 19	73 / 23	75 / 24	70 / 21	68 / 20	68 / 20	65 / 18	60 / 15	52 / 11
Min temp (°F / °C)	34 / 2	37 / 3	45 / 7	52 / 11	60 / 15	61 / 16	61 / 16	60 / 15	57 / 14	50 / 10	45 / 7	39 / 4
Precipitation (in / mm)	2.6 / 65	1.9 / 48	2.3 / 58	1.5 / 38	2.1 / 54	5.7 / 147	16.2 / 415	15.0 / 385	7.6 / 195	1.8 / 45	2.7 / 70	0.9 / 24

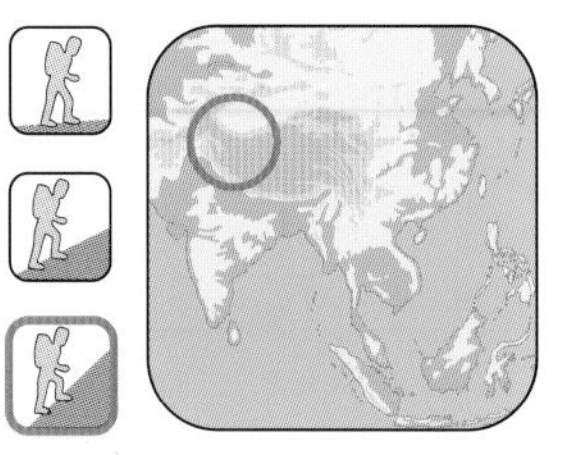

the KARAKORAM:
Snow Lake and the Hispar La
PAKISTAN

Richard Gilbert

THIS OUTSTANDING TREK CROSSES THE MIGHTY RANGE OF THE KARAKORAM BY TAKING THE LINE OF WEAKNESS CAUSED BY THE BIAFO AND HISPAR GLACIERS. THROUGHOUT THE TREK YOU ARE SURROUNDED BY SOME OF THE HIGHEST PEAKS IN THE HIMALAYA AND THE SCENERY IS NEVER LESS THAN AWE-INSPIRING.

In many places the route follows lush ablation valleys on the side of the glaciers between the lateral moraines and the mountainsides. This is a tough trek only to be attempted by fit and determined walkers with considerable mountain experience, and it should be realised that there are no escape routes. The mid-point of the trek traverses the edge of the historic Snow Lake, an elevated plateau of ice at 16,400 feet (5,000m), set amid towering peaks which remained unexplored until 1937.

It is a long and difficult journey to reach the start of the trek at Askole, and delays must be expected. It is advisable to join a party organised by one of the big trekking companies who have access to local facilities, guides and porters. In spite of the severity of this trek, the rewards are unsurpassed, and it is one that should not be missed by the adventurous and experienced trekker.

itinerary

•DAY 1 11 miles (17.5km)

Askole to Namla Brangso

The campsite at Askole is in a clearing of mud and grass overhung by trees, 1 hour's walk before the village is reached. The path winds up the terraced hillside through fields of maize and barley, which are expertly irrigated from mountain springs. Chickens and goats wander at will through the village, and the Balti inhabitants will look at you with curiosity. The stone houses have turf roofs which are popular grazing sites with the goats.

The path becomes dry, dusty and rocky as it contours the steep hillside overlooking a thundering river, white with sediment. The source of this river is the Biafo and Baltoro glaciers, and after a few miles you reach the junction of the

↑ *The huge Biafo glacier runs deep into the heart of the Karakoram, and the first half of the trek climbs the full length of the glacier to the high pass known as the Hispar La. Towering peaks overlook the glacier, some rising to over 23,000 feet (7,000m). Meltwater runnels, crevasses and seracs must be negotiated by trekkers on this serious and demanding route.*

two glaciers and turn left, scrambling up the moraines on the south side of the Biafo icefall.

Camp is in the ablation valley at the uninhabited grazing site of Namla Brangso, which looks across the Biafo glacier to the peak of Bullah (20,650 feet/6,294m).

•DAY 2 **5 miles (8km)**

Namla Brangso to Mango

This is a fairly short day to aid the acclimatization process since you are already above 11,400 feet (3,500m). The edge of the Biafo glacier is strewn with ice blocks and boulders. Karakoram boulder fields are unstable and no boulder is to be trusted. It is all too easy to fall and suffer cuts and bruises on this section of the trek. Once you are on the glacier proper, the difficulties lessen, but there is no marked path and you must make the best way ahead yourself. Another level campsite, off the ice on some sand, is found at Mango, where a side valley comes in.

•DAY 3 **10 miles (16km)**

Mango to Baintha

This is a long and arduous day over mixed ground. It is advisable to leave camp by 6:00 A.M. so that you can traverse some of the snowfields before they start to melt.

The Biafo glacier at this point is 3 miles (5km) wide, and the edge is a complex mixture of rubble, gravel on bare ice (which is lethal), and seracs. It is necessary to weave round crevasses and jump meltwater rivers, but you should still have time to enjoy the spectacular peaks that completely surround you.

Crossing over to the north side of the glacier, you reach a maze of ice blocks so high that it is easy to lose your bearings but, finally, you arrive at a lush ablation valley and excellent campsite at Baintha. Here, there is a clear stream, grass, flowers, and you may even see ibex grazing.

•DAY 4 **Rest day**

This is the last oasis for many days, until you are well down the Hispar glacier. It is important to consolidate your acclimatisation – for you are now at 13,868 feet (4,227m) – to recover your strength, wash your clothes, and reorganise. In addition, there is no firewood beyond Baintha and the porters need to bake bread and make chapatis for the days ahead.

•DAY 5 **8 miles (13km)**

Baintha to Napina

The trek continues along the ablation valley until it peters out on to a loose and steep mountainside. Then it is back to the glacier for some reasonably level walking and some enticing views up a side valley to the north, to the rock spires of Latok and the tremendous peak of Baintha Brakk (The Ogre), first climbed by Chris Bonington and Doug Scott in 1977. You leave the Biafo glacier at a side valley called Napina, where there is a rough and stony campsite.

•DAY 6 **6 miles (9.5km)**

Napina to Biafo Glacier

More glacier work as you approach Snow Lake. The southern side is dominated by the beautiful, twin-peaked Sosbun →

key

route of trek

road

peak

campsite/hut

lodge/hotel

provisions

The lateral moraines of the Hispar glacier have to be traversed after reaching the Hispar La at the top of the Biafo glacier, to reach the terraced fields of Hunza.

walk profile

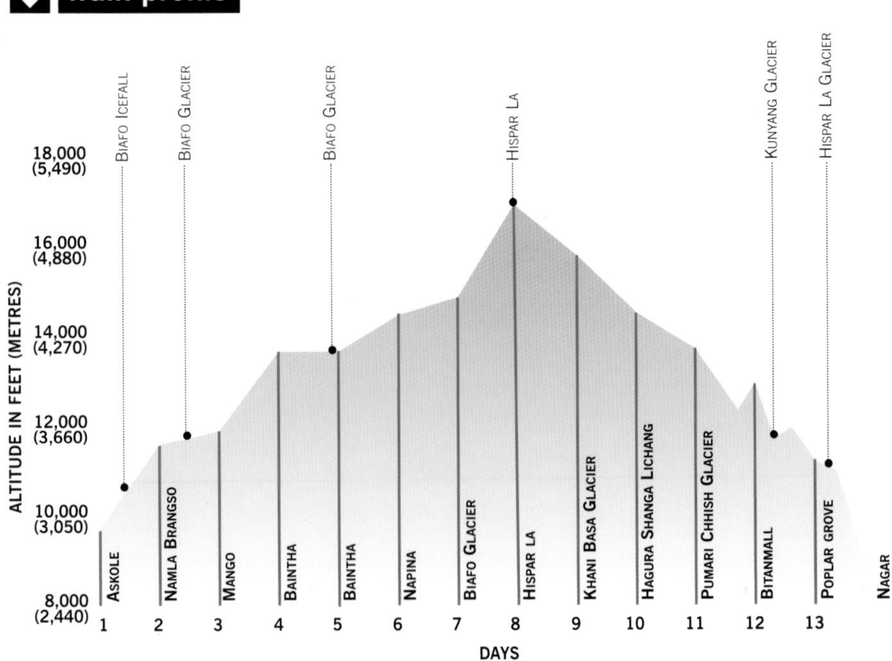

Brakk, 21,040 feet (6,413m). The glacier steepens and becomes badly crevassed, but an ascent of a boulder field brings you to a perch overlooking the Biafo glacier, which stretches away to distant horizons, and the Sim Gang glacier coming in from the east. It also provides a tantalising view of the edge of Snow Lake.

•DAY 7 6 miles (9.5km)

Biafo Glacier to Hispar La

An early start is advised for the long and exhausting ascent to the Hispar La (16,900 feet/5,151m). Deep snow lies over the ice, and it is essential to rope up for safety. Height is slowly gained and you can soon see over the vast expanse of Snow Lake with its myriad island peaks, the major proportion of which are still unclimbed. It is easy to imagine this could be the home of the legendary Abominable Snowman.

The slopes leading up to the Hispar La are steep and seemingly unending, but eventually you arrive at the summit of the pass and can look down the contorted upper reaches of the Hispar glacier. This is the point of no return. Pitch your tent in the soft snow and try to get a good night's sleep, in spite of the roar of avalanches pouring down from the mountains that overlook the camp on the south side.

•DAY 8 8 miles (13km)

Hispar La to Khani Basa Glacier

Going west down the Hispar glacier from the Hispar La, you lose height quickly. Frozen snow gives way to soft, deep snow with yawning crevasses, and you will have several

Near the summit of the Biafo glacier, chaotic moraines must be crossed. The overnight campsite is dominated by the magnificent rock spire of Sosbun Brakk, which soars into the heavens.

anxious moments before the slush gives way to hard, grey ice running with water.

Just before the Khani Basa glacier comes in from the north, you leave the ice of the Hispar glacier for a campsite on stony ground. From the tent entrance you can look across the glacier to the long, unbroken line of peaks making up the Balchish Group.

Having lost 2,600 feet (800m) from the Hispar La, you should have regained your appetite and be sleeping much better.

•DAY 9 6 miles (9.5km)

Khani Basa Glacier to Hagura Shanga Lichang

An early rise is necessary because the crossing of the Khani Basa glacier involves the fording of several big glacier torrents and these are at their lowest level at dawn. In distance this day is a short one, but it takes much time, effort and care to negotiate the icefall of the side glacier, the rivers, and an extremely loose and thin traverse line above the ice of the Hispar glacier. The most dramatic sight of the day is the view up the Khani Basa Valley to the towering peak of Kanjut Sar, at 25,460 feet (7,760m).

Camp is at another stony site at Hagura Shanga Lichang, where you will be heartened to see thyme and other flowers bravely flowering among the boulders.

•DAY 10 5 miles (8km)

Hagura Shanga Lichang to Pumari Chhish Glacier

Another similar day's descent – up and down mountainous and loose moraines, across side glaciers, and through torrents of brown, silt-laden water. But the last hour is through a grassy ablation valley, with the aroma of thyme, sage and mint in the air and yaks grazing peacefully. On my crossing, the cooks found wild rhubarb for a rhubarb crumble and custard supper.

•DAY 11 6 miles (9.5km)

Pumari Chhish Glacier to Bitanmall

This day starts with a crossing of the chaotic Pumari Chhish glacier and then a very nasty ascent of a mud cliff beyond the ice in order to regain a traverse line. The exceptionally →

Looking across the legendary Snow Lake, where footprints of the Yeti or Abominable Snowman were first reported in 1937.

steep slopes of mud and conglomerate have nicks cut to provide footsteps, but the position is exposed and precarious, and any slip would be disastrous.

The compensation for this difficult section is a view up the side valley to Khunyang Chhish peak, one of the Karakoram giants at 25,761 feet (7,852m).

A long, grassy stretch leads on to an excellent campsite at Bitanmall, 12,795 feet (3,900m), where there is a group of primitive huts used by shepherds in the summer months.

•DAY 12 **13 miles (21km)**

Bitanmall to poplar grove

Yet another side glacier, the Kunyang glacier, is crossed early on, and another, even steeper and more committing, mud cliff is climbed before an ablation valley, green with juniper and sage brush, runs down toward the snout of the Hispar glacier.

A long and tortuous section of jumbled boulders must be crossed to reach the foaming river emerging from the ice below the Hispar glacier. One mile (1.5km) downstream, a narrow suspension bridge crosses the river and a zigzag path climbs to Hispar village, where you are likely to be besieged by excited children.

The jeep track from Hispar to Nagar has been cut in many places, but at least it provides a reasonable path down the valley. After another 2 hours a fine campsite is reached beside a grove of poplars.

•DAY 13 **11 miles (17.5km)**

Poplar grove to Nagar

Follow the old jeep track as it winds its dusty way down the valley toward Nagar. On our crossing we stopped at the tiny summer settlement of Huru. We sat in the shade of apricot, fig and walnut trees and waited for jeeps to arrive to take us to Nagar and then on to Karimabad. If the Nagar–Huru road is down, it is another 4 hours' hot trekking to Nagar.

You have now completed one of the world's longest and most demanding treks and can bask in a well-deserved glow of satisfaction.

factfile

OVERVIEW

This magnificent mountain trek crosses the Central Karakoram, linking the regions of Baltistan and Hunza in northern Pakistan. Including time for acclimatisation, the actual trek will take 14 days (allowing 1 day for bad weather or sickness), but another 7 days should be set aside for travelling from and to Islamabad. The nature of the glacial terrain changes quite dramatically year by year necessitating diversions, but the length of this trek is somewhere between 90 and 100 miles (145–160km).

Start: Askole, a tiny village 7 hours by jeep from Skardu.

Finish: Nagar, a village 1½ hours from Karimabad.

Difficulty & Altitude: This is a long and demanding trek, which crosses the entire length of the Biafo and Hispar glaciers. Together, these glaciers make up the longest continuous glacier system in the world outside the polar regions. Near the top of the pass (the Hispar La at 16,900 feet/5,151m), the trek crosses the famous Snow Lake, a vast area of unbroken snow first explored by Shipton and Tilman in 1937. The trek is serious and committing. There are no escape routes and emergency rescue from the Hispar La would be extremely difficult to organise. It is essential to hire a local guide and porters, or to join one of the trekking companies that organise crossings of the Hispar La.

ACCESS

Airports: Islamabad International Airport. Small airstrips at Skardu and Gilgit.

Transport: Several flights a week link Islamabad with Skardu but they operate only in good weather. Travelling to Skardu by road takes 2 days, or longer if the Karakoram Highway is washed away by floods. Likewise, the jeep road from Skardu to Askole is likely to be cut by landslides, enforcing long detours on foot with all your luggage. The return to Islamabad can be either by bus from Karimabad (36 hours) or by bus to Gilgit and a flight, weather permitting, to Islamabad.

Passport & Visas: Passports and tourist visas are required for all visitors. Bring several

viburnum

photocopies of your passport and visa to hand to the authorities at the checkpoints.

Permit & Restrictions: None required at the time of writing.

LOCAL INFORMATION

Maps: Swiss 1990 1:250,000 Karakoram Sheets 1 and 2; AMS U-502 series 1: 250,000 Sheets NJ-43-14 and NI-43-3; Leoman Maps 1:200,000 Sheet 2 Karakoram.

Guidebooks: *Pakistan Handbook*, Isobel Shaw (Moon); *Pakistan Trekking Guide*, Isobel Shaw (Moon); *Pakistan, A Travel Survival Kit*, John King (Lonely Planet).

Background Reading: *Blank on the Map*, Eric Shipton (Baton Wicks).

Accommodation & Supplies: Hotels in Skardu and Karimabad, otherwise camping. The K2 Motel in Skardu is very helpful in arranging jeeps, guides, and porters. Staple local food, such as flour, eggs, chickens and cooking oil, is available in Askole and Nagar, but lightweight dried rations should be brought into Pakistan for the trek.

Currency & Language: The Rupee (Rps). Traveller's cheques and credit cards will not be accepted in the hill villages. Most guides and hoteliers speak English.

Photography: Photographs of bridges, radio stations and military installations are not allowed. No restrictions in the mountains.

Area Information: The latest information on conditions on the roads and in the mountains can be obtained from the Shalimar Hotel, Islamabad, or the K2 Motel, Skardu. There are tourist information centres and trekking agencies in Islamabad, Skardu, and Gilgit.

TIMING & SEASONALITY

Best Months to Visit: July to September. Earlier than July the crevasses in the glaciers are likely to be covered in soft snow, which makes them particularly dangerous.

Climate: The weather in the Karakoram is unpredictable and storms may blow up at any time. However, the monsoon rains do not reach the Central Karakoram, resulting in dry and desert-like foothills. On the trek you are likely to experience extremes of temperature from over 85°F (30°C) in the day to 15°F (-10°C) at night.

HEALTH & SAFETY

Vaccinations: Polio, tetanus, typhoid, hepatitis A; anti-malarials should be taken.

General Health Risks: It is extremely easy to pick up a debilitating stomach infection resulting in chronic sickness and diarrhoea. Below glacier level drink only bottled or treated water. Once you have reached the Biafo glacier, you should be free from infection. Take Dioralite (rehydrant salts) and Immodium (anti-diarrhoea). It is vitally important to gain height slowly and to drink large volumes of fluid to prevent Acute Mountain Sickness. If you suffer from persistent headaches, shortness of breath or vomiting, it is essential you return to lower ground as quickly as possible.

Politics & Religion: Pakistan is a mainly Muslim country. Discretion in dress is important, particularly for women.

Crime Risks: Carry money and passport in a money belt and never leave a rucksack unattended.

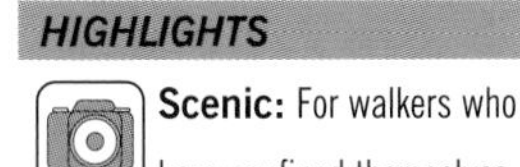

Scenic: For walkers who have confined themselves to lesser hills, the experience of the Karakoram is breathtaking. Giant peaks, some glaciated, some bare rock spires, thrust skywards on each side of the Biafo and Hispar glaciers. There is a constant rumble of avalanches from the steep faces accompanied by clouds of powder snow – but the route taken by the trek is free of avalanche danger. It is awe-inspiring to look into the icy depths of a crevasse from a snow bridge and see the green ice walls disappearing into a seemingly bottomless abyss. In several areas, though, particularly on the sides of the glacier, the enormous pressures have thrown up huge blocks of ice, as large as houses which could topple at any time. After nearly 2 weeks of struggling over some of the most wild, remote and rugged territory to be found anywhere, the green, irrigated, terraced fields around Nagar provide a welcome and startling change.

Wildlife & Flora: The vast expanses of ice mean few animals and little flora can survive in the Hispar La. The lower you are, the more likely you are to see short woody plants such as sage and viburnum. But one living creature that many who have been to the Karakoram believe to live there is the Yeti or Abominable Snowman. In 1937 Tilman and Shipton explored the Snow Lake and described footprints in the snow that were "...8 inches in diameter, 18 inches apart, almost circular and without sign of toe or heel".

→ Erecting a tent on the Hispar La. At 16,900 feet (5,150m), this is the highest point of the trek.

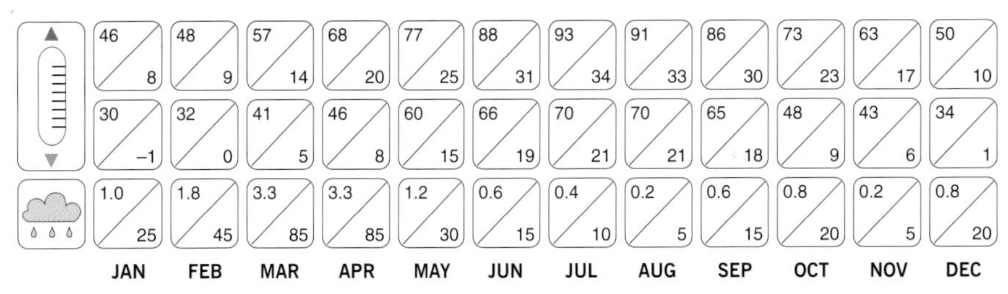

	JAN	FEB	MAR	APR	MAY	JUN	JUL	AUG	SEP	OCT	NOV	DEC
Max temp (°F / °C)	46 / 8	48 / 9	57 / 14	68 / 20	77 / 25	88 / 31	93 / 34	91 / 33	86 / 30	73 / 23	63 / 17	50 / 10
Min temp (°F / °C)	30 / -1	32 / 0	41 / 5	46 / 8	60 / 15	66 / 19	70 / 21	70 / 21	65 / 18	48 / 9	43 / 6	34 / 1
Precipitation	1.0 / 25	1.8 / 45	3.3 / 85	3.3 / 85	1.2 / 30	0.6 / 15	0.4 / 10	0.2 / 5	0.6 / 15	0.8 / 20	0.2 / 5	0.8 / 20

→ *Trekkers follow the fixed ropes up the granite slabs of the great whaleback en route to Kinabalu's summit.*

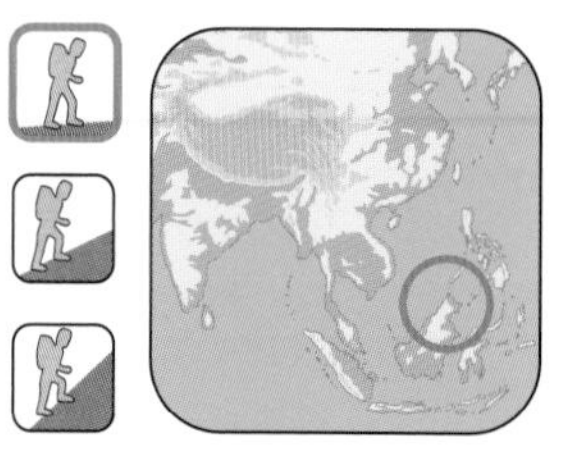

KINABALU

SABAH, MALAYSIA

John Cleare

APPARENTLY FLOATING BETWEEN EARTH AND SKY, KINABALU CAN BE SEEN FROM FAR OUT ON THE SOUTH CHINA SEA. AN ISOLATED MOUNTAIN, IT REARS HIGH ABOVE THE TANGLED HILLS OF NORTH BORNEO, A HUGE WHALEBACK OF GRANITE, ITS UNDULATING SUMMIT PLATEAU BRISTLING WITH PINNACLES, ITS STEEP FLANKS HUNG WITH TROPICAL FOREST.

The trek described is a short, steep, two-day journey to Kinabalu's summit, which, at 13,455 feet (4,101m), is the highest point between Western Australia and China. This regular route, virtually that taken on the first ascent by Hugh Low in 1851, follows the wide, blunt South Ridge up through the forest until it fades into the bare rock of the great whaleback. The large draughty hut – the Laban Rata Resthouse – stands not far below this point. As the forest gives way to rock walls and terraces, the route becomes more complicated, though easy to follow, and this section is normally climbed in the dark. The final stage ascends to the wide saddle on the summit crest – reached at dawn – above Low's Gully, the great feature of the northern flank of the mountain. Low's Peak, the actual summit, is a small, blocky pyramid and rather steeper.

→ *At around 7,000 feet (2,130m) on the southern slopes of Kinabalu, you pass through a verdant tropical cloud forest.*

•DAY 1 6 miles (9.5km)

Kinabalu Park H.Q. to Laban Rata Resthouse

A good road with six hairpins climbs the initial 2½ miles (4km) to the Power Station road head at 6,150 feet (1,870m), the actual start of the hiking trail. Usually, it will be possible to cover this section in motorised transport. Otherwise, several footpaths avoid all or parts of the road; these are likely to prove more strenuous than hiking the road itself but more worthwhile for those interested in flora and fauna. At the road head, an elaborate gateway opens onto a short descent to the fine Carson's Falls, from where the trail starts the climb in earnest. There are several flights of steps and the odd short ladder, and, because of the heavy rainfall, short sections of trail may be washed out.

The route passes four simple shelters, and, after some 2½ hours, reaches Carson's Camp at 8,860 feet (2,700m). The steepest part of the day's ascent is completed.

The forest now is lower and less dense, and there are groves of giant heather. Soon, the bizarre walls and towers of the Kinabalu summit plateau rise over the trees; eventually the odd hut appears, and the Laban Rata Resthouse (10,760 feet/3,280m) comes into sight in a clearing on the steep hillside. This is where you check in for the night. Most walkers usually take 5 or more hours to reach Laban Rata.

The view from here out across the Crocker Ranges is worth the hike in itself, and before dark it is worth making a short reconnaissance of the route ahead. Wise trekkers will also determine what time other parties plan to set off in the morning as, in the dark, parties of slow amateur hikers will cause traffic jams on narrower sections of the route.

•DAY 2 2 miles (3km)

Laban Rata Resthouse to Low's Peak

It is usual to start this section in the small hours by torchlight, aiming to arrive on the summit at sunrise. Several easy ladders and flights of steps surmount vegetation-hung craglets before a fixed rope is encountered, traversing a →

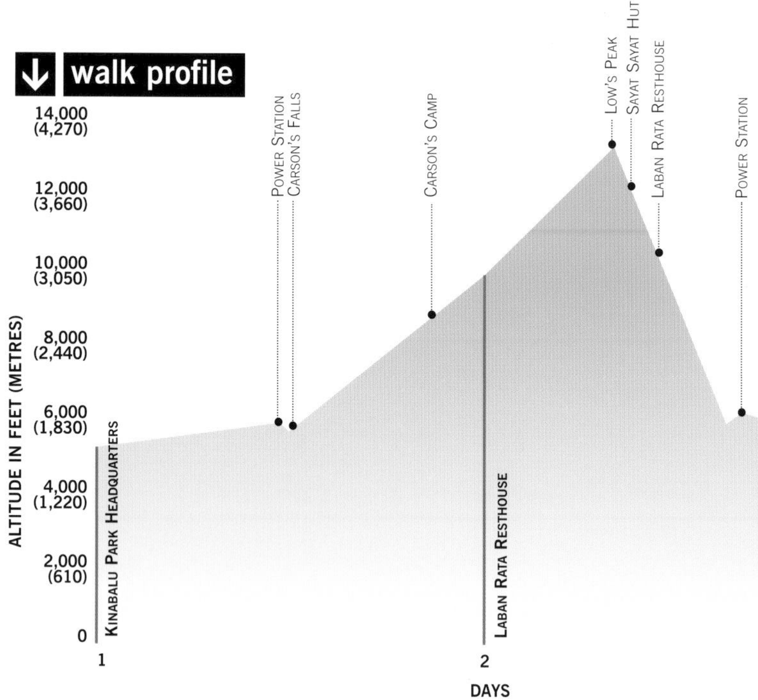

diagonal ledge across a granite slab. After 1½ hours or so, a bouldery path along the base of a steep rock wall leads to the locked Sayat Sayat Hut at 12,030 feet (3,670m). From here to the top you will be on rock the whole way. Indeed, the now continuous fixed ropes start almost immediately.

Cross a stream and, following the rope, climb on to the first of the slabs, the base of which apparently overhangs space. The rock scenery hereabouts is quite surreal, with acres of bare granite curling upwards beneath the twin, twisted towers of the Donkey's Ears. But the angle soon eases, and one just pads up and up, the thick white rope leading the way, right to the sharp summit of the small, blocky pyramid that is Low's Peak. Your time is likely to be under 3 hours from Laban Rata, although many hikers may take 5. Sunrise this close to the Equator is around 5:30 to 6:00 A.M.

The descent follows the fixed rope, but in daylight it is possible to explore the plateau if one's guide is confident of your capability. Ugly Sister Peak (13,220 feet/4,030m) is another easy, if very exposed, summit close above the route; and the view from the wide saddle between the two tops into the shadowed depths of Low's Gully, the huge abyss that splits the northern flank of the mountain, is awesome.

Depending on the weather and how long you stay high to admire the scenery, you will return to Laban Rata for brunch before trundling back down the trail to park HQ, where a certificate will await you to prove that you made it. By evening you can be back in Kota Kinabalu.

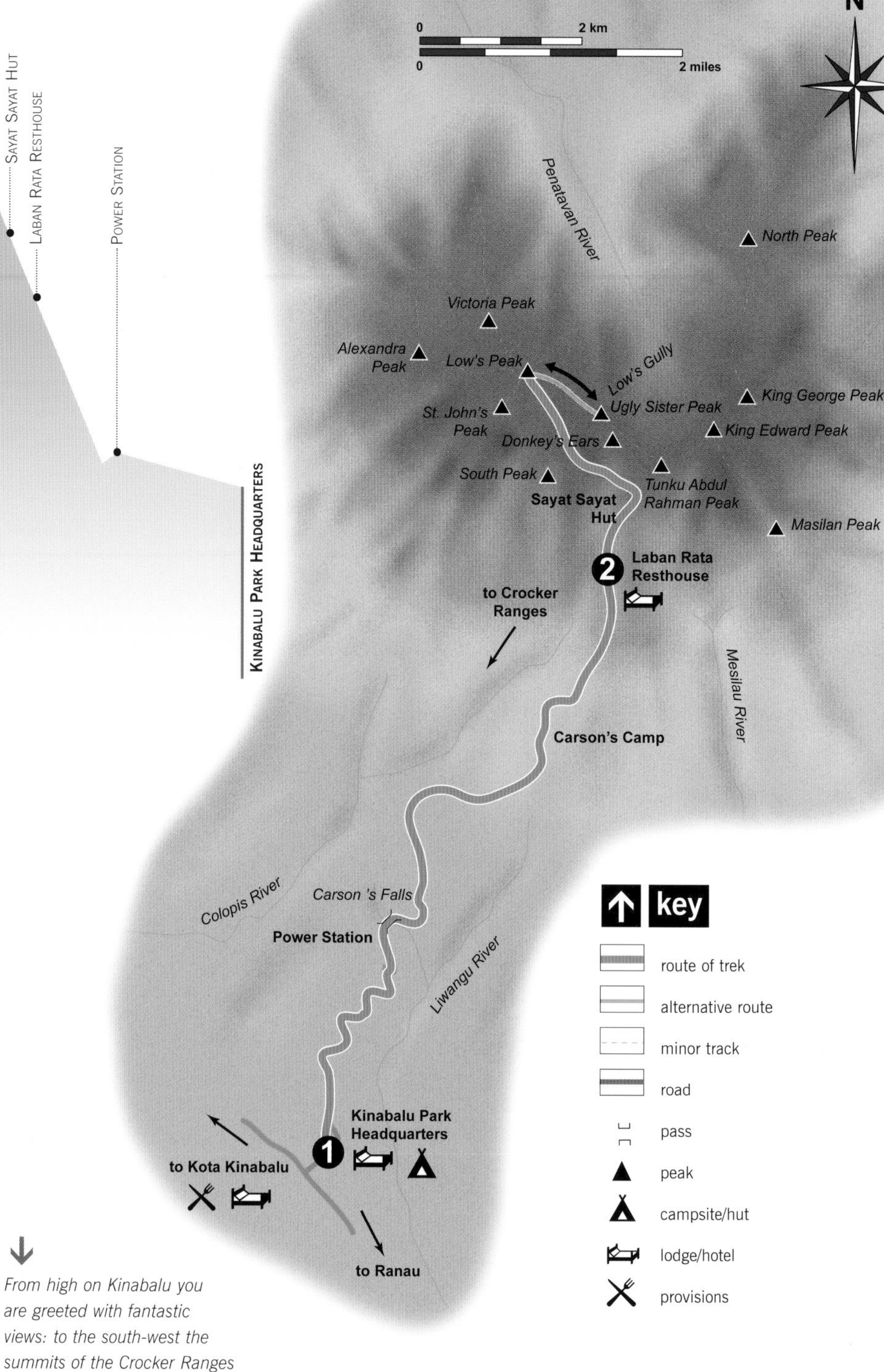

From high on Kinabalu you are greeted with fantastic views: to the south-west the summits of the Crocker Ranges rise like waves above a sea of monsoon cloud.

OVERVIEW

This 2-day ascent of North Borneo's legendary Mount Kinabalu is a short, steep journey to the summit. Traversing an easy route across serious terrain is always satisfying, and many will wish to return to explore more of Kinabalu or to sample the other challenging routes that traverse it.

Usual Start/Finish: Kinabalu Park Headquarters.

Difficulty & Altitude: The regular route is in no way technical. A well-maintained and consistently steep trail climbs through tropical forest to a large and comfortable rest house. Granite slabs lead some 1,500 feet (450m) up the wide whaleback to the summit (13,455 feet/4,101m). This rocky section of the route is equipped with over a mile of continuous fixed rope, to ensure that summiteers are unable to lose their way.

ACCESS

Airports: International airport at Kota Kinabalu.

Transport: Daily buses run from Kota Kinabalu to Ranau via Park HQ (a 2-hour trip). Charter minibuses, taxis and rental cars are readily available.

Passport & Visas: Visitor visas are issued on arrival.

Permits & Restrictions: A permit for the climb must be obtained from the park headquarters, where you must register in and out. It is mandatory to hire an authorised local guide. Porters can also be hired to carry extra food or equipment. Environmental conservation is taken seriously, and collecting any flora, fauna or natural object is strictly forbidden.

LOCAL INFORMATION

Maps: Sketch map at about 1:35,000 scale included in the Kinabalu Park leaflet listed below; Directorate of Overseas Surveys (D.O.S.) 1:50,000 aerial survey.

Guidebook: *Kinabalu Park* – an extremely useful leaflet published by Kinabalu Park authorities.

Background Reading: *Kinabalu, Summit of Borneo*, edited by K.M. Wong and A. Phillipps (The Sabah Society/Sabah Parks) – the definitive monograph of the mountain.

Accommodation & Supplies: There are numerous hotels and guesthouses in Kota Kinabalu; the park operates a series of chalets and hostels in the vicinity of the park headquarters, and it is possible to camp. On the mountain the park maintains the Laban Rata Resthouse, an alpine-style hut with accommodation for some 60 people and restaurant service. Smaller chalets close by provide extra accommodation. The higher Sayat Sayat Hut is a simple one-room building that is kept locked (available to expeditionary parties by arrangement). Drinks and meals can be purchased at the park headquarters and at the Laban Rata Resthouse. If you are planning a normal 2-day ascent, it is unnecessary to carry more than snacks.

Currency & Language: The Malaysian Ringgit (RM) or "Malay Dollar" of 100 sen. Bahasa Malay is the national language, but English is taught in schools and widely spoken.

Photography: No restrictions.

As you descend past Laban Rata Resthouse, the path passes through giant heather forest.

Area information: Tourist Information Office in Kota Kinabalu; Sabah Parks Office, P.O. Box 10626, 88806 Kota Kinabalu, Sabah, Malaysia (tel.: 00 60 88 211 652).

TIMING & SEASONALITY

Best Months to Visit: The sunniest, driest period is February to May, the wettest October to January. Nevertheless, bad weather can occur at any time.

Climate: Sabah enjoys a humid equatorial climate with high temperatures that vary little throughout the year. Temperatures become gradually cooler the higher one climbs; at Laban Rata, temperatures range between 34°F (2°C) and 50°F (10°C), while ice may be encountered around the summit. The characteristic daily weather pattern gives a clear dawn, cloudy morning and rain in the afternoon. The summit area can be cold, wet and inhospitable.

HEALTH & SAFETY

Vaccinations: Typhoid and tetanus are routine. A polio booster and gamma globulin against hepatitis B are recommended. Consult your doctor about current malaria precautions.

General Health Risks: Hypothermia is a serious risk for those without appropriate warm and weatherproof clothing, but precautions should also be taken against the sun. Rabies is present in Borneo, so keep dogs at a distance. Kinabalu is high enough for altitude sickness, and the ascent swift enough for headaches and nausea to occur.

Politics & Religion: Malaysia is a Muslim country, but in Sabah there are also Buddhist and Christian communities. Local customs should be observed and appropriate dress worn. Smoking is disapproved of, and in some public places may even result in fines.

Crime Risks: Low.

Food & Drink: Drink bottled water only, and avoid raw food, unpeeled fruit, unwashed salads and local dairy produce.

HIGHLIGHTS

Scenic: Kinabalu's scenery is unique, its views are stupendous, and Borneo's culture is intriguing.

Wildlife & Flora: The plant life of the park ranges from tropical rain forest, through montane forest to the stunted, moss-draped cloud forest of the higher levels and the hardy alpines of the very highest zones. The park is home to 150 varieties of orchid, many kinds of nepenthes (carnivorous pitcher plants) and the rare, parasitic Rafflesia, whose flowers can span 30 inches (75cm). Deer, wild pig, honey bear, civet, loris, gibbon and the Kinabalu rat frequent the park, while the orang-utan is common, if secretive. There are some 300 species of bird, exotic insects and a wide variety of reptiles including snakes and tree frogs.

orang-utan

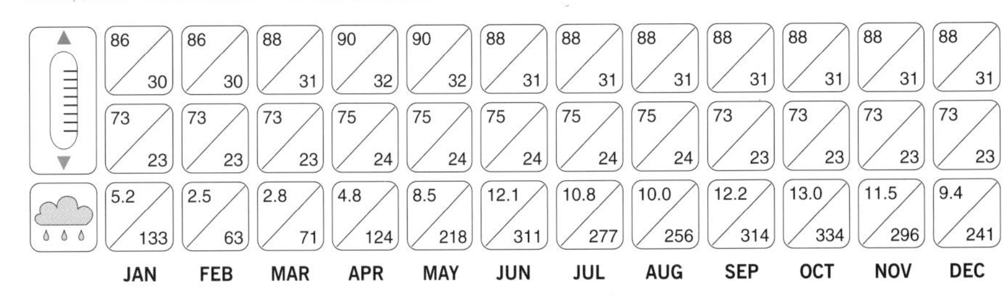

	JAN	FEB	MAR	APR	MAY	JUN	JUL	AUG	SEP	OCT	NOV	DEC
Max °F	86	86	88	90	90	88	88	88	88	88	88	88
Max °C	30	30	31	32	32	31	31	31	31	31	31	31
Min °F	73	73	73	75	75	75	75	75	73	73	73	73
Min °C	23	23	23	24	24	24	24	24	23	23	23	23
Precipitation (in)	5.2	2.5	2.8	4.8	8.5	12.1	10.8	10.0	12.2	13.0	11.5	9.4
Precipitation (mm)	133	63	71	124	218	311	277	256	314	334	296	241

Africa

The magic that is Africa proves an irresistible attraction to the dedicated and experienced trekker. This continent is noted for its wealth of wildlife and huge open landscapes. The four treks selected will take you from the tribal lands of South Africa to the Islamic world of Morocco on the continent's northern tip. In between, there are two mountains to be conquered: the dormant volcano of mighty Kilimanjaro and the impressive Mount Kenya.

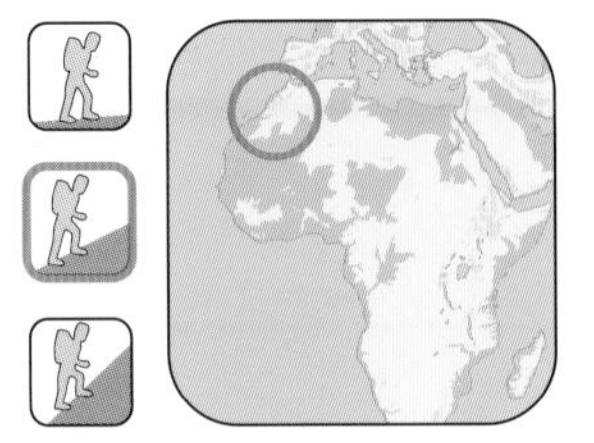

the TOUBKAL CIRCUIT

MOROCCO

Kev Reynolds

STARTING ON THE NORTHERN SLOPES OF THE HIGH ATLAS MOUNTAINS, WITH DJEBEL TOUBKAL – THE HIGHEST MOUNTAIN IN NORTH AFRICA – AS ITS HUB, THE TREK MAKES A CIRCUIT OF THE MASSIF BY WAY OF A SERIES OF HIGH PASSES LINKING THE VALLEYS THAT RADIATE FROM TOUBKAL ITSELF. THE TREK CULMINATES IN AN ASCENT OF TOUBKAL.

Trekking on these high trails, you will meet Berber families crossing the mountains, their mules piled up with children and goods, the women spinning as they walk. And as you pass through one Berber village after another, you may well be invited to take mint tea. The trails are ancient mule tracks used by generations of Berbers for trade and communication, and although the mountains appear wild and formidable, the valleys have been settled for centuries. Where there are villages, hillsides are terraced and irrigated. Elsewhere the mountains are snow-draped in winter and spring, and arid and bare in summer. As each new pass is gained, so a fresh landscape appears ahead, and from the summit of neighbouring peaks a golden haze in the south hints at the distant Sahara. A journey through these mountains offers a unique experience in a land of undisputed charm.

itinerary

•DAY 1 8 miles (13km)

Imlil (or Aroumd) to Amssakrou

Whether you start the trek in Imlil or Aroumd will depend on personal preference or time of arrival at the road head. Imlil has more facilities; Aroumd is 1 to 1½ hours upstream. From either village the way leads to Tizi n'Tamatert (*tizi* means pass), at about 7,477 feet (2,279m). The mule track linking Imlil with Tachdirt is followed for some way down the eastern side of the *tizi* before breaking off on the descent to Ouaneskra, a classic Berber village almost completely ringed by walnut trees. Now heading west, a clear path contours the north flank of the Imenane Valley, passing through Tamegguist and Ikis before arrival at Amssakrou, among an impressive set of terraces.

•DAY 2 5 miles (8km)

Amssakrou to Oukaïmeden

The route climbs steeply out of the Imenane Valley, across gritty slabs to gain a saddle shaded by juniper trees. Views south hint at one ridge after another; later, a red valley far below is lined with villages, and away to the north is the plain that leads to Marrakech. The trail becomes a ledge heading east, crosses a ridge, and makes a rising traverse of screes leading to another saddle overlooking Oukaïmeden (Ouka) and its grassy basin, at the head of which stands the impressive peak of Angour (11,864 feet/3,616m). The way down to the ski resort of Ouka passes a curious, adobe-style village, which seemingly belongs to a different century from that of Ouka's hotels and ski tows. Ask a guide to show you the local prehistoric rock carvings.

•DAY 3 6 miles (9.5km)

Oukaïmeden to Tachdirt

Crossing the 9,711-foot (2,960-m) Tizi n'Eddi south-east of Ouka is straightforward in summer, but deep snow lingers well into spring. The pass affords the most direct way between Ouka and Tachdirt. It's perfectly feasible to

Dwarfed by the landscape, trekkers work their way across the rocky mountainside en route to the ski resort of Oukaïmeden.

complete this route in 3 hours, so it's possible for strong mountain walkers with scrambling experience to climb the summit of Angour. On the south side of the *tizi* the trail forks – mules usually take the right branch, and the alternative is the Angour option. Tachdirt has a refuge, a shop, and a site for a few tents just beyond the village.

•DAY 4 **5 miles (8km)**

Tachdirt to Azib Likemt

East of Tachdirt, a trail climbs to Tizi n'Tachdirt and then down to Setti Fatma and the Ourika Valley – a recommended 2-day trek for another visit. This route, however, tackles some epic scree slopes to the south on the ascent to Tizi n'Likemt (11,663 feet/3,555m), then descends to the summer grazing hutments of Azib Likemt (*azib* means pasture). Irrigation channels line the path, and below the village grassy meadows form the valley bed. Before winter sets in the *azib* is closed, and the Berbers who spend their summers here cross the mountains to more hospitable villages.

•DAY 5 **8 miles (13km)**

Azib Likemt to Amsouzart

The fifth pass on the circuit is Tizi n'Ouraï. The way climbs into a hanging valley drained by the Assif Tinzart, a region of sweeping hillsides, bluffs, boulders and screes. With an old moraine arcing round a bend, the valley becomes less austere: low-growing cushion plants spatter the hillside, and there's broom, thyme and alpines specially adapted to the fierce environment provided by these mountains on the fringe of the Sahara. A gully divides a village of high-rise blocks of mud and stone. Beyond stretches a luxuriously fertile valley: terraced fields, walnut groves and streams. Waterfalls cascade through the lower hills, and the fragrance of vegetation replaces that of dust and scree. Down the valley lies the village of Amsouzart.

•DAY 6 **4½ miles (7.5km)**

Amsouzart to Lac d'Ifni

This stage only requires a morning's exercise as the following day's climb to Tizi n'Ouanoums is fairly demanding. Begin by climbing west out of the Tifnout Valley, passing small villages; the final village of Imheline is the best of all – hanging in tiers on the steep mountain wall, a waterfall spatters beside it, and maize terraces turn the hillside into a staircase of vegetation. The trail passes below the houses and rises again to cross a rib, beyond which lies Lac d'Ifni. Rimmed with savage mountain walls, and with a rocky valley to the west, Lac d'Ifni makes a →

At the foot of the gorge leading from Lac d'Ifni to Tizi n'Ouanoums, goats forage among the boulders.

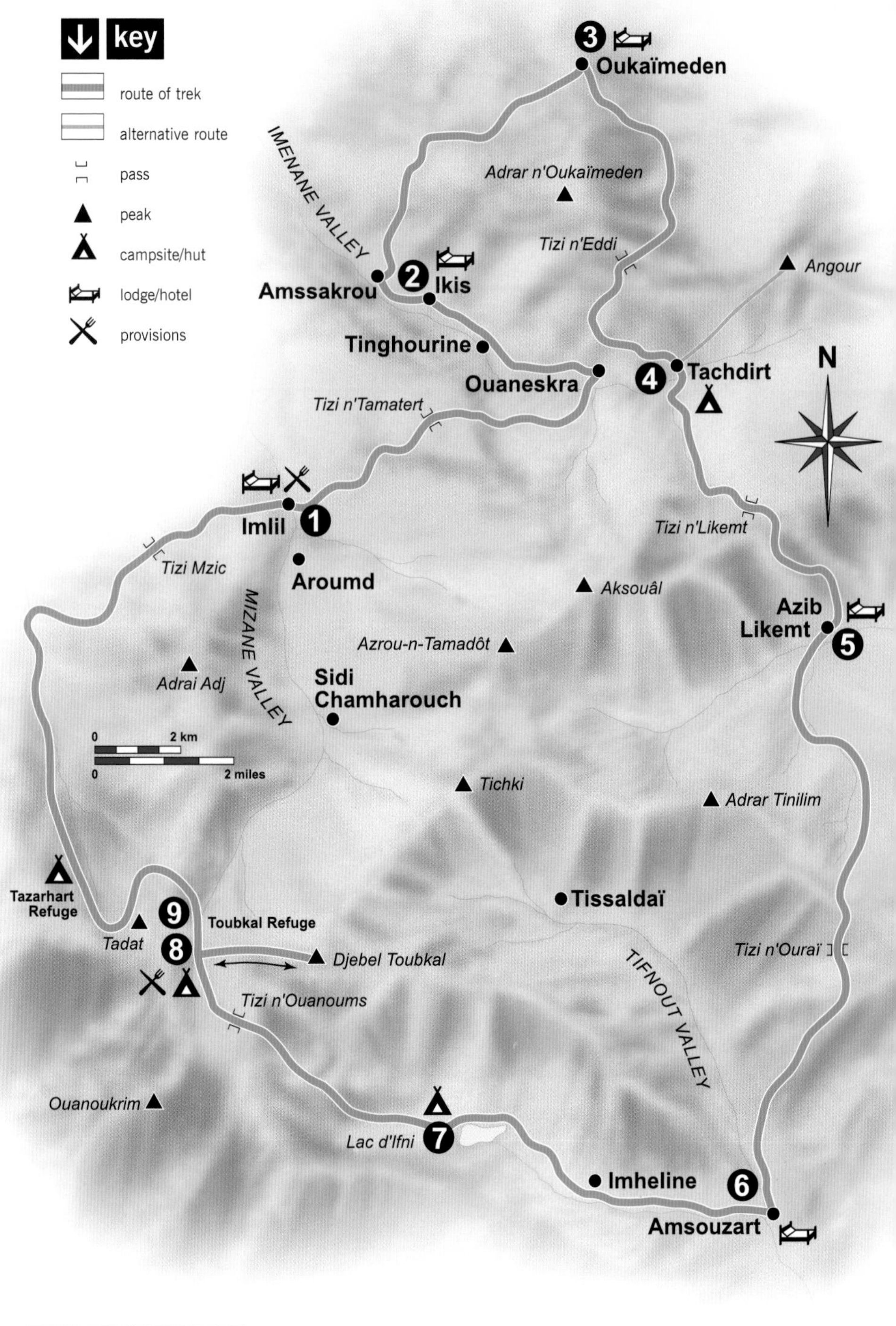

tranquil site for a lazy afternoon. There's no accommodation here, and your only choice is to bivouac along the stony shoreline.

•DAY 7 4½ miles (7.5km)

Lac d'Ifni to the Toubkal Refuge

The route to Tizi n'Ouanoums is the toughest so far. Leaving Lac d'Ifni, it enters a wild gorge dotted with gigantic boulders and a demanding trail. The path climbs under the looming cliffs of Toubkal, but on emerging at the narrow gateway of the pass, the view west across the upper Mizane Valley shows the long Ouanoukrim ridge with a group of pinnacles projecting from it. Below to the north stands the Toubkal Refuge, the main base for tackling Djebel Toubkal.

•DAY 8 4½ miles (7.5km)

Djebel Toubkal

Under normal summer conditions, the ascent of North Africa's highest mountain by the South Cirque (Ikkibi Sud) demands little more than a tiring uphill walk on scree. But in winter or spring this is a serious undertaking. The path begins upstream of the hut and soon rises to a mess of boulders, becoming a scree treadmill most of the way, until gaining the south-west ridge that plunges into shadowed depths on the far side. The summit is marked by a metal tripod.

•DAY 9 9½ miles (15km)

Toubkal Refuge to Imlil

The normal route down to Imlil follows a well-trodden path through the Mizane Valley, passing the pilgrim hamlet of Sidi Chamharouch and Aroumd. A more demanding alternative involves crossing Tizi n'Tadat, north of the refuge and gained by a steep gully, screes and rock ribs. The col opens onto a narrow ridge below the Tadat pinnacle, and the tough descent involves a traverse below Tadat to a series of gullies and screes that plunge down to the Tazarhart Refuge. It continues downstream on a good path into a colourful valley, through a gorge with water cascading from the cliffs. The way passes an *azib* and eventually crosses Tizi Mzic into the Mizane Valley at Imlil, after a long but unforgettable day's mountain travel.

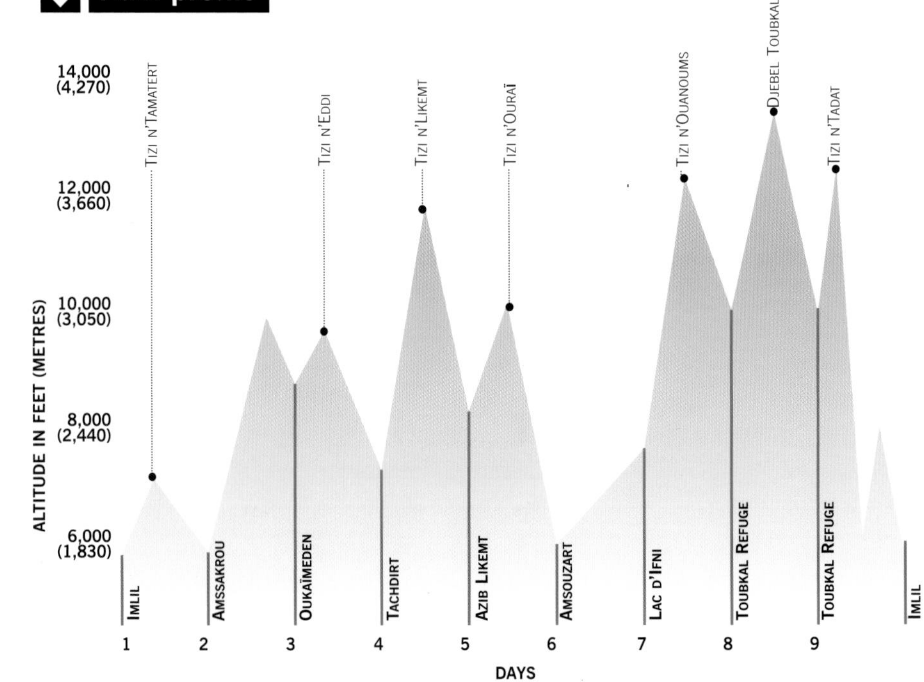

factfile

OVERVIEW

A 55-mile (88-km), 9-day clockwise circuit of Djebel Toubkal, the highest mountain in North Africa, crossing a series of high passes, visiting a number of remote Berber villages, and making the ascent of Toubkal.

Start/Finish: Imlil, Mizane Valley, the High Atlas Mountains.

Difficulty & Altitude: The basic circuit is moderately demanding, with some long ascents and descents over rough ground. The final stage should be attempted only by experienced mountain walkers; a straightforward alternative is offered. There are several passes in excess of 11,000 feet (3,500m), and the summit of Djebel Toubkal measures 13,671 feet (4,167m).

ACCESS

Airports: Marrakech (44 miles/71km to the north).

Transport: Groups are advised to use a minibus taxi from Marrakech to Imlil.

Passport & Visas: Passports required by all visitors.

Permits & Access: Visitors may not stay in the country for longer than 3 consecutive months.

LOCAL INFORMATION

Maps: Carte du Maroc Oukaïmeden–Toubkal at 1:100,000. This is sometimes difficult to obtain – try major map stockists well in advance of a visit.

Guidebooks: *The Atlas Mountains*, Karl Smith (Cicerone Press); *Atlas Mountains Morocco*, Robin G. Collomb (West Col).

Background Reading: *The Rough Guide to Morocco*, Ellingham, McVeigh & Grisbrook (Rough Guides); *The Great Walking Adventure*, Hamish Brown (Oxford Illustrated Press).

Accommodation & Supplies: Trekkers should be self-sufficient in food and cooking fuel. Some supplies are available in Imlil and one or two other villages, but it is better to stock up with provisions in Marrakech. On the circuit there are four mountain huts and a hotel, but most accommodation will be in village houses, sleeping on the rooftops in summer. Be prepared to camp.

Currency & Language: Dirham (dh), made up of 100 centimes. In the Atlas, villagers speak the Berber dialect, but many speak French also.

Photography: No restrictions.

Area Information: National tourist board (ONMT) offices: Australia: c/o Moroccan Consulate, 11 West Street North, Sydney NSW 2060 (tel.: 00 61 2 22 49 99); UK: 205 Regent Street, London W1R 7DE (tel.: 00 44 171 437 0073); USA: 20E 46th Street (Suite 1201), New York NY10017 (tel.: 001 212 557 2520); Morocco: Délégué Régional du Tourisme, Place Abdelmoumen Ben Ali, Marrakech. Imlil's C.A.F. refuge is also a useful source of information.

TIMING & SEASONALITY

Best Months to Visit: May to September are the prime trekking months for this circuit.

Climate: In July/August, at around 5,250 feet (1,600m), temperatures vary from 58 to 85°F (14 to 30°C), and the mountains should be free from snow. The High Atlas receives heavy winter snowfall with temperatures ranging from 23 to 55°F (-5 to 13°C). Precipitation is at its heaviest in November, February, and March.

HEALTH AND SAFETY

Vaccinations: Precautions against tetanus, typhoid, cholera, polio, and Hepatitis B are advised.

General Health Risks: Stomach upsets with diarrhea are fairly common. In the summer months heat stroke is possible in the foothills and out of the mountains. On the ascent of Djebel Toubkal trekkers should be aware of altitude sickness. There is no mountain rescue service, but in the event of a serious accident a helicopter may be sent from Marrakech. Huts have first-aid boxes and stretchers. Make sure you have adequate insurance.

crimson-winged finch

Special Considerations: Beware of camping close to streams in inclement weather as flash floods can occur. Take great care descending and crossing Atlas screes. Scorpions are found in the mountains – their sting is poisonous, but they're usually harmless unless provoked.

Politics & Religion: Morocco is a democratic monarchy which, while facing the challenge of radical Islam, has managed to remain largely peaceful. Morocco is an Islamic state.

Crime Risks: The greatest risk out of the mountains are pickpockets, and various scams carried out by touts who latch on to tourists in Marrakech. In the mountains there are no untoward concerns.

Food & Drink: For most of the trek it is likely you will cater for yourself, with foodstuffs brought in from Marrakech. In village houses a *tajine* (meat and vegetable stew) is often served. All stream water should be treated before drinking. Mint tea is offered in villages, and it is customary to drink three glasses (no more, no less).

Simple houses stacked one upon another on the steep mountainside form the attractive village of Imheline.

HIGHLIGHTS

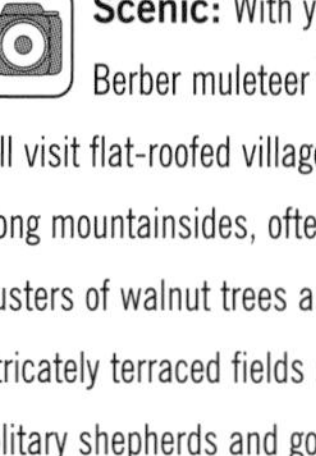

Scenic: With your Berber muleteer you will visit flat-roofed villages tiered along mountainsides, often with clusters of walnut trees and intricately terraced fields nearby. Solitary shepherds and goatherds tend their flocks, ranging the hillsides in search of good grazing.

Wildlife & Flora: The most unusual mammal in the region is the elephant shrew, a mouse-like creature with a trunk for a snout. The Moorish gecko is often seen spreadeagled on the houses. Two birds unique to the mountains of North Africa, and seen in the Toubkal region, are Moussier's redstart and crimson-winged finch. Wildflowers include wild gladioli, daffodils of the hooped-petticoat variety,and orchids.

temperature and precipitation

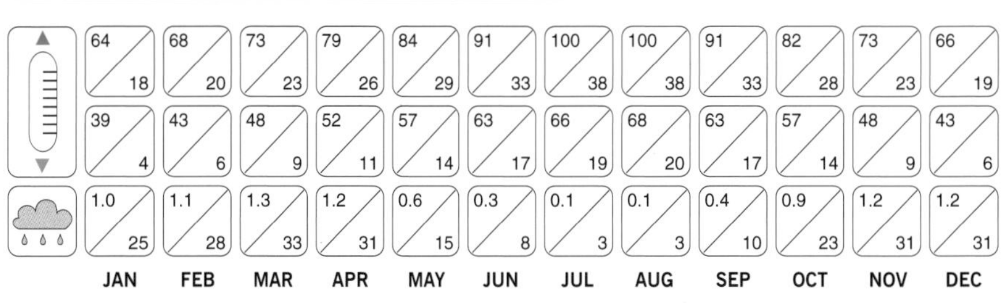

	JAN	FEB	MAR	APR	MAY	JUN	JUL	AUG	SEP	OCT	NOV	DEC
Max temp °F / °C	64 / 18	68 / 20	73 / 23	79 / 26	84 / 29	91 / 33	100 / 38	100 / 38	91 / 33	82 / 28	73 / 23	66 / 19
Min temp °F / °C	39 / 4	43 / 6	48 / 9	52 / 11	57 / 14	63 / 17	66 / 19	68 / 20	63 / 17	57 / 14	48 / 9	43 / 6
Precipitation in / mm	1.0 / 25	1.1 / 28	1.3 / 33	1.2 / 31	0.6 / 15	0.3 / 8	0.1 / 3	0.1 / 3	0.4 / 10	0.9 / 23	1.2 / 31	1.2 / 31

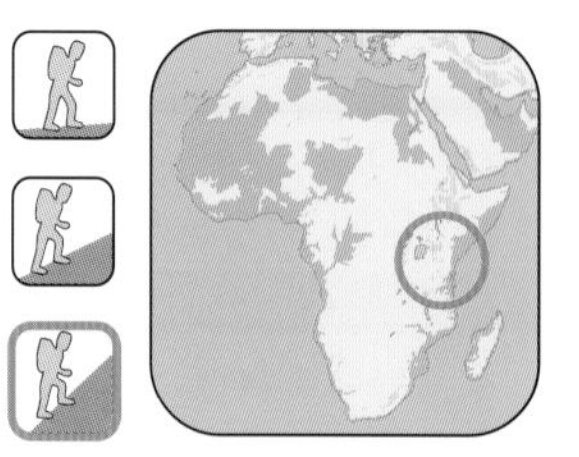

across the ROOF of AFRICA:

A Traverse of Kilimanjaro

TANZANIA

Bill O'Connor

THIS TREK IS A TRAVERSE ACROSS "THE ROOF OF AFRICA". THE VAST BULK OF KILIMANJARO, THE HIGHEST MOUNTAIN ON THE AFRICAN CONTINENT, IS 200 MILES (330KM) SOUTH OF THE EQUATOR IN TANZANIA. ITS PERFECT VOLCANIC CONE DOMINATES THE SKYLINE, RISING TO 19,340 FEET (5,895M) ABOVE THE SURROUNDING PLAINS AND GAME RESERVES.

"Kili" is a popular mountain (it is not in fact a single peak, but a vast complex of volcanic cones and cores spread over an area 38 miles/61km long by 25 miles/40km wide), and attracts around 15,000 tourists annually. Of course, not all of these people climb the mountain, and most that attempt it follow what is known as the Coca-Cola trail, the standard route from Marangu village and the Kilimanjaro National Park headquarters. But the traverse of Kili described here combines several established routes to provide the finest non-technical route to the summit. You can expect to see few if any other trekkers until you reach the mountain's rim, and it is a trek full of variety and interest, passing through five distinct ecological zones. And, apart from its sheer beauty, a major benefit of the route is that it allows for proper acclimatisation before the final summit push. Tough in parts, it's a trek that fills you with a sense of adventure and achievement.

itinerary

•DAY 1 6½ miles (10km)

Machame Gate to Machame Huts

The park offices are east of the trail and it is necessary to sign in. A four-wheel drive vehicle will be able to continue for several more miles/kilometres, to a point where the track through the forest runs out. From here the path breaks right, beneath a dense canopy of podocarpus trees hung with vine-like lianas and through an understorey of tree ferns and nettles. The sometimes muddy path climbs, steeply in places, a well-defined ridge bounded by a stream. The walking, although easy, can be hot, as "jungle" envelops you in intoxicating, green, tree-filtered sunlight. In light gaps in the forest look out for brightly coloured balsam flowers and red gladioli. As you near the hut site, the forest thins

The landscape of northern Tanzania is dominated by the distinctive flat-topped mountain of Kilimanjaro, a popular challenge with travellers to East Africa.

and gradually gives way to giant heather as the ridge crest narrows and you emerge at the Machame Huts. These circular, corrugated tin huts (Uniports) are, sadly, in a terrible condition. It's usual to let porters sleep in the huts and to camp near by. Water can be found by descending steeply to the north-west. Keep a lookout for hyrax, which live around the huts, and for leopard, which eat them.

•DAY 2 4½ miles (7.5km)

Machame Huts to Shira Hut

As you traverse the moorland zone, the path follows a narrowing rib of lava, winding its way through giant heather, groundsel plants, soft-petalled lobelia and aromatic ever-lasting flowers. On the left, the hillside falls steeply to a densely wooded valley until finally at its head the path bends left (west), ascending a final wall of volcanic rock in a series of steep steps. Pass beneath a tiny waterfall (fill your bottle), cross a second stream and continue to the high ground and open moorland of the Shira plateau. In places the ground is scattered with fragments of black volcanic glass (obsidian). Leopard and African hunting dogs have been sighted along this section of trail. Eventually, the vague path leads north-west to the dilapidated Shira hut. Fortunately there are plenty of spaces for camping close to the hut and water can be had nearby. The evening views of the Shira Peaks are often spectacular as the sun sets and afternoon clouds settle into the forest below. North-east across the plateau, the summit snows and southern glaciers can be seen burnished by evening light.

•DAY 3 6½ miles (10km)

Shira Hut to Barranco Hut

This is a good acclimatisation day, following the maxim of walking high and sleeping low. For although the Barranco Hut at 12,795 feet (3,900m) is only about 330 feet (100m) higher than Shira Hut, the route in between climbs

The higher you trek, the more barren the landscape becomes: to reach Uhuru Peak a lava desert must be crossed.

to 14,764 feet (4,500m) at the Lava Tower before eventually descending to the Barranco.

Striking east from the hut over gently rising moorland, the Shira Trail connects with the South Circuit Path (marked with red paint blobs), which leads more or less east-south-east. All the while, views of the rarely visited Credner and Penck Glaciers dominate the skyline. The trail traverses two broad valley depressions to an outstanding rock pinnacle called the Shark's Tooth. At the Lava Tower Hut, the trail divides. Stay on the main trail, going east then south round the Lava Tower. Once on the east side of the tower, you will join up with the South Circuit Path. This is an austere landscape, and the altitude will not go unnoticed as you climb out of steeper hollows.

The path eventually winds its way down a remote valley, cresting the ridge above the gorge of the Great Barranco before descending through giant groundsel to the Barranco Hut. Here you will find good camping places and water, and be able to enjoy the awesome views of the Breach Wall and hanging ice of the Heim and Kersten Glaciers.

•DAY 4 **5 miles (8km)**

Barranco Hut to Barafu Huts

The trail descends north to the stream and crosses it following a distinct trail toward the Barranco Wall, a 990-foot (300-m) rock step visible from the campsite. The path now zigzags with surprising ease up this well-prepared path, with extensive views of the Western Breach and Breach Wall. Once up the Barranco Wall, the path levels and you traverse east across a desolate landscape of volcanic ash and scree. Hanging high above you to the north are the great southern glaciers of Kilimanjaro: Heim, Kersten and Decken. In traversing eastward, the trail climbs in and out of numerous small valleys, and below the Decken glacier a meltwater stream has cut a deeper valley and provides the last water for 2 days. The South Circuit Path eventually meets the Mweka Ridge Path and the Barafu Huts are signposted. The way to the huts is up a broad, rock-strewn ridge and the 1½-mile (2-km) walk can seem endless. Camping spots are limited but adequate. Trekkers who have not yet acclimatised should spend another day around the Barranco before going higher.

•DAY 5 **12 miles (19.5km)**

Barafu Huts to Horombo Huts

This is a long and strenuous day above 13,000 feet (4,000m), and it should begin in the early hours of the

walk profile

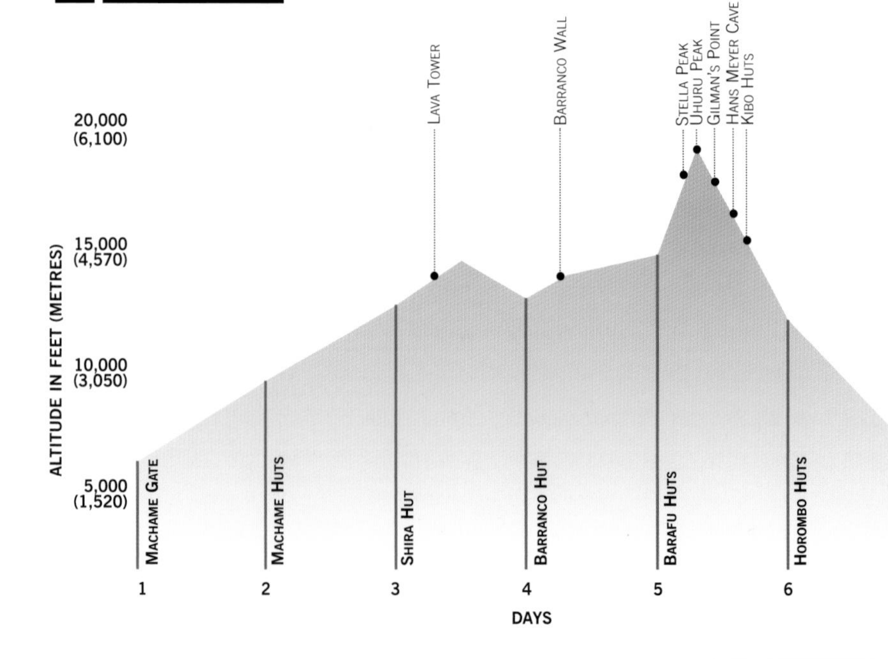

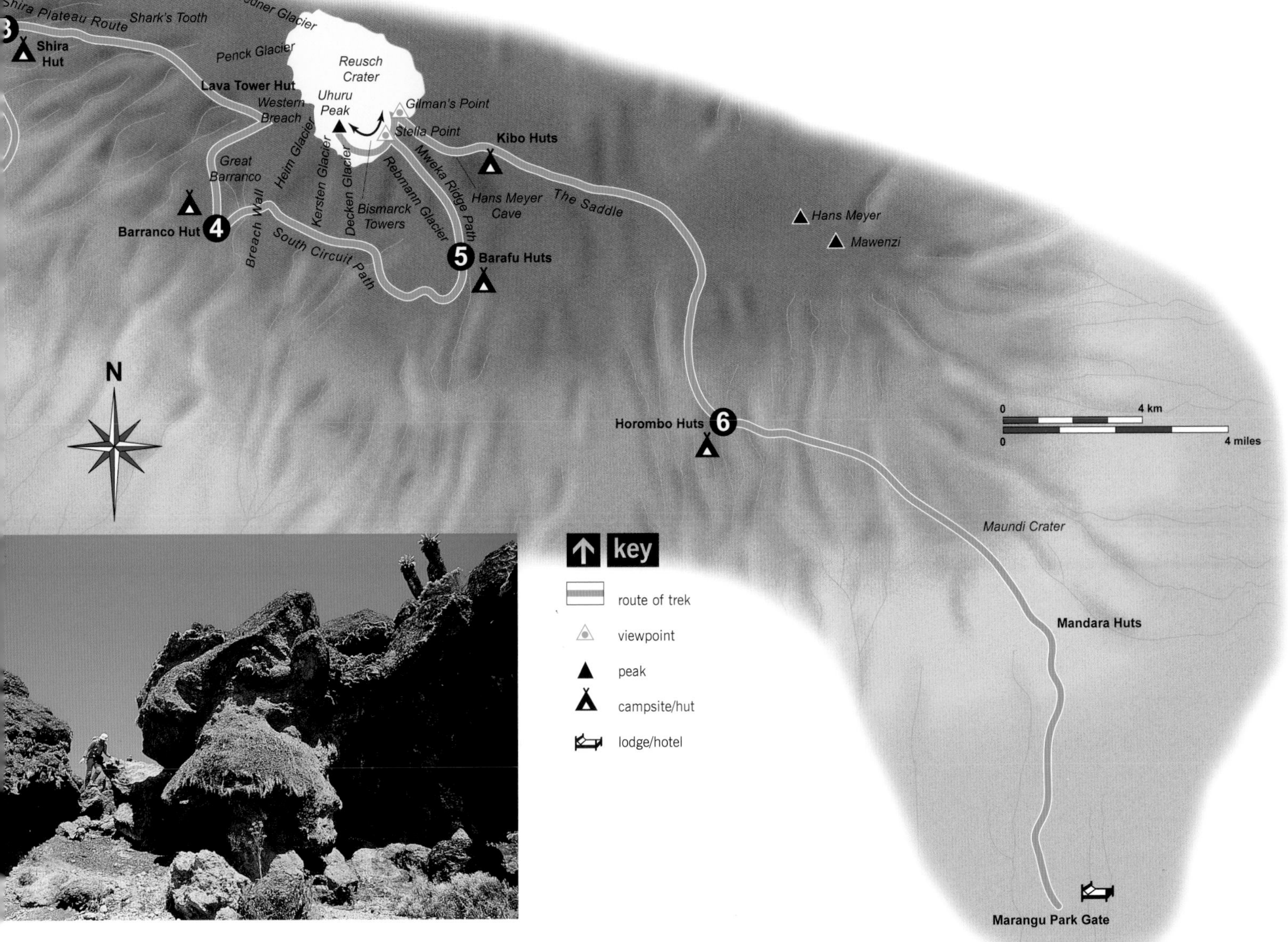

The South Kibo Circuit takes "Kili"-climbers on a scramble through the bizarre, amorphous forms of lava towers.

morning so that you reach Stella Point (19,000 feet/5,800m) on the crater rim, at, or soon after, sunrise. The hike is steep but without technical difficulty. It follows the ridge above the huts, steeply at first, and then up a monotonous, hopefully frozen, scree to the right of the Rebmann Glacier. At Stella Point, you can look down into the crater and across to the stepped eastern ice fields and darker rim of the main Reusch Crater and still-smoking ash pit. Uhuru Peak lies to the north-west and is reached by hiking round the rim. Finally, after 6 to 7 hours, you will arrive at the "Roof of Africa", marked by a small monument, some flags, and cairns of stones left by others. If you reach the summit, you can congratulate yourself on being among the 40 per cent or so of those that attempt the mountain to conquer it. The first recorded ascent of Kilimanjaro was made on October 5, 1889 by Hans Meyer and Ludwig Purtscheller. They planted the German flag on the summit and named it Kaiser Wilhelm Spitze, in honour of their king. The views are magnificent. The crater is adorned with stepped ice cliffs and fast-melting glaciers, and beyond that the vastness that is Africa.

But reaching the summit is only half a success. You must now descend by retracing your steps to Stella Point and continuing round the crater rim to Gilman's Point (18,635 feet/5,680m), with the jagged crest of Mawenzi beckoning to the east. The crater rim is not at all flat; in fact, the path rises and falls over several small pinnacles including Hans Meyer and Elvida Point before skirting the Bismarck Towers and climbing from the crater side to the rocky step of Gilman's Point (marked by a cairn and sign). Having arrived, you can look forward to a fast descent down the soft scree slopes of the "normal route" into thicker air and your porters waiting at the Kibo Huts (15,420 feet/4,700m). En route you will pass Hans Meyer Cave (16,900 feet/5,150m), where the angle relents and →

→ *The unrelenting African sun beats down on trekkers as they pass through the eerie giant heather above Machame Huts.*

a well-marked track leads to a plethora of huts. You can stay here, but after a rest and something to eat most people opt to continue to the relative luxury and lower altitude of the Horombo Huts, 3,300 feet (1,000m) descent and 7 miles (11.5km) away – but it's mostly downhill!

Refreshed, rehydrated, and having shaken the gravel from your boots, you can begin the magnificent walk across the rock-strewn lava desert, known as the Saddle, that stretches between Kibo and Mawenzi. Your legs may be tired but lungs and head will delight in the thick air. An increasing number of plants and everlasting flowers mark the start of the moorland zone and in no time at all you arrive at a strange collection of Scandinavian A-frame wooden buildings – the Horombo Huts (12,140 feet/3,700m). Most opt to stay in these bunkhouses, but there are places to camp.

•DAY 6 12 miles (19.5km)

Horombo Huts to Marangu Park Gate

In ascent, this part of the standard Marangu Route is normally done in 2 days, to allow for acclimatisation. In descent, it is easily covered in one day, with a lunch stop near Mandara Huts (8,850 feet/2,700m). Be sure to look back on the route you've covered, because the views of the mountain are striking. After days at high altitude, the walk through the moorland with its seemingly lush vegetation, including large-headed protea flowers, is a real treat. The well-marked trail drops south-east, crossing stream beds and traversing an area devastated by fire in 1997 close to the Maundi crater. This small, but perfectly formed volcanic cone is well worth a side trip.

The main trail is easily rejoined and the way, now through forest, is straightforward. The trail below the Mandara Huts is obvious as it files through the dense forest where you can enjoy an increasing variety of plants and wildlife. The main trail eventually becomes a wide dirt road, but a narrower trail on the right helps delay the return to civilisation, winding as it does through dense forest and past streams and pools. Eventually you will arrive at the Park Gate and the Park Headquarters to sign out and be issued with your summit success certificate.

↓ factfile

OVERVIEW

This is perhaps the finest high-level traverse in Africa, and certainly the best non-technical route on Kilimanjaro. Over 46 miles (75km), it links several well-established high-level trails: the Machame and Shira trails, the South Circuit Route, and the final part of the Mweka Ridge, known as the Barafu, which leads to the summit rim and Uhuru Peak. From the summit the trek follows the edge of the crater to Gilman's Point to descend the Marangu Route. In all, it's a wonderful journey, and an opportunity to see the little-visited southern glaciers of Kilimanjaro.

Start/Finish: Most begin this trek in the village of Marangu by staying in a lodge or hotel near the park headquarters. From here a taxi ride will take you to the start of the trek close to Machame village, at the edge of the forest. Moshi is the nearest large town.

Difficulty & Altitude: The week-long trek is relatively strenuous, with daily ascent over 3,300 feet (1,000m) and 6 hours of walking being the norm. The route is also at altitude, much of it over 13,123 feet (4,000m). The summit day is even longer and higher, but with more than half of it going downhill, it is readily achievable by fit walkers.

ACCESS

Airport: Kilimanjaro International Airport, between Arusha and Moshi.

Transport: It is possible to travel by bus from Moshi to Marangu very cheaply. A taxi will naturally cost a lot more (the price should be agreed before starting). From Marangu you will need a four-wheel drive vehicle to take you and your porters to Machame village and the start of the trek. This can be arranged by your outfitter.

Passport & Visas: Passports and tourist visas required by all visitors.

Permits & Restrictions: To climb the mountain you must work through a Tanzania outfitter and use a local guide and at least one porter to carry the guide's belongings while inside the park. In fact it is usual to have two or three porters per trekker. It is better to join a group and accept the fact that you will be part of a mini-expedition. There are numerous hotels around Marangu that arrange park reservations, porters, guides, food and transport to the start of the trek. Park fees have to be paid to the Park Headquarters at Marangu Gate.

LOCAL INFORMATION

Map: 1:50,000 Kilimanjaro map and guide, Mark Savage.

Guidebook: *Kilimanjaro & Mt. Kenya – a climbing and trekking guide*, Cameron M. Burns (Cordee).

Accommodation & Supplies: There are plenty of hotels and lodges in and around Marangu, Arusha, and Moshi. On the mountain it is usual to camp rather than use the huts, which are in a poor and dirty condition, apart from those on the Marangu route. All food should be taken with you as none is available on the trek. It can be bought in either Arusha or Moshi, although there is limited choice. It is normal for an outfitter to supply food and for it to be carried and cooked by porters.

Currency & Language: Tanzanian shilling. The official language is Swahili, but many of the waChugga, who live around Kilimanjaro, speak English and a smattering of other European languages.

Photography: On the mountain there are no restrictions, but be wary of taking pictures around airports and other public buildings, and of Tanzanian personnel in uniform.

Area Information: Chief Park Warden, P.O. Box 96, Marangu, Tanzania; Tanzania National Parks, PO Box 3134, Arusha, Tanzania.

TIMING & SEASONALITY

Best Months to Visit: Basically there are two trekking seasons in East Africa – from the middle of December to mid-March and from early June until mid-October. The busiest time is during the warmer dry season over Christmas.

giant forest pig

Climate: Daytime temperatures are usually quite warm, considering the altitude, and can reach 50°F (10°C) on the mountain, but in the early hours of the day it can drop to around 40°F (5°C), especially above 15,000 feet (4,570m). Cloudy days certainly feel cooler on the mountain so you still need to carry warm clothing and a comfortable sleeping bag.

HEALTH & SAFETY

Vaccinations: Yellow fever, hepatitis A, typhoid, polio, tetanus and meningococcal meningitis. You will also require anti-malarial tablets.

General Health Risks: Health insurance is essential. Those from northern latitudes will need to protect against the equatorial sun. A sun hat and sunscreen are essential.

Special Considerations: You will need to carry a comprehensive first-aid kit and be able to use it should the need occur. The mountain is high and you reach altitude quickly, so it is essential to acclimatise well and be on the alert for high-altitude ailments. In the case of altitude sickness, descending is the best cure. Better still, avoid it by taking time to acclimatise and going at a slow pace.

Politics & Religion: Tanzania is a stable, socialist country that has all the usual problems of Africa and the developing world.

Crime Risks: The crime rate is low, but normal security precautions should always be taken with valuables.

Food & Drink: You can always take your own food, but most outfitters provide good food for the mountain, including fresh meat and vegetables and the African staple *ugali* (cornmeal). Mangos and bananas abound. Breakfasts are usually a mixture of toast, porridge, eggs and fruit and, of course, tea, coffee, or squash. Water on or around the mountain is good, but precautions should always be taken around habitation or campsites in case of human contamination. It is essential to hydrate well.

scarlet-tufted malachite sunbird

HIGHLIGHTS

Scenic: Lush vegetation, thriving sisal and banana plantations, and bustling *shambas* (cultivated patches) line the routes to the mountain. The people that live on its flanks are the waChugga, a cheerful, friendly and independent people – the porters and guides are hard-working and, with their singing, make delightful company. On a clear day the dwindling summit snows and glaciers of Kilimanjaro glint in the equatorial sun as if burnished with silver.

Wildlife & Flora: On Kili there is always the chance of a rare sighting of larger mammals or smaller reptiles, and there are abundant birds, from colourful sunbirds and turraco to large raptors. The ever-changing vegetation as you climb through five distinct ecological zones never ceases to fascinate: the lower slopes (2,600–5,900 feet/800–1,800m); montane forest (5,900–8,850 feet/1,800–2,700m); heath and moorland zone (8,850–13,000 feet/2,700–3,960m); alpine desert zone (13,000–16,500 feet/3,960–5,000m); and the summit zone (16,500–19,340 feet/5,000–5,895m).

→ *The rich vegetation of the moorland zone includes the huge giant lobelia and senecio plants.*

↓ temperature and precipitation

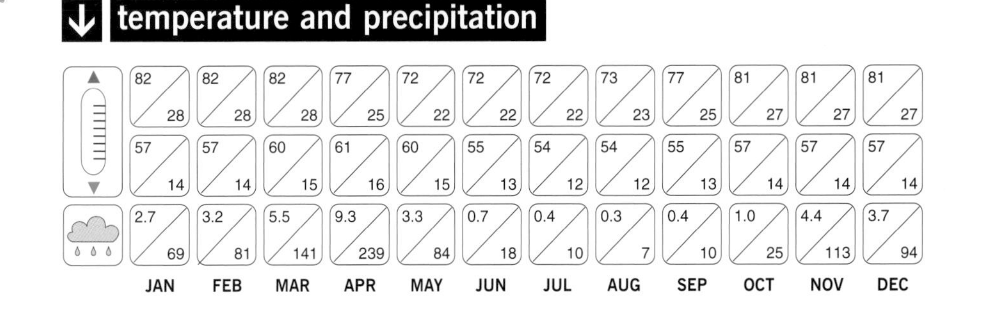

	JAN	FEB	MAR	APR	MAY	JUN	JUL	AUG	SEP	OCT	NOV	DEC
High (°F / °C)	82 / 28	82 / 28	82 / 28	77 / 25	72 / 22	72 / 22	72 / 22	73 / 23	77 / 25	81 / 27	81 / 27	81 / 27
Low (°F / °C)	57 / 14	57 / 14	60 / 15	61 / 16	60 / 15	55 / 13	54 / 12	54 / 12	55 / 13	57 / 14	57 / 14	57 / 14
Precipitation (in / mm)	2.7 / 69	3.2 / 81	5.5 / 141	9.3 / 239	3.3 / 84	0.7 / 18	0.4 / 10	0.3 / 7	0.4 / 10	1.0 / 25	4.4 / 113	3.7 / 94

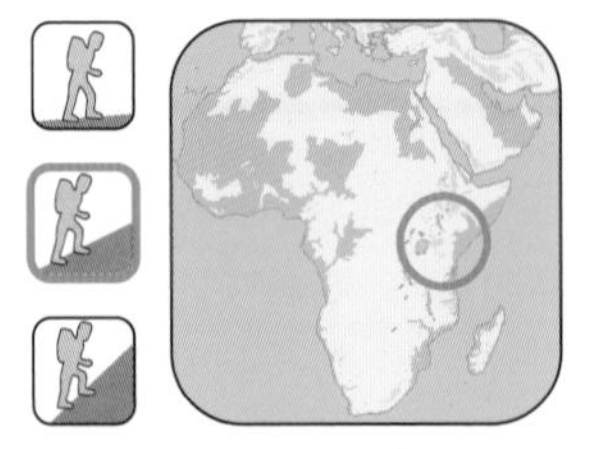

the MOUNT KENYA CIRCUIT

KENYA

Bill O'Connor

THIS TREK IS A WONDERFUL COMBINATION OF SEVERAL ESTABLISHED TRAILS THAT CONVERGE ON MOUNT KENYA FROM SEVERAL POINTS OF THE COMPASS. THROUGHOUT, IT OFFERS SPECTACULAR, EVER-CHANGING SCENERY: VISTAS OF VAST PLAINS BORDERING THE GREAT RIFT VALLEY, DENSE FOREST, DEEP GORGES, TUMBLING WATERFALLS, MOORLAND AND EVEN GLACIERS AND MOUNTAIN TARNS.

There are so many layers to a trek to Mount Kenya. Combined, the Chogoria Trail, the Summit Circuit Path, and the Naro Moru Trail provide the best the mountain has to offer in terms of diversity and interest. Chogoria is the best introduction to the mountain's intricate vegetation, wildlife and geology, and, as important, it will allow you to acclimatise and enjoy the experience to the full. Then you choose whether to traverse Mount Kenya by linking the Chogoria Trail with the Naro Moru Trail or to go the whole hog and include the Summit Circuit Path as well. Either way, you will be participating in a magnificent and unusual adventure.

•DAY 1 14 miles (22.5km)

Mount Kenya Forest Gate to Meru Mount Kenya Lodge

Only the hardy few opt to walk this section. Most prefer the dubious comfort of a vehicle for most if not all of the 14 miles (22.5km) to the Chogoria Park Gate. The walk, for those that do it, is through magnificent rain forest, dense with mutati, podocarpus, giant camphor, rosewood, oak, and seemingly endless tree species, as well as bamboo. The Meru Mount Kenya Lodge is an assortment of self-catering cabins or *bandas*. They are quite comfortable and have hot showers, a kitchen, small lounge and bedroom. You can reach them in a day from Nairobi. At an altitude of 9,850 feet (3,000m), it is wise to spend at least one night here, and, because it is outside the national park, there are no entry fees to pay.

•DAY 2 9½ miles (15km)

Meru Mount Kenya Lodge to Hall Tarns

On entering Chogoria Park Gate you will need to pay entrance fees. It is possible to take a four-wheel drive vehicle up the rough track to the road head about 4½ miles (7.5km) on, but the walk will do you good! Not far from the gate, take a path on the left which crosses a stream and heads north-west via the Urimandi Hut and the delightful Nithi Falls, (just south of the path) before joining the road head at about 10,500 feet (3,200m). From the road head, the trail crosses the stream to the south before turning west to ascend a well-marked ridge. You are now out of the woods and into grassy moorland, with scattered rosewood trees, head-high heath and whispering grasses. Bearing →

← *Even in the scorching heat of Africa, there are mountains high enough to retain snow on their peaks throughout the year.*

→ *From the moorland zone, where the cabbage-like flowers of the giant groundsel trees line the route, the views back down the mountain are vast.*

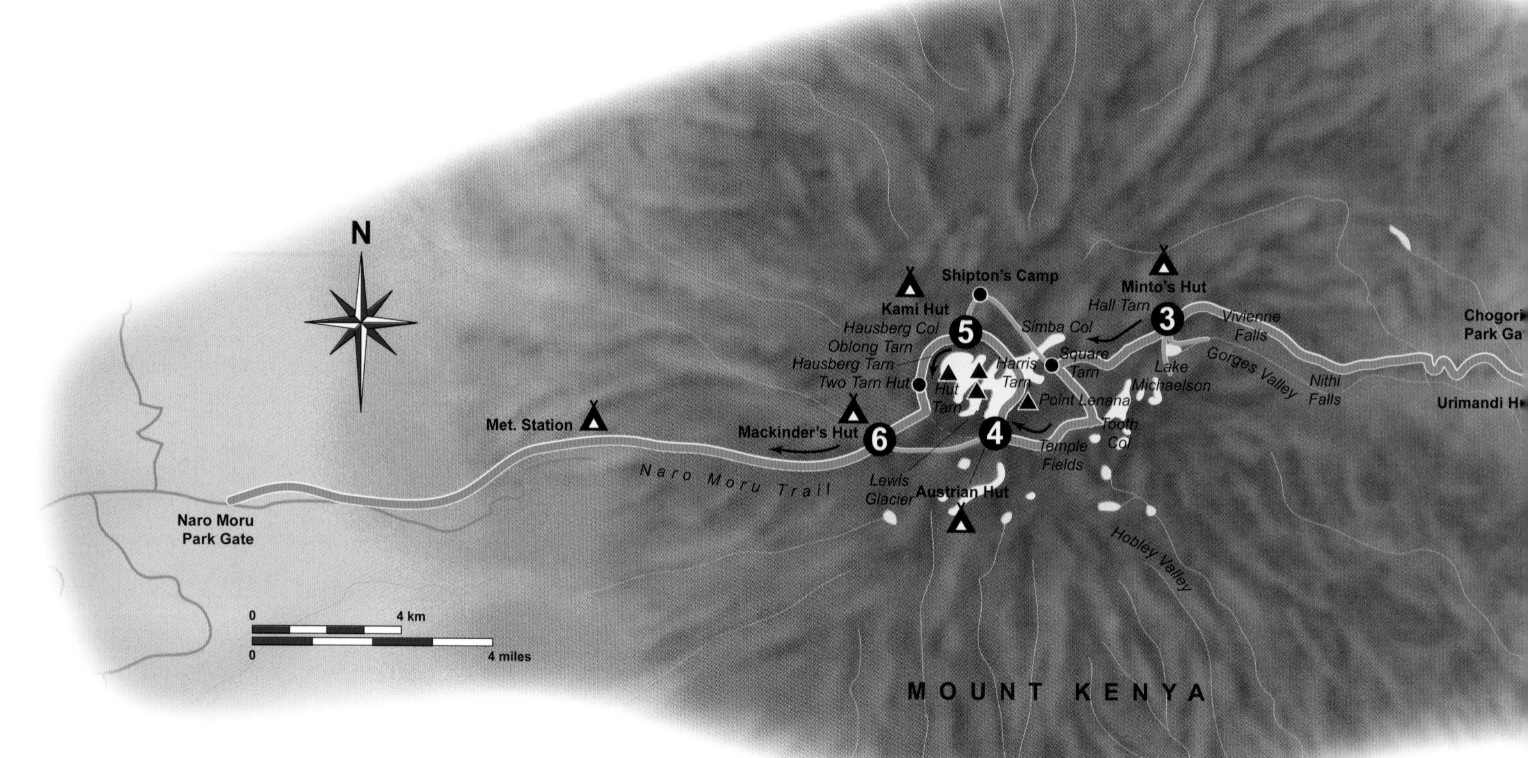

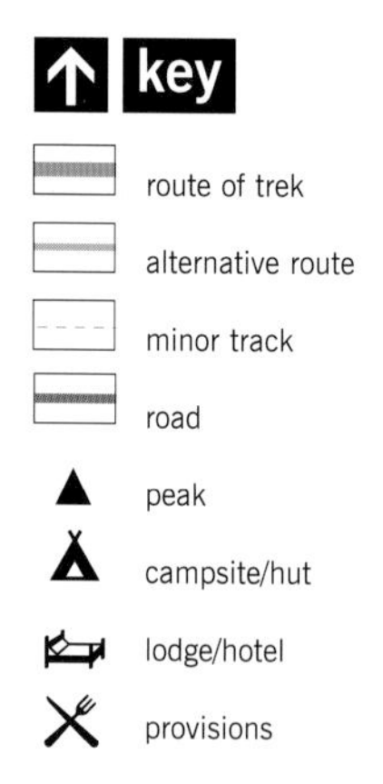

north-west, the trail borders the spectacular U-shaped trench of the Gorges Valley. Staying above the gorge, the trail turns west yet again through mountain moorland dotted with giant senecio and lobelia to reach the Hall Tarns and Minto's Hut at about 13,800 feet (4,200m). The hut is not recommended, but the camping around the tarns is. The landscape is a magnificent mountainous moonscape, and an extra day exploring the area is well worth taking. A descent into the gorge to visit Lake Michaelson and the Vivienne Falls will repay the effort, and the added acclimatisation will be of benefit later. Birds are a real feature of Mount Kenya: look out for malachite sunbirds, chestnut winged starlings, augur buzzard and Verreaux's eagle as you traverse its slopes.

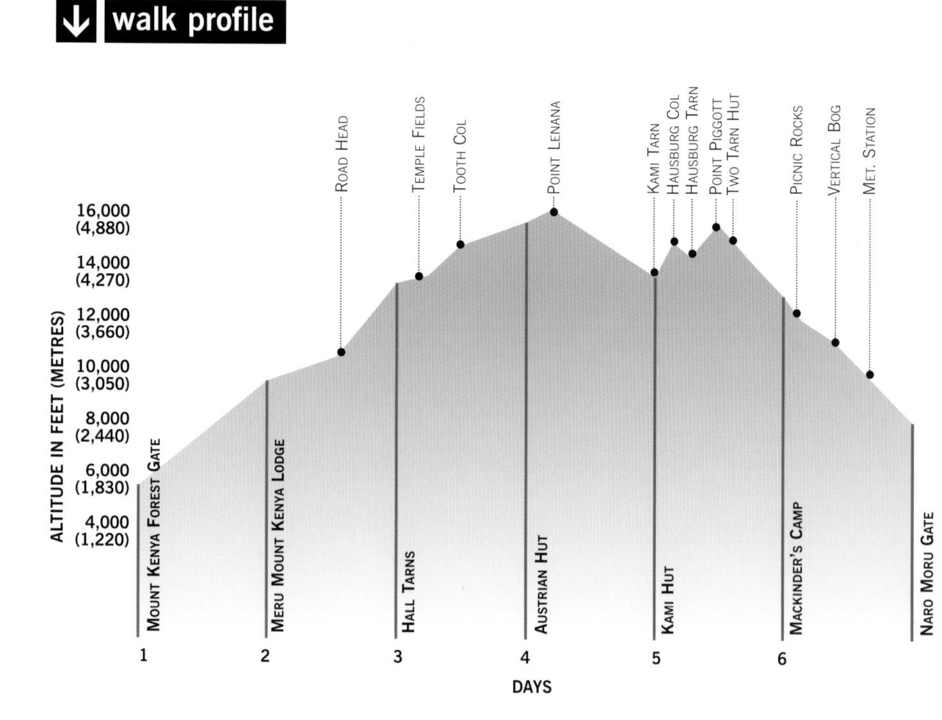

•DAY 3 6 miles (9.5km)

Hall Tarns to Austrian Hut

The trail now heads west over rough, boulder-strewn moorland to reach a flattish area known as Temple Fields, where the path divides to go either north or south following the Summit Circuit path (see Day 4). For those intent on climbing Point Lenana (1,519 feet/4,985m), the third-highest summit of Mount Kenya and a non-technical one at that, the option is to go south-west to the valley head, and then steeply to Square Tarn.

Temple Fields is a delightful spot, featuring the odd giant senecio and lobelia, along with everlasting flowers and decorative grass. From here, the trail climbs a sharp scree slope to Tooth Col, dramatically set amid volcanic towers. Ahead, it crosses the head of scenic, tarn-speckled Hobley Valley to reach Austrian Hut (1,575 feet/4,800m) close to the fast-retreating Lewis Glacier. This wooden hut is built on the site of the former Arthur Firmin Hut and is, sadly, in a sorry state. Water is available from the Curling Pond, close to the Lewis Glacier. What a setting this is: below, to the south,

↑ *Tarns of deep greens and blues provide a welcome break to the rocky landscape that greets you as you climb higher.*

you look over the dark spread of the forest and beyond that to the vast expanse of highland and African plain. If you are lucky and conditions are right, you might just be able to pick out the massive bulk of Kilimanjaro far off in Tanzania.

The ascent of Point Lenana from Austrian Hut is quite straightforward, although you may have to tread on a little ice or snow. At 1,519 feet/4,984m, it is the third-highest summit on Mount Kenya. The easy ascent follows the rocky ridge to the right of the Lewis Glacier, staying away from ice and gravel. The final rocky block is best overcome at its north-west corner.

•DAY 4 **2½ miles (4km)**

Austrian Hut to Kami Hut

For those intent on going full-circle on Mount Kenya, the Summit Circuit Path is the only option. Although it can be completed in 1 day, 2 are preferable, if only to enjoy at leisure different aspects of the mountain. It would have been possible to join the Summit Circuit by going north from Temple Fields, but that would have meant climbing Point Lenana from Mackinder's Camp by a long and tedious scree slope. Much better to climb it first and retrace your steps to Square Tarn, then go north to Simba Col. Alternatively, ascend Lenana by the Southwest Ridge and descend the North Ridge to Harris Tarn (about 15,500 feet/4,725m), continuing north-west to Kami Hut and tarn. This makes an ideal place to camp and split the circuit, rather than going to the lower Shipton's Camp. From the tarn there are spectacular views of the north flank of the mountain as well as of the outlying peaks of Terere and Sendeyo.

VARIATION: 2-hour walk. Austrian Hut to Mackinder's Camp. Those who don't wish to complete the Summit Circuit can reach Mackinder's camp directly from Austrian Hut by descending a seemingly endless scree slope, first toward Tilman Peak to the south and then steep and dustily westwards into the upper Teleki Valley and Mackinder's Camp. From there, follow directions for Day 6.

•DAY 5 **4½ miles (7.5km)**

Kami Hut to Mackinder's Camp

The Summit Circuit trail continues west of the Kami Hut tarn, zigzagging up scree through an austere landscape to Hausberg Col (15,062 feet/4,591m) before dropping steeply to Hausberg and Oblong tarns (about 14,534 feet/4,430m). The route now climbs south-west to crest a ridge (15,308 feet/4,666m), between Point Pigott (16,263 feet/4,957m) and Arthur's Seat. Ahead, the route descends a series of rock ledges to go between Nanyuki and Hut Tarn to reach the appropriately named Two Tarn Hut (14,730 →

feet/4,490m). From here, there are impressive views of Batian's West Face and the hanging ice of the Tyndall Glacier. Skirting Hut Tarn, the route drops into the upper Teleki Valley, where the Naro Moru Trail ends at the Ranger's Post and, slightly lower down, Mackinder's Camp (13,780 feet/4,200m). From here there are magnificent views of Mount Kenya's twin peaks, Batian (17,057 feet/5,199m) and Nelion (17,021 feet/5,188m), and the sabre slash of the icy Diamond Couloir falling from hanging Diamond Glacier and romantically named Gate of the Mists.

With the circuit complete it's time to get off the mountain. A full day will take you down the Naro Moru track to the Naro Moru Gate and the park exit.

• DAY 6 12½ miles (20km)

Mackinder's Camp to Naro Moru Gate

The trail descends eastwards, following the valley floor, or you can follow the ridge to the south, leading through a moorland dotted with giant plants to reach Picnic Rocks (12,300 feet/3,750m). The canopy of the forest awaits below, but before that there is the final highlight of Mount Kenya. The path ahead falls more steeply to cross a section of waterlogged moorland known as the Vertical Bog. If you are lucky it will be dry, but it's more likely that you will lunge from tussock to tussock and emerge with mud-coated legs. Soon enough, the path enters a deep forest and leads to the Met. Station (10,000 feet/3,050m) and the road head – look out for buffalo, monkey and even lion en route. Here, there is a campsite and bunk-houses. The Naro Moru Gate is still 6 miles (9.5km) down a rough, sometimes impassable, four-wheel drive track. Pre-arranged transport could meet you either at the campsite or at the park gate. The Mount Kenya National Park headquarters are the end point of a satisfying achievement.

factfile

OVERVIEW

This magnificent 6-day, 49-mile (78km) trek traverses Mount Kenya by linking two of its most famous trails – the Chogoria and Naro Moru – with a high-level circuit of the main peaks, including an ascent of Point Lenana, the mountain's third-highest summit.

Start: Mount Kenya Forest Gate.

Finish: Naro Moru Gate.

Difficulty & Altitude: This is a strenuous trek at high altitude, with 6 hours' walking and an ascent of 3,300 feet (1,000m) a day not uncommon. It is usual to use local porters to carry food and equipment.

ACCESS

Airport: Nairobi.

Transport: The Nairobi–Nanyuki road and the Mount Kenya ring road lead to Chogoria Village, on the east side of the Mount Kenya National Park. Local buses (*matatu*) travel north on the Nairobi–Nanyuki road daily, but for a small party it is easier to hire a taxi from Nairobi to the Chogoria National Park Gate. You should expect to take a day travelling to the park. From Chogoria Village a dirt road leads to Mount Kenya Forest Gate – the entrance to the National Park and the start of the trek.

Passport & Visas: Full passport required. A visa is required for all overseas visitors except British Commonwealth citizens.

Permits & Restrictions: On the mountain a daily park and camping fee is applicable.

LOCAL INFORMATION

Map: Mount Kenya 1:50,000 map and guide, Wielochowski and Savage.

Guidebook: *Kilimanjaro & Mt. Kenya – a climbing and trekking guide*, Cameron M. Burns (Cordee).

Backbround Reading: *No Picnic on Mount Kenya*, Felice Benuzzi – a classic account of an ascent of Point Lenana by three Italian Prisoners of War; *Trekking in East Africa*, David Else (Lonely Planet).

hyrax

Accommodation & Supplies: Nairobi is an international city with every kind of accommodation. Many trekkers opt to camp close to or at the point of entry into the National Park. Accommodation can be had in Chogoria, but the Chogoria Transit Motel, south of Chogoria village, is highly recommended. If you haven't

Rain forest is only one of the many landscapes you will trek through on this ascent of Mount Kenya. The diversity of the areas covered in the trek requires a sensible range of clothing.

organised things in advance, the manager can arrange porters, guide, cook, a vehicle to the park gate, and even food. A few people opt not to hire a porter, preferring to remain self-sufficient for the whole trek.

Currency & Language: The Kenyan shilling. Beware bogus money changers and counterfeit money. Swahili is the official language, although most Kenyans speak in their tribal tongue. Many Kenyans speak English. On the mountain, porters from Chogoria will speak the Meru language.

Photography: No restrictions.

Area Information: Mount Kenya National Park. P.O. Box 69, Naro Moru, Kenya (tel.: 00 254 171 2383).

TIMING & SEASONALITY

Best Months to Visit: The two trekking seasons in East Africa are from the middle of December to mid-March and from early June until mid-October; the warmer dry season over Christmas is the busiest time.

↑ *Hall Tarns, halfway up Mount Kenya, is not only an idyllic place to set up camp; there are also some good side trips to be made in the immediate area.*

Climate: Daytime temperatures are usually warm, reaching 50°F (10°C) on the mountain, but in the early hours of the day they can drop to around 40°F (5°C), especially above 15,000 feet (4,570m). An undoubted highlight as you traverse and circle "the mountain", is the transition from summer to winter. Remember, Mount Kenya is all but astride the Equator and, depending on the season, one side of the mountain can, quite literally, be enjoying the northern hemisphere's summer while the opposite side is snowy as if in the depths of the southern winter.

HEALTH & SAFETY

Vaccinations: Current immunisation requirements for East Africa include yellow fever, hepatitis A, typhoid, polio, tetanus and meningococcal meningitis. You will also require anti-malarial tablets.

General Health Risks: Health insurance is essential. Those from northern latitudes will need to protect against the equatorial sun. A sun hat and sunscreen are essential.

Special Considerations: On the mountain, high-altitude ailments are common: you will need to carry a comprehensive first-aid kit and be able to use it should the need occur – help could take days, not hours, to get to you at some points on the trek. There is a risk of altitude sickness if you go too high too quickly. Descending is the best cure. Better still, take time to acclimatise – as they say in Swahili, *"pole, pole"* (slowly, slowly).

Politics & Religion: Post-colonial Kenya has a stable government and democracy. More than other East African countries, Kenya is a multicultural/multiracial nation made up of numerous tribal groups, dominated by the Kikuyu.

Crime Risk: Personal crime seems to be increasing in Kenya and especially in Nairobi, which has around 2 million inhabitants. Robbery and violence are common. The best advice is to not go out after dark alone, stay away from the downtown districts, and don't display your wealth – however meagre! On the mountain, things are usually safe.

Food & Drink: Nairobi has some great restaurants that serve everything from antelope to zebra. "Up country", on the other hand, food is not so good, unless you are staying in an upmarket lodge. Boiled or deep-fried vegetables and meat or the occasional stew served with ugali (cornmeal) is standard. Fruit and vegetables, however, are plentiful, but should be washed and peeled if eaten raw. For the mountain you will need to take all your food with you. Just about anything can be bought in Nairobi, including tinned and dried food. On organised treks, the agency or outfitter will invariably provide food. In Nairobi and on the mountain the water is generally good, but be careful of contamination from others if your supply is at a well-used campsite. The safe rule is boil, filter or in some way purify it if you are uncertain. Remember, local porters can drink water that would make you dance the "Kenyan Quickstep" because they have built up immunity. One thing is for sure: you will need to hydrate well.

Verreaux's eagle

HIGHLIGHTS

Scenic: East Africa is a vast landscape that throbs with life, a colourful land full of contrast and big skies. The first view of Mount Kenya, an extinct volcano's core, is magical; it rises like a giant tooth from the surrounding plains, a remnant granite monolith that once plugged a long-gone volcanic pyramid.

Wildlife & Flora: Once you are on the mountain, the transition through differing vegetation zones as you climb toward the peak is an unforgettable experience: cultivated plantations give way in turn to luxuriant montane rain forest, a bamboo belt, a zone of open hypericum and hagena woodland, and then open moorland with patches of giant lobelia, senecio and everlasting flowers. Each zone has distinct fauna – in the forest, elephant, buffalo, monkey, forest antelope, hog, even lion, are not uncommon; higher on the mountain the hyrax – a tiny marmot-like creature, closely related (apparently) to the elephant – will often visit your camp. Don't be surprised if you see buffalo or elephant, or even leopard. Birds are everywhere and are quite stunning in their variety and form.

↓ temperature and precipitation

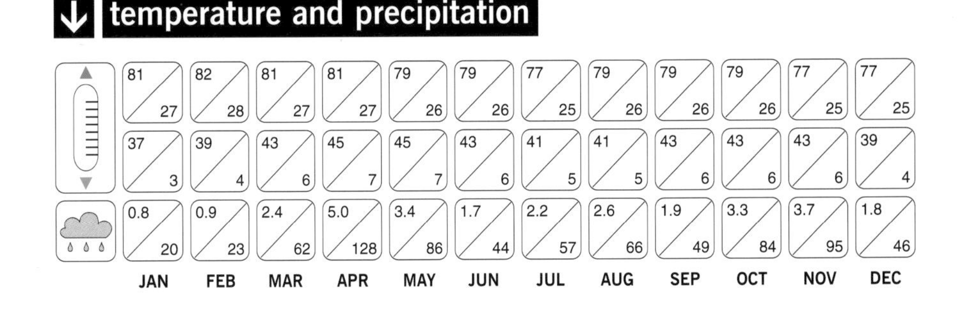

	JAN	FEB	MAR	APR	MAY	JUN	JUL	AUG	SEP	OCT	NOV	DEC
Max (°F / °C)	81 / 27	82 / 28	81 / 27	81 / 27	79 / 26	79 / 26	77 / 25	79 / 26	79 / 26	79 / 26	77 / 25	77 / 25
Min (°F / °C)	37 / 3	39 / 4	43 / 6	45 / 7	45 / 7	43 / 6	41 / 5	41 / 5	43 / 6	43 / 6	43 / 6	39 / 4
Precipitation (in / mm)	0.8 / 20	0.9 / 23	2.4 / 62	5.0 / 128	3.4 / 86	1.7 / 44	2.2 / 57	2.6 / 66	1.9 / 49	3.3 / 84	3.7 / 95	1.8 / 46

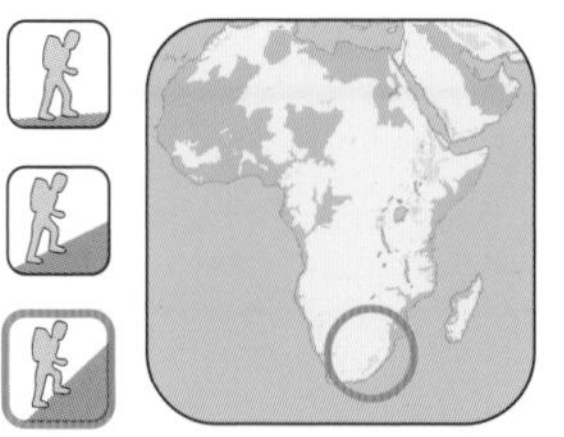

the NORTHERN DRAKENSBERG TREK

SOUTH AFRICA/LESOTHO

Tom Hutton

THE ZULUS CALLED THEM QUATHLAMBA, "BARRIER OF UPRIGHT SPEARS", A MOST FITTING DESCRIPTION OF THIS FORMIDABLE BASTION OF ROCK. RISING UP FROM THE FLATLANDS OF THE TRANSKEI, THE DRAKENSBERG, OR DRAGON MOUNTAINS, FORM THE BORDER BETWEEN THE SOUTH AFRICAN PROVINCE OF KWAZULU-NATAL AND THE MOUNTAINOUS KINGDOM OF LESOTHO.

This trek concentrates on the Northern Berg, which is generally accepted as the area north of Cathedral Peak. Nowhere is the escarpment more dramatic, and nowhere would the Zulu name seem more appropriate. Each time the route meets the escarpment edge there is a new vista: huge basalt pinnacles and treacherous knife-edge arêtes that penetrate the void between the cliff tops and the softer green lands of the Low Berg, way below.

The trail starts clearly, as many day walkers climb past the Sentinel and over the chain ladder to the Ranger's Hut on Mont-aux-Sources. But, from the Tugela Falls, the going becomes remote and the crossing enters a real wilderness. The route follows the general line of the escarpment but makes regular forays into Lesotho, where it follows river valleys and rock bands over an austere, rolling plateau. Each day ends back on the escarpment, watching a fiery sunset from the shelter of a cave or hollow.

itinerary

• DAY 1 **7 miles (11.5km)**

Sentinel car park to Ifidi Cave

Permits can be bought in the office at the car park, where you should leave details of your route. A good path runs from the gates up a series of zigzags, toward the huge tower of the Sentinel. Halfway, there are some incredible views over the Amphitheatre to the Eastern Buttress and the Devil's Tooth. The path traverses beneath the face of the Sentinel and then contours to a narrow zawn (a gap in the cliffs). Two chain ladders penetrate the steep rock faces and lead on to the plateau above. The path continues south-east, where it picks up the Tugela River, which leads to the Tugela Falls, one of the highest waterfalls in South Africa.

↑ *Storm clouds build over the escarpment, adding atmosphere to the already imposing rock scenery. From Mponjwane the views are to the north and the crossings end of Cathedral Peak. At times, the whole route can be seen; the distinctive shape of the peaks and pinnacles make them easy to identify against the skyline.*

The trail follows the escarpment south-east, along the Amphitheatre, past the daunting spires of the Inner Tower and the Devil's Tooth, to the Icidi Pass. A more sheltered alternative is to head south and pick up the Khubedu River, which leads easily back onto the escarpment north of the Icidi Pass. The pass is easily recognised by the two Icidi pinnacles and the cave is a shallow overhang, above the pass.

•DAY 2 7 miles (11km)

Ifidi Cave to Rwanqa Cave

This is one of the most breathtaking sections of the whole trek, with sharp spires and pinnacles thrusting skywards from the main rock walls. The trail continues southwards, drops into a valley and climbs back to the Icidi Pass. It then runs behind the Icidi Buttress and crosses another tributary of the Khubedu River. It's worth heading back to the escarpment at Mbundini for some great views of the pinnacles of Madonna and Her Worshippers. From Mbundini, the route crosses above Fangs Pass and drops into another valley. A gentle climb follows and leads back onto the escarpment at Rwanqa Pass, marked with cairns. A short, steep scramble down the pass leads you to the cave.

•DAY 3 9 miles (14.5km)

Rwanqa Cave to Mponjwane Cave

The climb back up the pass isn't as daunting as it looks. Once back on the escarpment, a track leads down to a river. Follow the river toward its source and the escarpment edge at Pins Pass. The trail hugs the Lesotho border and links the summits of a distinct ridge that overlooks the Mnweni Cutback. Keeping to the high ground, the route swings east and then north round the cutback. Back on the edge, the spectacles come thick and fast. First the Twelve Apostles and then the wonderfully named Eeny, Meeny, Miny and Mo. Heading east from the pass, the walk drops into the Senqu Valley, the source of the Orange River. From the river, a steady climb follows, ending back on the escarpment overlooking the Mponjwane Tower. The Mponjwane Cave, probably the most spectacular of the traverse, is reached by heading south from the tower, round the back of a raised rocky section. A trail of cairns leads back on to the escarpment, where a short scramble downward sgains a ledge. The cave is on the right, overlooking the saddle to the tower.

•DAY 4 9 miles (14.5km)

Mponjwane Cave to Twins Cave

Sunrise seen from Mponjwane is one of the route's many highlights, and worth setting your alarm for. A faint path leads southwards to the Rockeries Pass. This can be followed past the Rockeries themselves before descending to the Senqu River to walk downstream to a confluence with the Kokoatsoan River, east of some waterfalls. Follow the Kokoatsoan back up towards the Saddle and the Nguza Pass. A steep ridge heads south from the river's source. This should be climbed to 11,150 feet (3,400m) and then descended by traversing round the hillside to meet the Ntonjelana Pass. Follow the escarpment in a south-easterly direction to the majestic cluster of the Cathedral Peak range. →

The day ends at one of the biggest Drakensberg caves, Twins, further down the Mlambonja Pass. The Bell and Cathedral Peak are the high points of a rocky peninsula that stands proud of the main escarpment. Twins Pass, well-marked with cairns, is south-east from here. After a short but difficult descent, a saddle is reached, where a track comes in from the left and leads steeply down to Twins Cave.

↑ *The lush vegetation of the Mlambonja Pass comes as a refreshing change from the stark austerity of the plateau above. The Mlambonja Buttress and the Cathedral Peak Range, bathed in afternoon sunshine, make a stunning backdrop.*

•DAY 5 **7 miles (11.5km)**

Twins Cave to Cathedral Peak Hotel

This is a day of descending, but starts with a short climb back up to the Mlambonja Pass. The descent follows the river, and is marked with small cairns in the more obscure sections. The lush green vegetation and colourful flowers of the pass come as a refreshing change after 5 days on the stark wilderness of the plateau above. At about 6,550 feet (2,000m), the path splits – the righthand track is generally accepted as the better of the two. This contours for a while but then climbs slightly to get a clearer line. A long zigzag section leads back down to river level, and the going eases as it ducks out of the Mlambonja Wilderness Area and crosses more open ground. The final section threads through grasslands and picks up a waymarked track back to the hotel and civilisation.

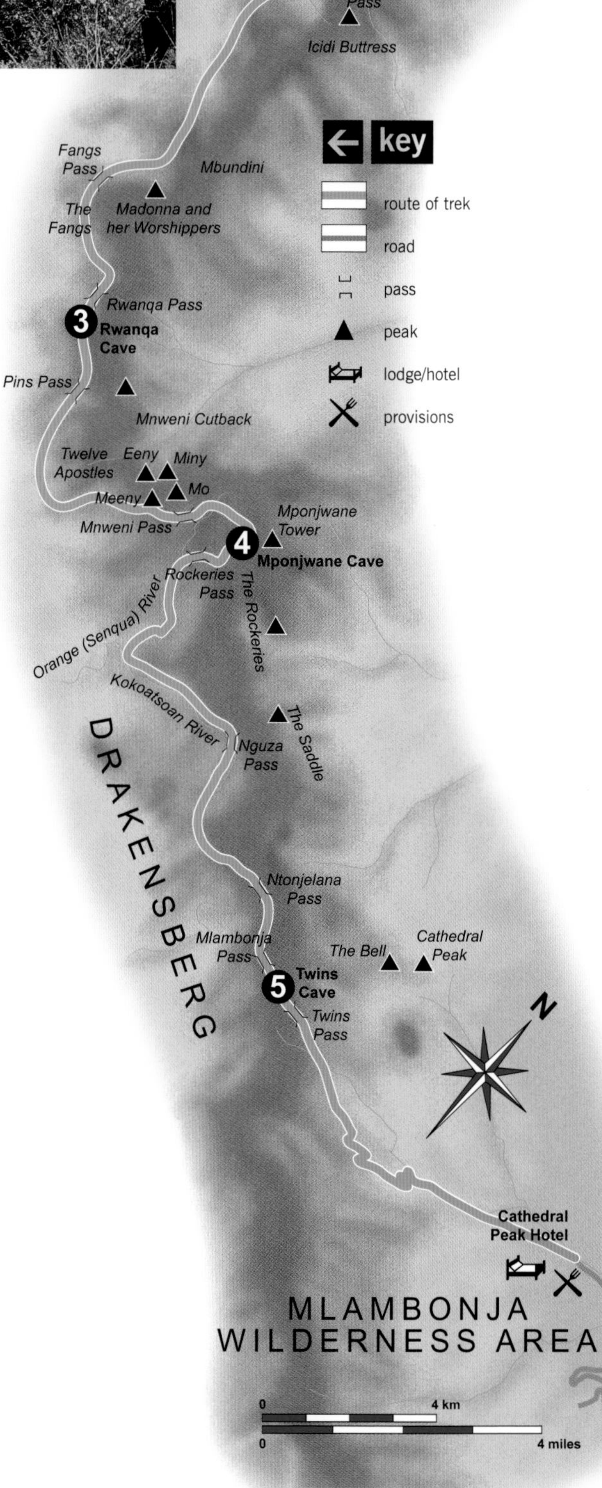

walk profile

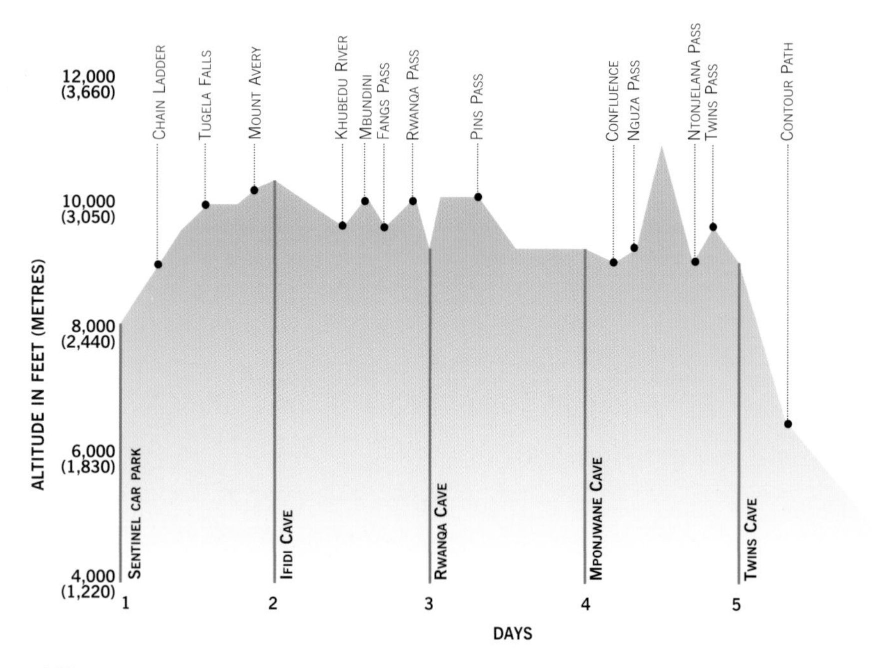

factfile

OVERVIEW

A high-level traverse of the northern Drakensberg escarpment. The route, which is approximately 40 miles (65km) long, straddles the border between South Africa and Lesotho, taking 5 strenuous days to complete.

Start The Sentinel car park, Royal Natal National Park.

Finish: The Cathedral Peak Hotel.

Difficulty & Altitude: This is a high-mountain walk in one of the more remote areas of the Drakensberg Range. The major part of the route is at an altitude of around 9,800 feet (3,000m) and the going can be quite tough and the days long. Once on the escarpment, there is little in the way of established paths and good navigation skills are essential for a safe crossing. In full winter conditions, the traverse would become a serious undertaking. It is certainly worth considering hiring a guide.

ACCESS

Airports: International Airport at Johannesburg. Durban is also close (internal transfer).

Transport: There are a number of options: many backpack lodges and hotels will arrange transfers and drop-off/pick-up services. You could hire a car at the airport and drive to the area. Pick-ups, etc. could then be arranged from there.

Passport & Visas: Passports (no visas) required for all visitors. But this is an ever-changing situation, so check before leaving. The route makes numerous crossings into Lesotho, so a passport should be carried throughout the trek.

Permits & Restrictions: No access restrictions, but there are a number of small fees payable, including one to use the toll road through the Qwa-Qwa National Park. Permits are needed to walk and camp in the park.

LOCAL INFORMATION

Maps: 1:50,000 Natal Drakensberg Park Hiking Series No. 1 Rugged Glen to Mweni and No. 2 Cathedral Peak.

Guidebooks: *Drakensberg Walks*, David Bristowe (Struik); *South Africa, Lesotho and Swaziland* (Lonely Planet).

Background Reading: *Barrier of Spears*, Reg Pearse (Howard Timmins); *Mountain Odyssey*, J. Clarke and P. Coulson (Macmillan).

berg adder

Accommodation & Supplies: South Africa has become littered with backpack lodges, which offer good quality bed and breakfast at very reasonable prices. The route finishes at the Cathedral Peak Hotel, a well-situated, upmarket establishment. The nearest decent-size town to the start is Harrismith. Provisions and supplies for the trek can be acquired there.

Currency & Language: The Rand. Mainly English.

Photography: No restrictions.

Area Information: KwaZulu Natal Nature Conservation Services (KZN NCS) P.O. Box 13053, Cascades, 3202, KwaZulu Natal (tel.: 00 27 331 845 1002).

TIMING & SEASONALITY:

Best Months to Visit: April, May, June or September, October, November.

Climate: Summers in the Drakensberg can be very hot and very wet. The winters are much drier, but there is always a chance of precipitation, which will probably take the form of snow on the high ground. In spring and fall the day-time temperatures are ideal (between 60°F/15°C and 70°F/20°C), but at night will frequently drop below zero.

HEALTH & SAFETY

Vaccinations: None required, but polio and tetanus are recommended, as are typhoid and hepatitis A. There may be malaria in some parts of KwaZulu-Natal.

General Health Risks: Sunscreen and a hat are essential.

Special Considerations: This is a long trek in a very remote area and it's essential to be completely self-sufficient.

Politics & Religion: Once a political hotbed, South Africa has become much more stable since the democratic elections of 1994, and now welcomes tourism with open arms. This is a multiracial society and, as such, many different religions are practised, varying according to region and race.

Crime Risks: In cities such as Johannesburg, crime rates are high. In the Drakensberg, it is quite common to come across small groups of Basotho (people of Lesotho) who use the mountain passes to cross into South Africa. In general they are peaceful and mind their own business, but walkers with all their gear appear considerably more wealthy than they are. Good advice is to walk in groups of three or more.

Food & Drink: As the emphasis is on self-sufficiency, it's really down to what can be carried. Most common types of food are available in the supermarkets. The water on the escarpment is safe for drinking.

HIGHLIGHTS

Scenic: The real highlight of this trek is the landscape. The escarpment and adjoining pinnacles are absolutely awesome. Africa is justifiably famous for its sunsets, and they are particularly dramatic when viewed from the shelter of a cave high up on the escarpment.

Wildlife & Flora: Despite the stark nature of the plateau, there are many species that manage to scrape a living out of this harsh environment. The birds are the most spectacular. A keen eye might spot black eagles, cape vultures, rock kestrels, and even the rare, bearded vulture or lammergeier. Bucks, such as the grey roebuck and, king of the antelopes, the mighty eland, can survive at this height, as can baboons and rock hydraxes. Reptiles include the puff adder, berg adder, and various geckos and skinks. The plant life is more interesting in the Low Berg and the descent route is coloured by ericas and proteas.

The views from the escarpment never disappoint. From the Icidi Pass, the unique patterns of Low Berg far below are visible.

temperature and precipitation

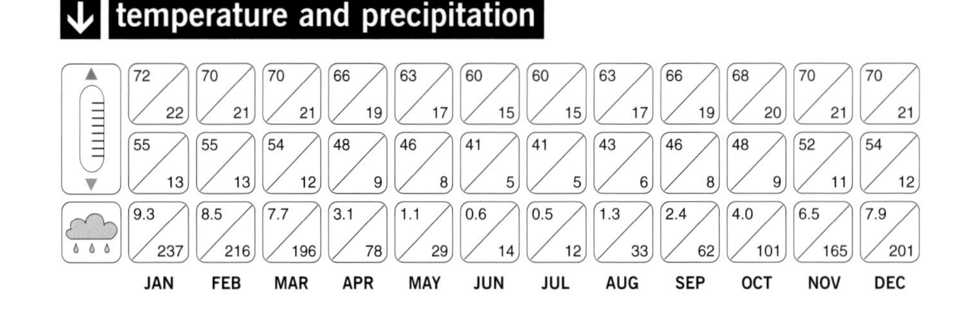

	JAN	FEB	MAR	APR	MAY	JUN	JUL	AUG	SEP	OCT	NOV	DEC
Max °F	72	70	70	66	63	60	60	63	66	68	70	70
Max °C	22	21	21	19	17	15	15	17	19	20	21	21
Min °F	55	55	54	48	46	41	41	43	46	48	52	54
Min °C	13	13	12	9	8	5	5	6	8	9	11	12
Precipitation (in)	9.3	8.5	7.7	3.1	1.1	0.6	0.5	1.3	2.4	4.0	6.5	7.9
Precipitation (mm)	237	216	196	78	29	14	12	33	62	101	165	201

Australasia

The amazingly diverse ecology and environment of the Southern Hemisphere are well represented by both New Zealand and Australia. Scenically splendid New Zealand is covered by two treks, one from the active volcanic region of North Island, and one along the seashore of South Island. Those chosen from Australia and Tasmania contrast tropical jungle with the bushwhacking delights of the lake and mountain wilderness of a World Heritage site.

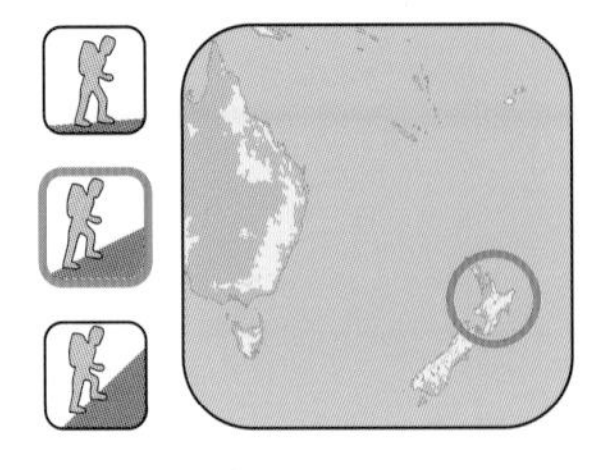

Tongariro Northern Circuit

the RING OF FIRE

NORTH ISLAND, NEW ZEALAND

Judy Armstrong

THIS IS ONE OF NEW ZEALAND'S GREAT WALKS, IN A NATIONAL PARK OFTEN FORGOTTEN BY WALKERS. TONGARIRO NATIONAL PARK IS THE HOT HEART OF THE NORTH ISLAND AND HAS DUAL WORLD HERITAGE STATUS, ON BOTH NATURAL AND CULTURAL GROUNDS.

Created in 1894 as the country's first national park, Tongariro is a volcanic wonderland with active cones and thermal hot springs, tussock grassland and the country's only desert.

The Tongariro Northern Circuit includes the Tongariro Crossing, regarded as the best one-day walk in the country because of the unique craters, steaming lakes and volcanic formations along the route. It uses a hut system which, in true Kiwi tradition, is basic but comfortable; all the huts on this circuit have staggering views of mountains, including the active (and often snow-covered) volcanoes of Ruapehu and Ngauruhoe. The easy-to-follow route takes in rimu and beech woodland, crashing waterfalls, golden tussock and turquoise lakes. It ducks past volcanic debris, lava blocks the size of cars, steaming craters and ground too hot to touch. It attracts plenty of Kiwi trampers and a trickle of tourists who have heard its praises sung...this is easily the best classic walk in the North Island, and one of the best in New Zealand.

itinerary

•DAY 1 5 miles (8km)

Whakapapa Village to Mangatepopo Hut

The trek – or tramp, as Kiwis call it – starts at Whakapapa Village, site of the Whakapapa Visitor Centre. Walkers sign in here, buy hut passes, state intended routes, and check out the weather forecast.

The track wanders north-east through tussock grass and a few stands of beech, diving in and out of moist native bush. Its main users are day-walkers on forays to the spectacular Taranaki Falls; this is signposted right at a junction, while Tongariro Northern Circuit heads straight on. The walking is fairly level, across the western flank of the great cone of Mount Ngauruhoe. The Maori people hold it sacred, and legend tells of a slave thrown into the blazing crater to appease Ruamiko, the volcano god – look out for the plumes of gas and steam that often spiral from its summit.

The path deteriorates at the Taranaki Falls junction: scores of small streams pouring off Ngauruhoe have gutted the track, and the black slop of soil held together by the innocent gold of tussock makes for entertaining walking. The day's destination is the Mangatepopo Hut, a 24-bunk building with water tank and stove, well-positioned above the Mangatepopo Stream and below the towering cone of Ngauruhoe. The smaller volcano Pukekaikiore nudges between the hut and its taller sibling.

•DAY 2 8 miles (13km)

Mangatepopo Hut to Ketetahi Hut

If Ruamiko is smiling, the walker is in for a treat, covering the Tongariro Crossing with the peak of Ngauruhoe as an extra bonus for the super-fit. The path heads east up the Mangatepopo Stream, climbing gently and dodging a succession of old lava flows – huge chunks of sharp, grey rock which give an indication of the erupting volcano's power.

A spur track to Soda Springs soon crosses the stream to the left. The sulphurous springs can be smelled long before they can be seen – it takes about 15 minutes as a round

↑ *The impressive view from the ridge above Red Crater takes in the ice-cold Emerald Lakes. To the right of the picture are lava flows leading down to the desert; to the left is a track across Tongariro's flanks, toward Ketetahi Hut.*

← *Native woodland, known as "bush" by New Zealanders, is a feature of the first and last days of this trek. Trees are predominantly rimu and beech, and house a variety of noisy native birds.*

trip. The rocks below the springs are coloured golden by iron oxide from the fallout of volcanic ash.

The main path climbs up lava blocks to the saddle between Ngauruhoe and Mount Tongariro. From here, if the weather is with you, the hour is early, and the cone is free of snow, it's feasible to add an attempt on Ngauruhoe to the day's equation. The 3- to 4-hour round trip is the experience of a lifetime, scrambling up steep scree slopes to the lip of the crater, an ascent of nearly 2,300 feet (700m). Ngauruhoe should be treated with caution, and should not be approached when the crater shows vigorous activity. The last major eruption occurred in 1975, producing a plume between 7 and 8 miles (11 and 13km) high. As the cloud collapsed, ash bombs and large blocks crashed down the slopes, leaving great sheets of debris on the lower slopes. Unfortunately, the chances of all conditions being good enough to climb this seductive peak are slim.

Back at South Crater, poles indicate the way forward. The crater is a huge, walled amphitheatre with a pool of meltwater often glistening in the explosion pit. As the path crosses the flat crater bottom then climbs steeply up a lava ridge, the feeling of being a stranger in an alien land intensifies. To the right is Red Crater: the impact is enormous, with blood-red crater walls, billowing sulphur-laden steam, and the feeling of exposure on the high ridge.

The path drops down toward Emerald Lakes, brilliant-green pools of water that provide a vivid contrast to the red and grey of the crater. Despite the steam vents spouting all around, the lakes are ice-cold. The visual circus continues all the way to the Ketetahi Hut – over Central Crater, past Blue Lake, round the flanks of North Crater, and down rough red rocks to the hut.

•DAY 3 **9½ miles (15km)**

Ketetahi Hut to New Waihohonu Hut

Ketetahi Hut is within spitting distance of natural hot springs (390–480°F/200–250°C) which are fed by an underground reservoir. But the springs are privately owned by the Ketetahi Trust, and trampers are asked to respect the Maori tapu, or sacredness, of the area and should not bathe in or visit the pools.

This day's route retraces the track up the eastern reaches of Mount Tongariro past Blue Lake, a remarkable expanse of blue-tinted water in an inactive crater. Its Maori name slips neatly off the tongue: Te wai-whakaata-o-Te Rangihiroa, meaning Rangihiroa's Mirror. Chief Te Rangihiroa is said to have explored the Tongariro volcanoes in about 1750.

→

walk profile

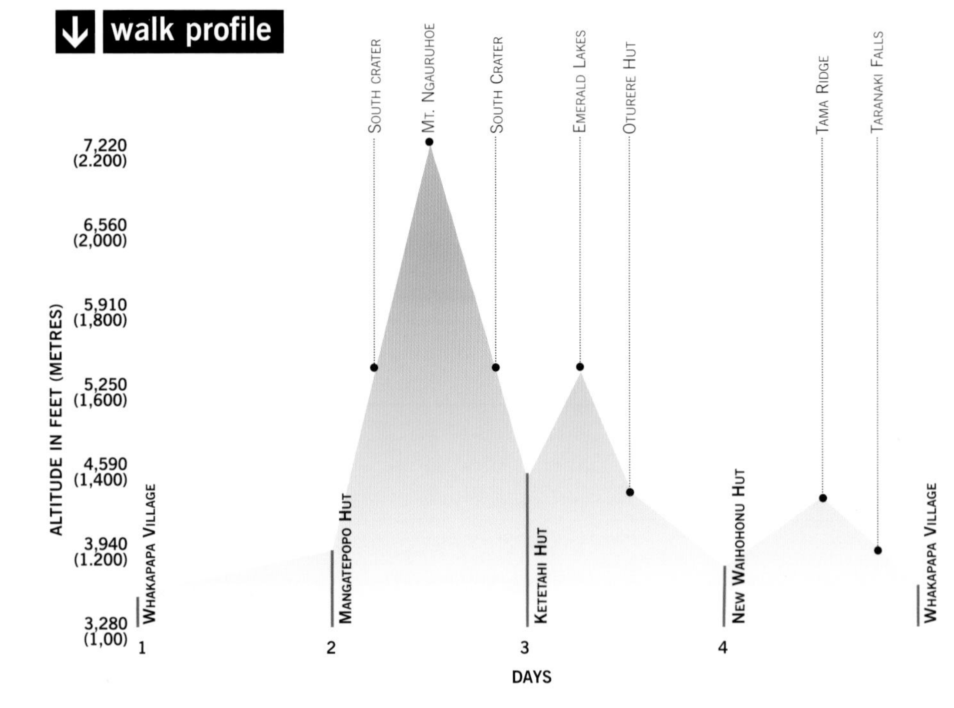

The path finally peaks at the junction below Emerald Lakes and dives down a solid wave of lava to cross New Zealand's only desert – a silent, eerie place where the wind whispers through honeycomb rock and kicks grey sand over telltale footprints, jagged lava, and struggling flora. Soon the Oturere Hut appears behind a giant lava block.

Now the way swings south-west through open country, descending straight towards Mount Ruapehu, where there is much evidence of its recent spate of eruptions.

It's a fair distance to the immaculate New Waihohonu Hut, perched in beech woodland in a clearing above a stream. The views of Ruapehu along the way are spectacular.

•DAY 4 10 miles (16km)

New Waihohonu Hut to Whakapapa Village

The day begins with a descent, and almost immediately there is a junction. Trampers taking in a section of the Around The Mountain Track branch off here, to cross the rain-shadowed slopes of Ruapehu to the Rangipo Hut. This is set in a barren landscape of red-brown sand and ash, a unique area resulting from 2 million years of volcanic eruptions which pumped out oceans of pumice and destroyed all vegetation. A short diversion in the same direction involves a scramble over river rocks to reach the Ohinepango Springs which bubble up from beneath an old lava flow.

Back at the junction the main track heads west. It soon passes the Old Waihohonu Hut – the oldest building in the park, it has corrugated iron walls filled with pumice as primitive insulation against winter storms. The hut, renovated in 1997 and now preserved by the Historic Places Trust, is no longer used for accommodation.

The track back to Whakapapa Village climbs in and out of streams, up and over lumpy grassland, through beech woodland and spidery tussock. The views are breathtaking: Ngauruhoe's perfect cone on one side, and the snow-capped summit of Ruapehu on the other. The Taranaki Falls interrupt the mountains as the Wairere Stream hurls itself down a 65-foot (20-m) rock face into a boulder-ringed pool, casting a fine mist over dusty trampers.

key

- route of trek
- minor track
- road
- peak
- campsite/hut
- lodge/hotel
- provisions

factfile

OVERVIEW

Tongariro Northern Circuit winds around three sacred volcanoes: Ngauruhoe (active, 7,503 feet/2,287m); Tongariro (extinct, 6,453 feet/1,967m); Ruapehu (active, approximately 9,176 feet/2,797m – volcanic activity is changing the shape of the upper reaches). The route passes through craters and volcanic debris, tussock grassland, desert and native woodland. The route is some 32½ miles (52km) and takes 4 to 6 days to complete. It can be extended by joining the Around The Mountain Track at the Waihohonu Stream junction.

Start/Finish: Whakapapa Village, Tongariro National Park.

Difficulty & Altitude: The route has some stiff climbs and rocky descents, mixed with a gravel desert and sometimes boggy tussock. Walkers should be fit and prepared for bad weather. The South Crater on the Tongariro Crossing should not be attempted in foul weather. The highest altitude reached, if you climb it, is Mount Ngauruhoe at 7,503 feet/2,287m.

ACCESS

Airports: Auckland and Wellington.

Transport: Taupo is at the geographical hub of the North Island, midway between Auckland and Wellington. Long-distance buses arrive in Taupo with daily connections to Turangi. Shuttle services are available from here to National Park township, stopping at Whakapapa Village on the way. The nearest train station is at National Park. Cars can be left at Ohakune and Whakapapa Visitor Centre.

Passport & Visas: USA and Canadian citizens are given a 3-month visitor's permit; UK passport holders get 3 months.

Permits & Access: From October to June a Great Walk pass is required for the huts; camping is not permitted within 1,600 feet (500m) of the tracks for environmental reasons.

LOCAL INFORMATION

Maps: Tongariro National Park (Infomap 273–4) 1:80,000 is sufficient for this walk; Tongariro is also covered by four 1:50,000 Topomaps, T19, T20, S19, S20.

Guidebook: *Tramping in New Zealand*, Jim DuFresne (Lonely Planet).

Background Reading: *Volcanoes of the South Wind*, Karen Williams (Tongariro Natural History Society, P.O. Box 2421, Wellington, NZ) – a field guide to the volcanoes and landscape of Tongariro National Park.

Accommodation & Supplies: Accommodation at Whakapapa Village (expensive Chateau Grand; budget Skotel; and a motorcamp). Turangi, National Park, and Ohakune have a wide choice of hotels, motels, campsites, and cabins, as well as grocery stores. Food is available at Whakapapa village, Ohakune, National Park township and Turangi. No supplies are available on the tramp.

Currency & Language: The New Zealand dollar ($NZ). New Zealand has two official languages: English, which is most widely spoken, and Maori.

Photography: No restrictions.

Area Information: The best place to organise your tramp is at the Turangi Information Centre (tel.: 00 64 7 386 8999). Whakapapa Visitor Centre (tel.: 00 64 7 892 3729; fax 00 64 7 892 3814) can supply maps, information about walks and huts, plus track and weather conditions. Websites: www.nztb.govt.nz (New Zealand Tourist Board); www.doc.govt.nz (Department of Conservation); www.nzalpine.org.nz (The New Zealand Alpine Club).

TIMING & SEASONALITY

Best Months to Visit: The most popular and safest time is December to March, when there is the greatest chance of clear weather and the tracks should be clear of snow. In winter it is a full alpine adventure, requiring ice-axes and crampons.

Climate: The mountainous terrain means unpredictable weather patterns, and the western slopes of all three volcanoes can experience sudden periods of heavy rain and snow as late as December. Winds can reach gale force on the ridge, and it rains frequently.

Grey sand, jagged lava and honeycomb rock feature in New Zealand's only desert.

HEALTH AND SAFETY

Vaccinations: None required.

General Health Risks: Medical attention is of the highest standard and is reasonably priced, but trekkers should have adequate medical insurance. Sunburn is a real issue in New Zealand, so keep covered up, wear good quality sunglasses and a hat.

Special Considerations: The Whakapapa Visitor Centre warns trekkers of any volcanic activity.

Politics & Religion: No concerns.

Crime Risks: Violent crime is uncommon, but cars left unattended for significant periods are a target for thieves – hide your valuables, or store them elsewhere while you're away walking.

Food & Drink: Tap water is clean and safe to drink, but boil or treat hut water before drinking.

HIGHLIGHTS

Scenic: Active volcanoes, boiling mud pools, fumaroles and sulphurous pits are all part of the Tongariro experience. The Maori people believe the volcanoes to be sacred and this belief saved the tussock grassland around Tongariro from becoming a giant sheep-grazing paddock. In 1887 Chief Horonuku Te Heuheu Tukino IV presented the area to the Crown and New Zealand's first national park was formed.

Wildlife & Flora: Much of New Zealand's native flora and fauna is not found anywhere else in the world. Native woodland, or "bush", houses an incredible variety of birds and ferns. Many alpine areas are home to the kea, a green parrot with a penchant for rubber and shiny metal.

temperature and precipitation

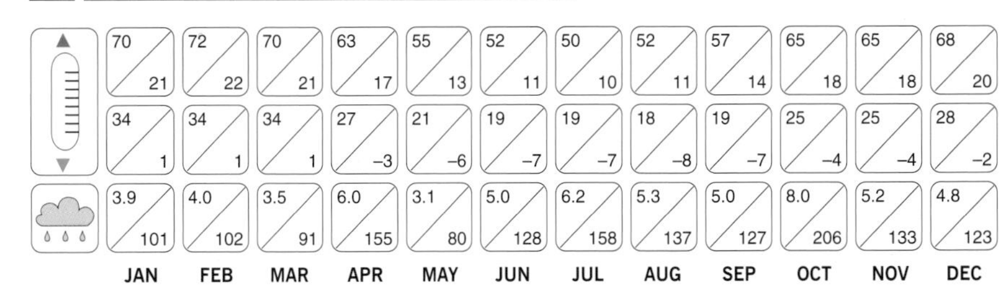

	JAN	FEB	MAR	APR	MAY	JUN	JUL	AUG	SEP	OCT	NOV	DEC
Max °F	70	72	70	63	55	52	50	52	57	65	65	68
Max °C	21	22	21	17	13	11	10	11	14	18	18	20
Min °F	34	34	34	27	21	19	19	18	19	25	25	28
Min °C	1	1	1	–3	–6	–7	–7	–8	–7	–4	–4	–2
Precip. in	3.9	4.0	3.5	6.0	3.1	5.0	6.2	5.3	5.0	8.0	5.2	4.8
Precip. mm	101	102	91	155	80	128	158	137	127	206	133	123

On the second day of the Abel Tasman Coastal Track, you must use this dramatic suspension bridge to cross the Falls River.

the ABEL TASMAN COASTAL TRACK

SOUTH ISLAND, NEW ZEALAND

Walt Unsworth

AT THE NORTHERN TIP OF NEW ZEALAND'S SOUTH ISLAND, BETWEEN GOLDEN BAY AND MOTUEKA, LIES THE ABEL TASMAN NATIONAL PARK. IT MEETS THE OCEAN IN A SERIES OF ROCKY BLUFFS, BETWEEN WHICH ARE GOLDEN, SANDY BAYS AND TIDAL ESTUARIES, FORMING THE BACKGROUND FOR ONE OF NEW ZEALAND'S MOST POPULAR LONG-DISTANCE WALKS, THE ABEL TASMAN COASTAL TRACK.

At just over 22,000 acres, the Abel Tasman National Park is New Zealand's smallest national park and was opened in 1942 thanks to the persistence of one woman, Perrine Moncrief, a Nelson conservationist who was determined to save this beautiful coast from the ravages of logging. It was Perrine who suggested it be named after the Dutchman Abel Tasman, the discoverer of New Zealand, who landed here in 1642. Inland, the park is thick bush with deep valleys and cave-riddled marble rock. The coast, which the route follows, is stunningly beautiful.

The well-maintained path to Anchorage Bay makes this section of the walk easy going for those trekking with heavy rucksacks.

itinerary

•DAY 1 7 miles (11.5km)

Marahau to Anchorage

From the National Park information centre at Marahau a long causeway crosses the estuary to an easy path covered in pine needles, which winds round the coast to Tinline Bay, then Stilwell Bay. There is a rough descent to the bay, which has a gorgeous sandy beach. Adele Island lies offshore and the coast here is known as the Astrolabe Roads, after a French corvette that explored it back in 1827. Frederick Carrington, an early surveyor, thought this part of the coast was the finest in New Zealand and there are times, when the light catches the water, that it is indeed breathtaking.

There is a climb out of the bay, then over a saddle to descend to Anchorage, where there is a hut (24 bunks). Those staying at the lodge, however, need to cross a low ridge to the next inlet, Torrent Bay, and walk round this (45 minutes) or, at low tide, wade the muddy bay and ford the Torrent River (25 minutes) – which can be exciting!

•DAY 2 6 miles (9.5km)

Anchorage to Bark Bay

After crossing Torrent Bay (this can only be done 2 hours either side of low tide) and continuing on a more inland track over a saddle between two attractive tree-filled valleys, the track leads down to the lush greenery of the Falls River, which is crossed by a suspension bridge. The going is a bit tougher than yesterday. After another climb there is a sidetrack to South Head, with views over Sandfly Bay. The main track descends to the Bark Bay Hut (25 bunks). There is no lodge at Bark Bay.

VARIATION: 2½ miles (4km). The Falls River walk. An interesting sortie into the rugged interior for about half a day can be made from Torrent Bay. A good track follows the Tregidga Creek past Cascade Falls to meet the Falls River about 15 minutes from the main falls, which consist of six leaps. Care is needed: it is hard, slippery going to reach a place where →

the falls can be seen, and the stream has to be crossed twice. There are intermittent waymarks. Only the bottom three sections of the falls are visible before it becomes impossible to venture further upstream. Return by the same route: it is dangerous to follow the Falls River downstream.

• DAY 3 **7 miles (11.5km)**

Bark Bay to Awaroa

You can cross Bark Bay at low tide or walk round the shore to climb up through beech woods to a saddle and then by the side of Long Valley to drop down to the beach again at Tonga Quarry. Granite from here was used in some of New

↑

The rough descent to Stilwell Bay on the first day of the trek leads to a wonderful beach. An early surveyor of the country deemed this area to contain some of the finest scenery in New Zealand.

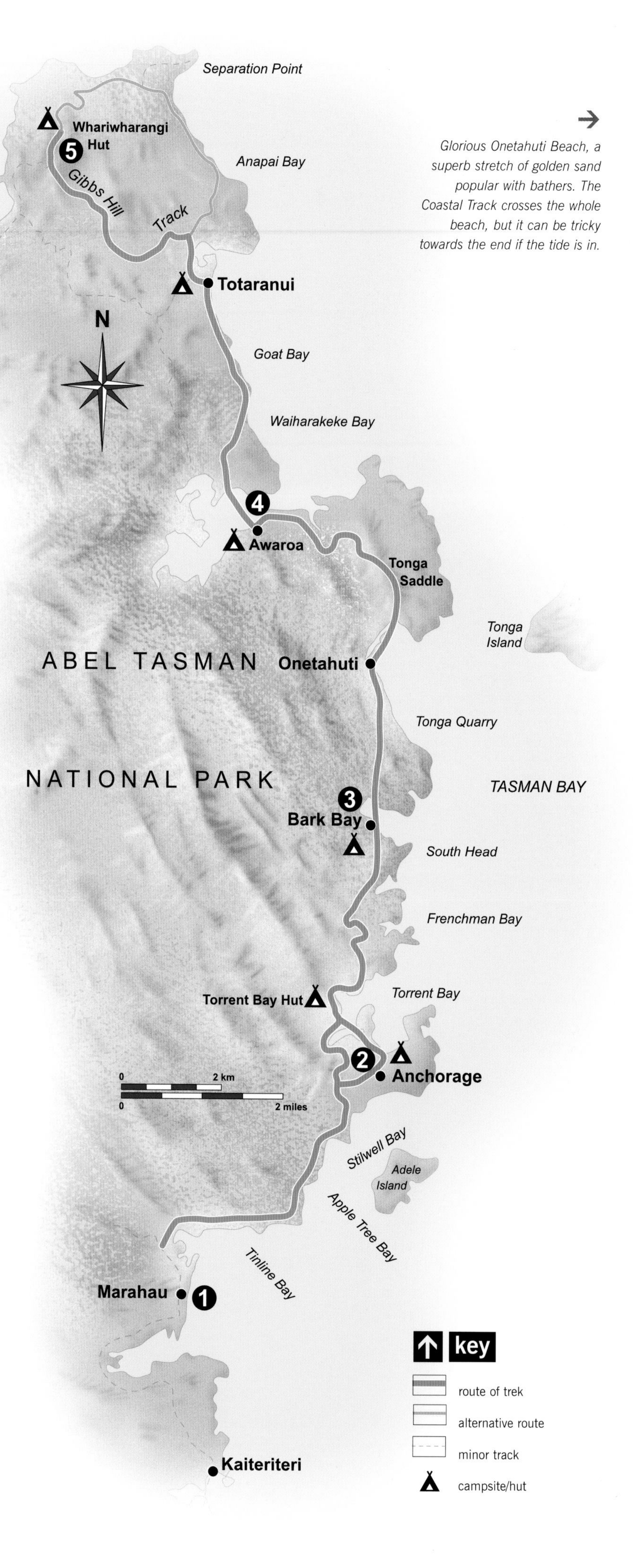

→

Glorious Onetahuti Beach, a superb stretch of golden sand popular with bathers. The Coastal Track crosses the whole beach, but it can be tricky towards the end if the tide is in.

walk profile

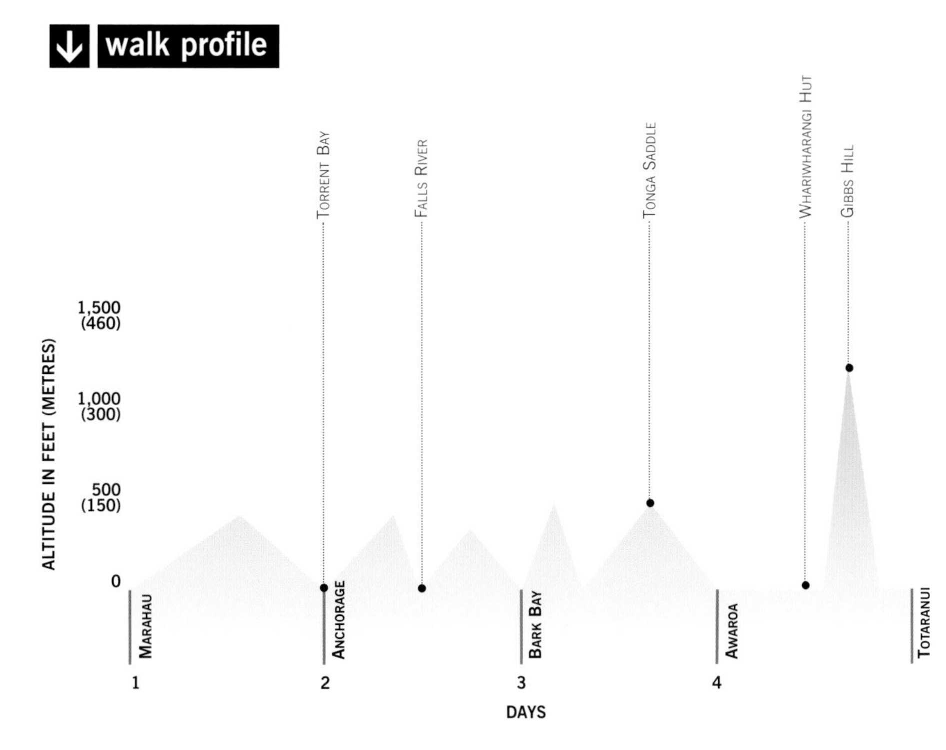

Zealand's public buildings at the end of the nineteenth century but there is little evidence of the industry now. Offshore is Tonga Island, a sanctuary for wildlife – including the little blue penguin, which has been wiped out on the mainland by feral stoats.

The track climbs again from the quarry, but before long it descends to the beach once more. This is glorious Onetahuti, a superb stretch of golden sand, very popular for bathing. Walk to the far end of the beach, where the Richardson Stream comes tumbling in. The last few paces are tricky at high tide because the wading can be very deep just here.

From the beach a good track climbs to the Tonga Saddle then down to Awaroa, the largest of all the inlets on this coast. The Awaroa hut (25 bunks) and Homestead Lodge are both on the near side of the inlet.

→

• DAY 4 3½ miles (5.5km)

Awaroa to Totaranui

This short day begins with the crossing of the inlet. This can only be done during the 2 hours either side of low tide when the sea retreats leaving firm sand and streams seldom more than knee-deep. Residents at Homestead Lodge can be taken across by boat during high water, but less fortunate travellers should note there is no way round the inlet. Once across, follow the path along the coast through a series of small but attractive bays to Totaranui, where there is a big campsite, ranger station and restroom. Buses and boats can be caught here for the return.

VARIATION: 4½ miles (7.5km) Totaranui to Whariwharangi Hut. There is no hut at Totaranui, but as an alternative to catching the boat back to Awaroa straightaway, you could carry on along the coast to the Whariwharangi Hut, a restored farm homestead housing 20 souls. Return next day to Totaranui by Gibbs Hill (1,329 feet/405m) on a waterless track of 3 hours or so, thus making a little circular tour.

extensions

Days 5–10 The Inland Track The trip can be extended by 3 to 5 days by returning along the Inland Track. You need Back Country Hut Tickets to use the four huts, and the track is more rugged than the coast, especially if you take the side track through the karst to Harwood's Hole. Boots are required. This is not a popular option: most of the huts hold just four people. Access is from Totaranui, Whariwarangi or even Awaroa; the first hut is Awapoto (holds 12), then Wainui (four), Moa Park (four) and Castle Rocks (four), ending back at Marahau. There are several variants, but the highlights are the karst (Harwood's Hole involves a 525-foot (160-m) abseil, should you be thinking of tackling it) and the tussock country of Moa Park.

factfile

OVERVIEW

A traverse of the Abel Tasman National Park coastline from Marahau to Totaranui, approximately 24 miles (38km). It takes 4 days to complete. Side excursions could add half as much again. At the end of the trek it is usual to return from Totaranui to Marahau by boat (there's a regular service of powerful launches along the coast) but for those who want to see more of the park by extending their walk there are several alternatives. You need to decide before you set out, however, whether you want the luxury guided tour or the economy backpack trip. On a guided tour you will be staying in comfortable lodges, being fed and watered; if you are backpacking you can use the basic huts of the Department of Conservation (D.O.C.), or camp. You need to either book for the lodges or get a D.O.C. hut/camp pass. If you decide to extend the trip by returning along the Inland Track, you need a separate Back Country Hut ticket to allow you use of the four huts.

Start: Marahau.

Finish: Totaranui.

Difficulty & Altitude: The walking is easy, but can involve wading. It can be done in trainers or sports sandals– preferable because there is quite a bit of

The trek ends at another fine beach, Totaranui, seen here from Skinner Point. Regular ferries call at this beach to take walkers back to Marahau.

beach walking and (unless you are very lucky) some sea wading, which would ruin boots (a spare pair of old trainers that you can throw away at the end of the walk is perfect). Climbs and descents are never more than 350 to 500 feet (100–150m), though there can be several in a day.

ACCESS

Airports: International flights to Aukland, Wellington, and Christchurch, with connecting flights to Nelson.
Transport: There is a bus service between Nelson and Marahau. There is also a ferry service. The Abel Tasman Seafaris (tel.: 00 64 3 527 8083) depart Marahau at 9:00 A.M. and 2:00 P.M., returning at 12:30 and 4:30 P.M. It takes about 2 hours to Totaranui (or return); the fare is $N.Z.28. There is also AT National Park Travel (tel.: 00 64 3 528 7801), whose boats leave at 9:00 A.M. and return at 12:15 P.M.
Passports & Visas: Passports are required but no visa for USA, UK, Commonwealth, and most European countries.
Permits & Restrictions: A Coastal Track hut or camping pass, as appropriate, must be obtained before entering the park, and a separate one for the Inland Track (available at outdoor shops, information centres, etc., in the region). Mountain biking is not permitted.

LOCAL INFORMATION

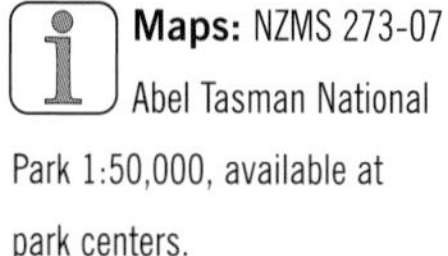

Maps: NZMS 273-07 Abel Tasman National Park 1:50,000, available at park centers.
Guidebook: *Classic Tramps in New Zealand*, Connie Roos (Cicerone).
Background reading: *A Park for all Seasons* (Department of Conservation); *The Enchanted Coast*, Emily Host (McIndoe).

The dramatic Falls River is one of the highlights of the Coastal Track. It is possible to make a fairly tough diversion from Torrent Bay to the falls on Day 2 of the trek.

Accommodation & Supplies: The Department of Conservation look after the park, the huts and campsites. The huts have bunks, heating and water but no cooking facilities. Accommodation cannot be guaranteed– it gets very crowded in season and stays are limited to 2 nights per hut. There are 22 equipped campsites. All trash must be packed out of the park. Awaroa Lodge: This private lodge and cafe has various types of accommodation, including family rooms. They have kayaks, rowboats and snorkels for hire; laundry, hot showers and pack storage; D.O.C. hut passes are also available. The restaurant is open to all. It is a good one-centre base for the park, and is connected by boat to everywhere along the coast.
Currency & Language: New Zealand dollars ($N.Z.). English.
Photography: No restrictions.
Area Information: Department of Conservation (D.O.C.), P.O. Box 97, Motueka, New Zealand (tel.:00 64 3 528 9117); Abel Tasman Tourist Accommodation Ltd, P.O. Box 72, Takaka, New Zealand (tel./fax: 00 64 3 528 8758).
Tours: 3-day and 5-day tours using the lodges are organised by Abel Tasman National Park Enterprises, Motueka Rd 3, Nelson, New Zealand (tel.: 00 64 3 528 7801; fax 00 64 3 528 6087). All meals and transport are included, and the service of a guide. Main luggage is taken from lodge to lodge by boat so you need carry only a day sack.

TIMING & SEASONALITY

Best Months to Visit: The walk is possible all year round. Facilities are probably best from September to June, when the track is busiest. Because of its fantastic beaches, the track is used by some 30,000 people a year.
Climate: This walk is noted for its subtropical climate.

HEALTH & SAFETY

Vaccinations: None required.
General Health Risks: All water must be boiled or filtered because of giardia in some streams. Insect repellent is needed to keep off the sandflies which are out in the day, not evening, and whose bite is very irritating. Wasps are a nuisance in late summer and fall.
Special Considerations: Tide tables, posted on all the huts, need to be studied before crossing estuaries, but be prepared to get your feet wet.
Lightweight raingear is useful, as is a sweater for cool evenings.
Politics & Religion: No concerns.
Crime Risks: Low.
Food & Drink: This is not a wilderness walk, and you are never far away from civilisation where food and drink can be found. But if you wish to enjoy the trek fully and avoid the towns, you must carry your food and cooking facilities with you. All huts have water supplies.

HIGHLIGHTS

Scenic: A wonderful track that takes you through native forest and along golden beaches. Don't miss the Astrolabe Roadstead along the first part of the walk, and, out of all the amazing beaches along the way, the Onetahuti beach stands out above the rest. If you get fed up with walking then experience the park from the coast by taking a boat trip. Sailings are subject to the weather and it is not uncommon to have to wade ashore, but it's worth it for the coastal landscape is fantastic.
Wildlife & Flora: The brown flightless weka and the blue pukeko bird may be seen along the trek. The islands are inhabited by fur seals and the blue penguin.

weka

temperature and precipitation

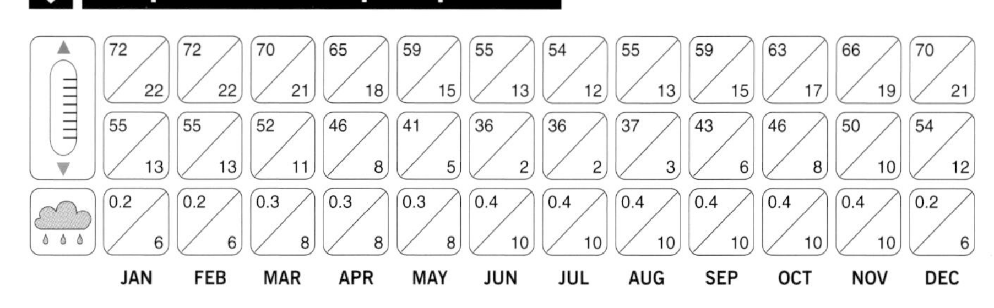

	JAN	FEB	MAR	APR	MAY	JUN	JUL	AUG	SEP	OCT	NOV	DEC
Max temp (°F / °C)	72 / 22	72 / 22	70 / 21	65 / 18	59 / 15	55 / 13	54 / 12	55 / 13	59 / 15	63 / 17	66 / 19	70 / 21
Min temp (°F / °C)	55 / 13	55 / 13	52 / 11	46 / 8	41 / 5	36 / 2	36 / 2	37 / 3	43 / 6	46 / 8	50 / 10	54 / 12
Precipitation	0.2 / 6	0.2 / 6	0.3 / 8	0.3 / 8	0.3 / 8	0.4 / 10	0.4 / 10	0.4 / 10	0.4 / 10	0.4 / 10	0.4 / 10	0.2 / 6

Duckboarding is often used on the Overland Track to protect the fragile, peaty terrain of this wilderness area.

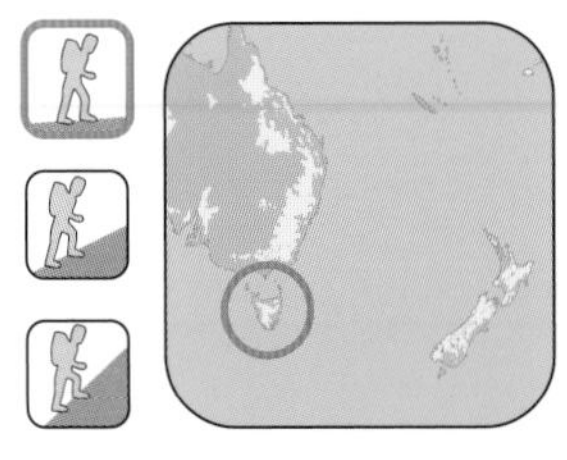

the OVERLAND TRACK

TASMANIA, AUSTRALIA

Alan Castle

THE MECCA FOR WALKING IN AUSTRALIA IS THE SMALL, BY AUSTRALIAN STANDARDS, ISLAND OFF THE SOUTH-EAST END OF THE CONTINENT, KNOWN AS TASMANIA. A CONSIDERABLE AREA OF THE ISLAND IS UNSPOILT WILDERNESS, FORMING THE TASMANIAN WORLD HERITAGE AREA.

There are no less than 17 national parks on the island, the jewel in the crown of which is the Cradle Mountain–Lake St. Clair National Park in the north-west of Tassie. The Overland Track, which traverses the full length of the park, from the imposing Cradle Mountain in the north to the huge alpine Lake St. Clair in the south, is considered by many to be one of the finest maintained walking trails in the world, providing a unique wilderness experience but one that is not too arduous and within the capabilities of most reasonably fit people. As a result the trail is a popular one, attracting walkers form all parts of the globe who are keen to experience what in our modern world is becoming more and more of a rarity – a landscape relatively untouched by man. People who trek the Overland spend around a week of their lives in a unique part of the world, and take away memories that they will cherish until their dying day. No one is untouched by this wild and dramatic landscape.

itinerary

•DAY 1 6¼ miles (10km)

Cradle Valley to Waterfall Valley Hut

The first few miles of the trail are the hardest, physically, of the entire standard route, so if you cope adequately with these then all should be well for the remainder of the trip. The track passes close by two picturesque glacial lakes, Crater and the larger Dove Lake, before climbing up to Marion's Lookout, a superb vantage point.

A descent to the tiny Kitchen Hut, an emergency shelter, allows the first optional detour of the walk to be tackled, provided the weather conditions are good – a signposted path leads to the summit of Cradle Mountain itself (1½ miles/2.5km return).

From Kitchen Hut the track descends gently before climbing to Cradle Cirque, an easy walk. A side track leads to the summit of Barn Bluff, a distinctive peak that dominates the whole area for many miles. If you intend to climb Barn Bluff (3 miles/5km return), then do check with the rangers before starting your walk, as the trail has suffered from considerable erosion in the past, and access may be restricted in the future.

Eventually the track descends steeply into Waterfall Valley. Follow a track to the right to reach Waterfall Valley Hut, your first night's destination.

•DAY 2 4½ miles (7km)

Waterfall Valley Hut to Windermere Hut

While at Waterfall Valley Hut try to make some time to visit Waterfall Valley (2 miles/3km return) where you can see the first of many spectacular waterfalls in this park. The Overland Track crosses open moorland; after 1½ miles (2.5km) the route reaches a side track off to the right, from where a recommended detour to isolated Lake Will (3 miles/5km return) can be made. Lake Will is fed by the Bluff River, on which will be found the Innes Falls and the start of a disused nineteenth-century track that led to the west coast across very remote country. Descend

to picturesque Lake Windermere, dotted with many small islands; a few minutes south of this Windermere Hut is located.

•DAY 3 **8¾ miles (14km)**

Windermere Hut to Pelion Hut

If you are feeling fit enough, the day can easily be split into two. In the morning, follow the Overland Track to the east of Mount Pelion West (a difficult peak to attempt), and then descend through dense eucalyptus forest to reach Frog Flats (2,362 feet/720m), the lowest point reached on the Overland. Climb through more forest to reach Pelion Hut. During the afternoon choose one or more detours: to Old Pelion Hut and Copper Mine (2 miles/3km return); to Lake Ayr (3¾ miles/6km return); or the ascent of Mount Oakleigh (5 miles/8km return) – this is the most rewarding of the three, but only attempt it in clear conditions.

•DAY 4 **5 miles (8km)**

Pelion Hut to Kia Ora Hut

In good weather conditions many trekkers will wish to include on today's itinerary an ascent of Mount Ossa, the highest peak in Tasmania (3 miles/5km return). It makes for a very rewarding and full day. Follow the Overland to Pelion Gap, the broad col between Mount Ossa to the south-west and Mount Pelion East to the north-east. Here, head west on the signposted track to Mount Ossa.

Return to Pelion Gap after the climb before descending to Kia Ora Hut. An alternative to Mount Ossa, particularly when cloud obscures the higher peaks, is to climb Mount Pelion East, a shorter and easier climb than Ossa, and equally rewarding (1½ miles/2.5km return).

•DAY 5 **5½ miles (9km)**

Kia Ora Hut to Windy Ridge Hut

Much of the route is in dense and attractive forest, first passing the very old Du Cane Hut – of considerable historical interest as it was used by some of the early trekkers to this region – and then reaching the junction for the path to the impressive D'Alton and Fergusson waterfalls on the Mersey River. The detour to these falls is about half a mile (1km) return. Farther on, another path (½ mile/1km return) leads to the Hartnett Falls, which should certainly not be missed; there are excellent views from the rim of the gorge near these falls. From this track junction, the trail climbs to the Du Cane Gap (3,511 feet/1,070m), beyond which a steep drop leads to Windy Ridge Hut.

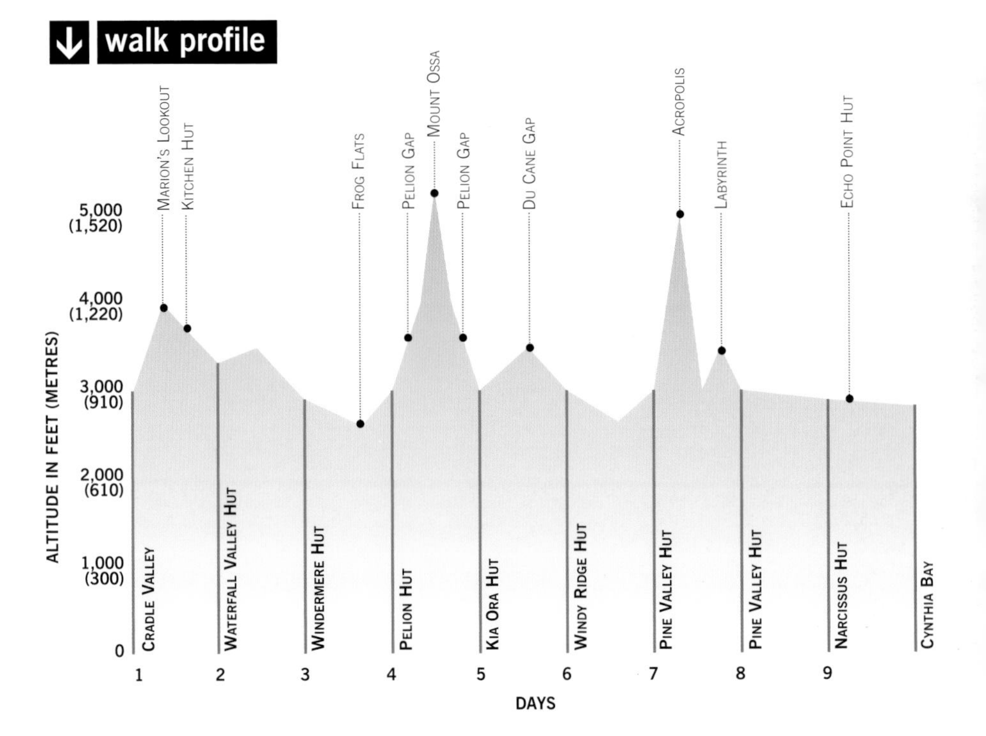

The prominent mountain peak of Barn Bluff is visible continuously for the first few days on the northern section of the Overland Track.

A trekker crosses the suspension bridge on the trail through Pine Valley. This bridge allows easy access to Pine Valley Hut and the remarkable Acropolis and Labyrinth areas.

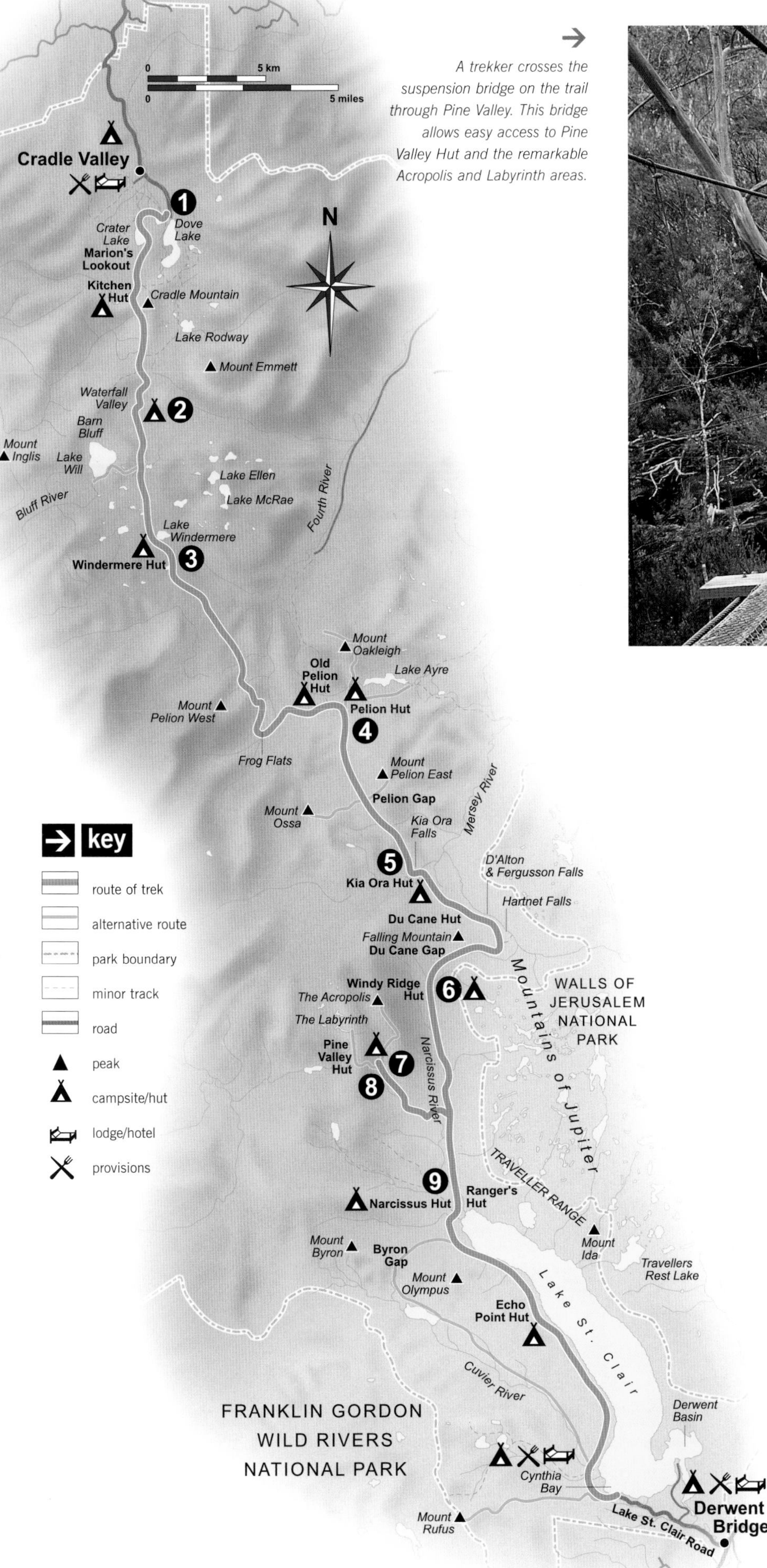

•DAY 6 6¼ miles (10km)

Windy Ridge Hut to Pine Valley Hut

An important decision needs to be made here. Provided the weather conditions are good, and the party still in fine fettle, with adequate food supplies still available, then an extra day should be spent in Pine Valley exploring the Acropolis and the Labyrinth – this area is scenically one of the best in the park and should not be missed (see Day 7).

Those who have had enough of the wilderness for now and yearn for a hot shower and the other comforts of civiliSation can easily reach Cynthia Bay today by walking the Overland Track to Narcissus Hut (5½ miles/9km) and taking the ferry across Lake St. Clair. Those heading for Pine Valley also follow the standard trail southwardS, but turn off to head north-westwardS on the signposted path to the Pine Valley Hut.

•DAY 7 7½ miles (12km)

Pine Valley Hut to the Acropolis and the Labyrinth, and back

A second night spent in Pine Valley Hut allows a climb to the Acropolis, an ascent that involves a little easy scrambling. An incredible view of seemingly countless mountain tarns and rocky summits awaits you at the top. The Labyrinth, as the name suggests, is an area dotted with →

The treescape and unique vegetation of the Cradle Mountain–Lake St. Clair National Park are reminiscent of the sets for Steven Spielberg's blockbuster film, Jurassic Park.

a maze of lakes; there are numerous possibilities for exploring this stunningly beautiful area before returning to the hut for dinner and a well-earned rest.

•DAY 8 **5½ miles (9km)**

Pine Valley Hut to Narcissus Hut

Return from Pine Valley Hut on the trail used on the outward route, to the path junction with the standard trail. Here turn right to head southwardS, crossing the Narcissus River on a long swing bridge to reach Narcissus Hut on the northern end of Lake St. Clair. Catch the afternoon ferry to Cynthia Bay or spend your last night in the wilds at Narcissus Hut.

•DAY 9 **10 miles (16km)**

Narcissus Hut to Cynthia Bay via the Lake Track

The lakeside path provides a good end to the trek, although views of lake and mountains are somewhat restricted. After about 3 miles (5km) you pass a last hut, Echo Point Hut, and your very last opportunity to spend a night in the park. There is an alternative route to Cynthia Bay via Byron Gap and the Curier River, but this is much longer and very wet and muddy, and as such cannot be recommended, although the views are certainly finer. Before leaving the area, a good day-walk from Cynthia Bay for the energetic is the ascent of Mount Rufus (4,643 feet/1,415m), for a final view over the peaks and plateaux of this most amazing of wilderness landscapes.

factfile

OVERVIEW

Trekking in Australia is known as "Bushwalking", and the Overland Track in Tasmania is the country's finest and best-known long-distance bushwalk trail, traversing the Lake St. Clair National Park. The basic trail is around 60 miles (95km) in length, but there are numerous detours and worthwhile excursions to lakes, mountains, and water-falls, so that those wishing to enjoy the wilderness experience to the full take 8 or 9 days to complete the route.

Start: Car park at the end of the public road, 4½ miles (7.5km) south of Cradle Mountain National Park Visitor Centre, close to the northern shore of Lake Dove.

Finish: Lake St. Clair.

Difficulty & Altitude: The Overland Track is neither a physically demanding walk nor one that requires a lot of bushwalking or mountaineering experience. The paths are on the whole well graded and well maintained, and there is only a moderate amount of ascent and descent. Navigation is generally straightforward. There are several very muddy sections, particularly on some of the side tracks. Any reasonably fit walker who is not afraid to "rough it" a little should have little trouble coping with the Track. Some of the optional detours from the main track are considerably harder than the Overland itself, often with steep paths on rough ground.

ACCESS

Airports: Tasmania international airport is at the capital, Hobart.

Ferry: *The Spirit of Tasmania* ferry makes the 14-hour journey between Melbourne and Devonport, on the north of Tasmania, three times a week in each direction.

Transport: Tasmanian Wilderness Travel (tel.: 00 61 3 6334 4442; fax: 00 61 3 6334 2029; e-mail info@taswildtravel.com.au) provide daily bus transport from Launceston and Devonport to the start of the Overland Track. Buses also operate from Lake St. Clair to Hobart, Launceston, Devonport and Strahan.

Passport & Visas: Passport with tourist visa for all visitors.

Permits & Restrictions: A Tasmanian National Park Pass is required. A range of passes is available, but the most suitable for the Overland Track is the Backpacker Pass, which is valid for 2 months and which also allows access to all of Tasmania's other national parks. The permit is available from the Park Visitor Centre, Tasmanian Travel Information Centres and on the internet.

LOCAL INFORMATION

Maps: 1:100,000 Cradle Mountain, Lake St. Clair National Park Map and Notes (Tasmap). The detail is quite sufficient for this easy-to-follow trail.

Guidebooks: *The Overland Track– A Walker's Notebook* (Tasmania Parks and Wildlife Service) – a pocket-sized, 80-page booklet, which can be ordered on the internet; *Cradle Mountain Lake St. Clair*, John Chapman and John Sizeman (Pindari Publications).

yellow gum

Background Reading: *Bushwalking in Australia* (Lonely Planet); *Welcome to the Wilderness* – a free trip-planning guide available on the internet and as a booklet from Tasmania Parks and Wildlife Service.

Accommodation & Supplies: Apart from the public, unmanned huts – basic, with communal bunks, no mattresses, and offering little more than overnight shelter – there are no facilities on the trail, and consequently walkers must be totally self-sufficient, carrying all food and cooking equipment for the duration of the trek. The park is a "fuel stove only" area; open camp and cooking fires are not allowed. Carry all food packaging and other rubbish out of the park. It is best to buy food supplies in one of the Tasmanian cities or large towns (Hobart, Launceston, or Devonport) rather than rely on stores locally. Be sure to take an extra 2 or 3 days' worth of supplies in case bad weather causes a delay. Hut space cannot be booked, and sleeping places are on a first-come, first-served basis, so it is essential to carry a tent, particularly during the main summer holiday period (Christmas to the end of January). It is advisable to camp in the areas adjacent to the huts. Note that Kitchen Hut, Ranger Hut and Du Cane Hut are for emergency shelter only. At the north end of the track, inside the national park, cabin accommodation is available at Waldheim. At Cynthia Bay there is backpacker accommodation and camping.

Currency & Language: Australian dollars ($Aus.). English.

Photography: Be sure to bring plenty of film, as none can be purchased en route.

Area information: The Department of Parks, Wildlife and Heritage, G.P.O. Box 44A, Hobart 7001, Australia (Information Officer tel.: 00 61 3 6233 6191). Websites: www.parks.tas.gov.au/natparks/; www.parks.tas.gov.au/natparks/.

TIMING & SEASONALITY

Best Months to Visit: December to April, i.e., during the summer months; conditions can be severe outside this period (and even within it). The track is at its most crowded from Christmas to the end of January. February is perhaps the best month.

Climate: Tasmania lies within the notorious "roaring forties" and, consequently, strong winds and rain are not uncommon, particularly in the wilderness mountain areas. Unlike much of Australia, Tasmania has all four seasons, winters (June to September) often being very harsh with low temperatures and considerable snow falls. Even in the summer months a trekker on the Overland Track can experience strong winds, driving rain, sleet, snow, low dense cloud and blazing sunshine – all within one day. Trekkers must come prepared with clothing suitable for all such eventualities. Temperatures can drop quite low at night, with frosts, even during the summer, so a good sleeping bag is required. The dangers of hypothermia should not be underestimated.

HEALTH & SAFETY

Vaccinations: None are required for entry into Australia.

General Health Risks: Australia has an excellent health service, but this will not be directly available to hikers on the Overland Track. A health and accident insurance policy is strongly recommended. An adequate

wallaby

first-aid kit is required to cope with minor emergencies during the time the hiker is on the trail, where medical help is not at hand. Be sure to carry plenty of insect repellent, and high-factor sunscreen and a sun hat.

Special Considerations: There are three species of snake on Tasmania, all of which are highly poisonous: the whip-snake, the copperhead and the tiger snake. The latter is the most deadly. Care should be exercised, but they are shy creatures who usually sense your presence before you are aware of them and move away quietly. They are very unlikely to strike a human unless they feel trapped or threatened; take care not to tread on one. Some walkers prefer to wear gaiters to lessen the chance of a strike puncturing the skin (gaiters also protect against the scrub and mud on the trail). Leeches are an unpleasant problem on the Overland, especially in the Frog Flats and Pelion Hut area. Use salt or a lighted cigarette to remove them from the skin; their effective anticoagulant sometimes means that a leech wound takes quite a time to dry up. Wherever possible use the toilet facilities provided at all of the overnight huts. Otherwise toilet sites must be at least 330 feet (100m) from water sources and campsites, and waste and paper must be buried to a depth of at least six inches (15cm). Be scrupulously clean with your personal hygiene.

Politics & Religion: No concerns.

Crime Risks: Low.

Food & Drink: Water purifying tablets are a good idea.

HIGHLIGHTS

Scenic: The mountain, lake and forest scenery is outstanding. There are high rocky mountains, numerous dramatic waterfalls, many tranquil glacial lakes, and the flora of the temperate rain forest is quite unlike that found in the northern hemisphere. From the summit of Tasmania's highest mountain, Mount Ossa (5,301 feet/1,617m), there are extensive views over half of Tasmania.

Wildlife & Flora: Tasmania abounds with wildlife, particularly in the remote national park areas. Wallabies and possums are common and are sure to be seen by trekkers, but there are several other marsupial and other mammalian species in the bush. Birdlife is particularly rich, with over 80 species inhabiting the highlands, several of which are endemic to Tasmania. The vegetation of western Tasmania is unique and very varied – in the valleys are beech and eucalyptus forests, and higher in the mountains will be found buttongrass moors and alpine flowers. The range of species is immense. In particular, the cushion plants – of which there are five species in Tasmania – are of great interest.

Narcissus Hut at the northern end of Lake St. Clair, where tired trekkers can opt to end their walk prematurely.

temperature and precipitation

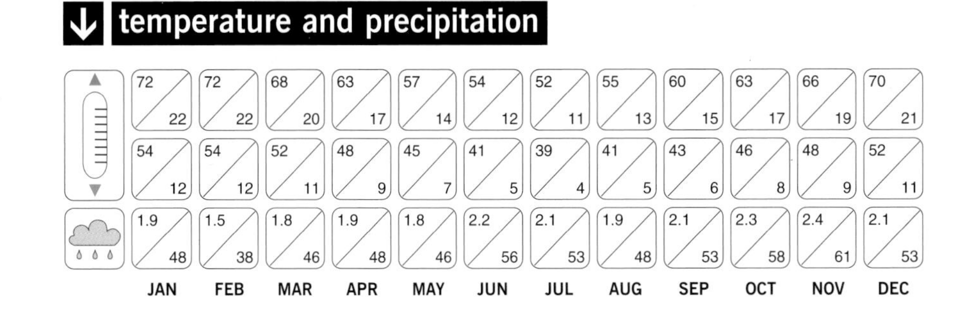

	JAN	FEB	MAR	APR	MAY	JUN	JUL	AUG	SEP	OCT	NOV	DEC
Max temp (°F / °C)	72 / 22	72 / 22	68 / 20	63 / 17	57 / 14	54 / 12	52 / 11	55 / 13	60 / 15	63 / 17	66 / 19	70 / 21
Min temp (°F / °C)	54 / 12	54 / 12	52 / 11	48 / 9	45 / 7	41 / 5	39 / 4	41 / 5	43 / 6	46 / 8	48 / 9	52 / 11
Precipitation (in / mm)	1.9 / 48	1.5 / 38	1.8 / 46	1.9 / 48	1.8 / 46	2.2 / 56	2.1 / 53	1.9 / 48	2.1 / 53	2.3 / 58	2.4 / 61	2.1 / 53

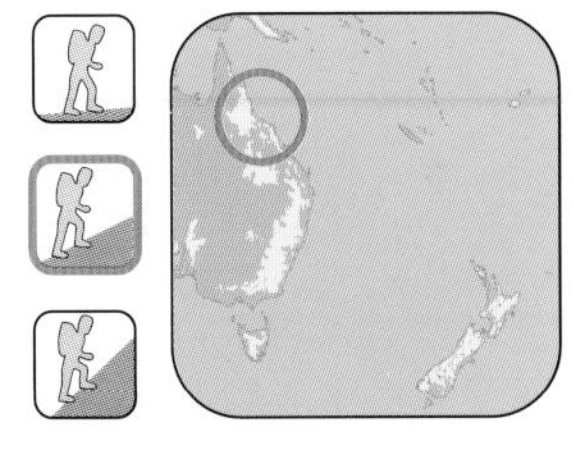

the THORSBORNE TRAIL

QUEENSLAND, AUSTRALIA

Alan Castle

NOT FAR OFF THE COAST OF NORTH QUEENSLAND, ACROSS A SMALL CHANNEL, LIES HINCHINBROOK ISLAND, AN UNSPOILT WILDERNESS ISLAND FAMOUS FOR THE RICHNESS AND VARIETY OF ITS HABITATS. IT LIES WITHIN THE GREAT BARRIER REEF WORLD HERITAGE AREA, AND IS AUSTRALIA'S LARGEST ISLAND NATIONAL PARK.

The Thorsborne Trail is named after the late Arthur Thorsborne, who, along with his wife, Margaret, spent a lifetime dedicated to the conservation of Hinchinbrook Island and a study of its abundant and varied wildlife. The trek follows the island's east coast and offers a unique chance to savour a tropical island wilderness, stopping for refreshing swims in creeks and water holes, and enjoying the abundance of wildlife in the rain forest, mangrove swamps, and saltpans. Trekkers must be completely self-reliant and self-sufficient, but the rewards are immeasurable. There are a number of creek crossings to negotiate, including a couple of tidal creeks, and the terrain is sometimes rugged. Nevertheless, the route is well waymarked, and, provided the weather conditions are not extreme, the trail should be within the capabilities of the average fit and well-prepared bushwalker.

itinerary

•DAY 1 4 miles (6km)

Ramsey Bay to Little Ramsey Bay

The ferry carrying Thorsborne Trail trekkers usually leaves Cardwell at 9:00 A.M., crosses Hinchinbrook Channel, and enters an area of mangrove swamp, cruising along a channel until it arrives at a landing point close to Ramsey Bay. Upon disembarking, walk along the boardwalk to arrive on the pristine sand of Ramsey Bay.

Walk south along the beach to enter an area of tall open forest. About halfway along this section there is an opportunity to take an hour's detour and climb Nina Peak, on a rather steep, rugged path that leads off the main trail at its highest point between Ramsey and Nina Bays. The spectacular view from the summit of Nina Peak is worth this little bit of extra effort. Return to the main trail on the same path.

The trail reaches Nina Beach, where there is a permitted camping area. Some may wish to spend the night here, but otherwise it makes an ideal spot for a prolonged lunch, rest and possible swim (but not during the box jellyfish season – see Factfile, page 185).

Rock-hopping around a rough headland makes up much of the next section, carefully following waymarks and cairns over Boulder Bay and on to Little Ramsey Bay, where you should arrive by mid-afternoon. This is the recommended site for your first night's camp, a gorgeous spot close to a freshwater lagoon. There's time to relax, cook and eat dinner, and perhaps have another cooling swim.

•DAY 2 6½ miles (10km)

Little Ramsey Bay to Zoe Bay

From the camping area, walk south along the beach to reach a tidal creek. This is usually fairly easy to cross, but on occasion can prove tricky. Continue to another small beach, from where you clamber over rocks to a larger sandy beach. From the other end of this beach, follow a gully up to the top of the ridge, from where a side

↑ *Backpackers stride across Ramsey Bay, taking their first steps on the Thorsborne Trail. The pristine sands of the many unspoilt beaches on Hinchinbrook Island offer a sense of space, tranquility and timelessness.*

← *Several of the creeks on this mountainous island have spectacular waterfalls, such as Zoe Falls. The falls are at their most dramatic during the wet season.*

path leads in to Banksia Bay. Descend on the main trail into Banksia Creek and continue to enter rain forest, open forest and mangroves.

After crossing North Zoe Creek (beware of estuarine crocodiles), there follows a section through extensive palm swamps where it is essential to follow the waymarks carefully. Wet feet will be the order of the day as there are many creeks and swampy areas to cross. And beware of spider's webs stretching across adjacent tree trunks, as if you walk into one you may find its large, unhappy owner on your rucksack or shoulder.

The trail eventually emerges onto the beach at Zoe Bay. The night's campsite is at the southern end of this beach. Alternatively, follow South Zoe Creek through rain forest to another camping area, where there is good fresh water. After a rest, do take some time to walk upstream to Zoe Falls to enjoy the cascade and huge, clear pool into which the water tumbles.

•DAY 3 **4¾ miles (7.5km)**

Zoe Bay to Mulligan Falls

The trail crosses South Zoe Creek downstream from Zoe Falls. Boulder- and rock-hopping follows, at one steep point crossing granite slabs with the assistance of a fixed rope. Be sure to look back for the fine views over Zoe Bay. South Zoe Creek is followed with further creek crossings (as always, take care and practise caution) to reach a height of 853 feet (260m), the highest point on the entire trail.

A descent eventually follows, during which there is a side track to Sunken Reef Bay. This short detour would provide an alternative campsite for the night or, if you left Zoe Bay early enough, it would make a great location for an extended lunch stop.

Diamantina is often one of the most difficult of the many creeks on the trail to cross; particular caution is necessary if the water level is swollen because of recent rains. Follow the markers diagonally across the creek. →

walk profile

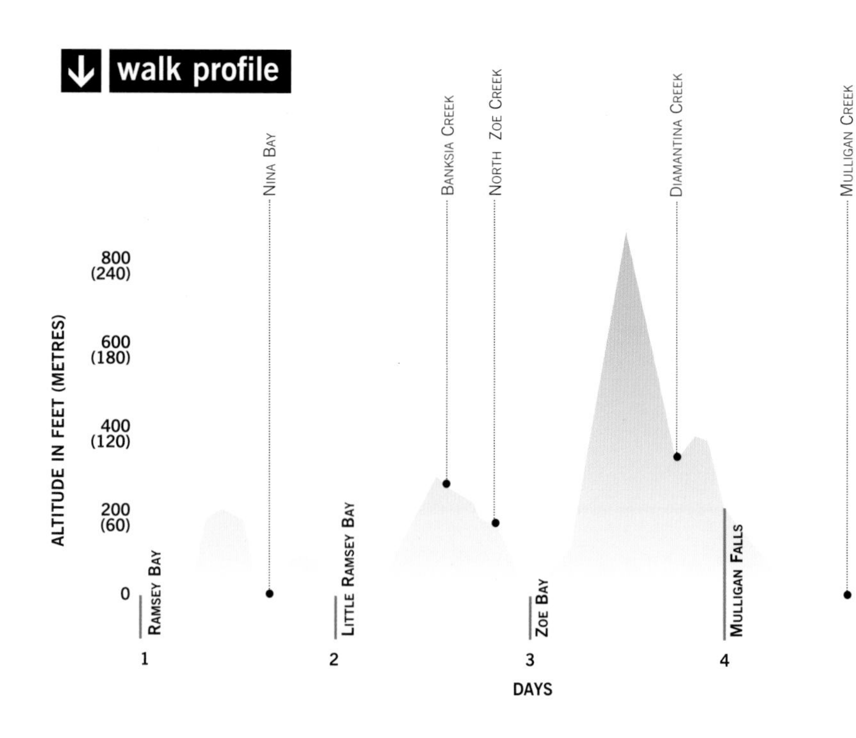

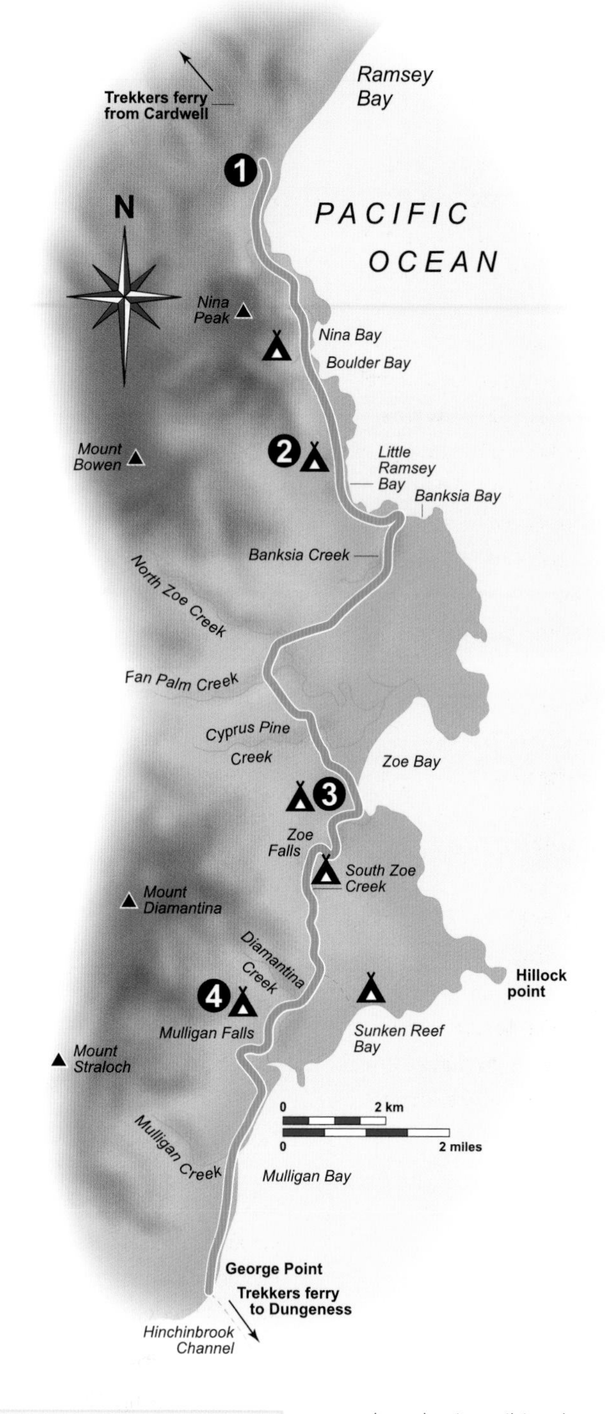

key

A short climb follows, after which the trail descends to Mulligan Falls campsite, where you must camp within the rain forest in the designated areas, which are located before the picturesque waterfall. You will no doubt use the large rock pool, which usually teems with fish, for yet another refreshing swim; this is a really splendid place to spend your last night on the island.

•DAY 4 **4¾ miles (7.5km)**

Mulligan Falls to George Point

The final day is relatively short, so there is little need to rise early, unless you wish to enjoy a last, early morning swim in the pool. Short the day may be, but your difficulties are certainly not yet over. The first part of the walk is through tropical rain forest, involving the crossing of as many as five more creeks. The trail emerges onto the huge beach of Mulligan Bay. A pleasant walk along the sand leads to the very last obstacle on the trail, the tidal Mulligan Creek. Attempt to cross this wide creek, which spills out across the beach, only at low to half tide; crossing at high tide is extremely dangerous. There remains just a short amble farther south along the beach to reach George Point. Here, you will await your ferry to Dungeness (from where the ferry company will arrange transport back to Cardwell), and reflect on what will almost certainly have been one of the most rewarding and exciting few days of your life.

factfile

OVERVIEW

A wilderness trek on tropical Hinchinbrook Island, off the coast of north Queensland. The island is Australia's largest island national park and is within the Great Barrier Reef World Heritage Area. Although the trail is only 20 miles (32km) in length, a minimum of 4 days is recommended. The trail passes over generally rough and varied terrain and involves many river (creek) crossings, so progress is often slow.

Start: Ramsey Bay, Hinchinbrook Island.

Finish: George Point, Hinchinbrook Island.

Difficulty & Altitude: The route is waymarked with small orange or yellow triangles. Considerable parts of the trail are rugged – there are many unbridged creek crossings to negotiate and two tidal creeks to wade. The ground in the rain forest is often very wet and muddy, as are the mangrove swamps. The wildlife can be extremely unpleasant and even dangerous. Nevertheless, most fit walkers with a taste for adventure should thoroughly enjoy the experience.

ACCESS

Airports: Cairns, 120 miles (190km) to the north.

Transport: Long-distance coaches run up and down the

North Queensland Coast daily.

Ferries: Hinchinbrook Island Ferries in Cardwell take walkers to Ramsey Bay at the northern end of the trail. Another ferry company takes walkers off the island from George Point, at the southern trail terminus, to Dungeness on the mainland, and from there provides transport back to Cardwell. Bookings for both ferries can be made through Hinchinbrook Island Ferries, 113 Bruce Highway, Cardwell, Queensland, Australia 4849 (tel.: 00 61 7 4066 8270; fax: 00 61 7 4066 8271; email: hinchinbrook@4kz.com.au).

Passport & Visas: Passport with tourist visa for all visitors.

Permits & Restrictions: Permits are required to walk the trail and to camp overnight at designated sites. To protect the fragile rain forest ecosystem and unspoilt coastline, the numbers of hikers on the trail are strictly limited. If you intend to leave the trail to climb adjacent mountains, notably Mount Bowen and Mount Diamantina, a special permit is required, available from the same office. Walking off-trail is very restricted. Access to the island will be restricted when it is deemed unsafe. Hinchinbrook Island is a "fuel stove only" area; fires are not allowed. Carry all rubbish off the island.

LOCAL INFORMATION

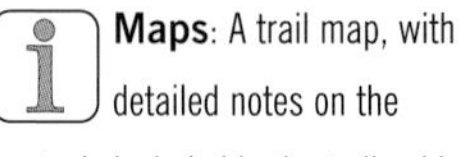

Maps: A trail map, with detailed notes on the route, is included in the trail guide leaflet (see below).

Guidebook: A trail guide leaflet containing all essential information is available from the Park Office in Cardwell.

Background Reading: *Hinchinbrook Island, the Land Time Forgot*, Arthur and Margaret Thorsborne (Lansdowne Publishing, Sydney).

Accommodation & Supplies: There is no accommodation along the trail. Trekkers must carry their own tents and may only camp at designated areas: Nina Bay, Little Ramsey Bay, Banksia Bay, Zoe Bay, Sunken Reef Bay, Mulligan Falls and George Point. The wardens at Cardwell Youth Hostel, where you can make all the necessary preparations for your trek, will keep safe any gear not needed on the trek. All food for the duration of the hike must be taken onto the island; adequate supplies can be purchased in Cardwell, although a wider choice is available in Cairns or Townsville.

saltwater crocodile

Currency & Language: Australian dollars ($Aus.). English.

Photography: No restrictions.

Area information: Queensland Department of the Environment and Heritage, Rainforest and Reef Centre, Shop 5, 79 Victoria Street, P.O. Box 74, Cardwell, Queensland; tel: 00 61 70 66 8601. Websites: www.gspeak.com.au/hinchinbrook/ www.gspeak.com.au/hinchinbrook/4day.htm

TIMING & SEASONALITY

Best Months to Visit: April to September.

Climate: North Queensland enjoys a year-round tropical climate, with high temperatures and humidity. There are only two "seasons", the "Wet" (generally from December to April) and the "Dry" (May to November). Cyclones and other tropical storms are not uncommon, particularly around March. Nighttime temperatures can sometimes be quite cool.

HEALTH & SAFETY

Vaccinations: None required.

General Health Risks: Australia has an excellent health service, but a health and accident insurance policy is necessary. An adequate first-aid kit is required to cope with minor emergencies where medical help is not at hand. Be sure to carry plenty of insect repellent, high-factor sunscreen, a sun hat, water-purifying tablets, and antiseptic cream. Feet will often be wet on the trail because of the creek crossings and the wet and humid conditions of the rain forest. Footcare should be practised to avoid infection.

Special Considerations: Dehydration and heat exhaustion are possible, so drink sufficient quantities of water. Apply liberal quantities of insect repellent and wear long-sleeved shirts and trousers to reduce the number of insect bites. Estuarine crocodiles inhabit some of the tidal lagoons and mangrove estuaries; they are extremely dangerous, so be vigilant when crossing creeks. There are several poisonous snakes on the island, and caution is required. Box jellyfish are extremely dangerous; swimming in the sea is not advised during October to May when they are present in the coastal waters of Queensland. Never attempt a river crossing at high tide, when the water levels are dangerously high, or the water very fast running.

Crime Risks: Low.

Food & Drink: Fishing from the freshwater streams is not permitted. Fresh food will rapidly spoil in this tropical environment, so take types of food that will not deteriorate. Water from mountain streams is considered safe to drink. Rats are present at most campsites; secure your food in the rat-proof boxes that should be found at each campsite (otherwise hang packs and food separately, and out of rats' reach).

HIGHLIGHTS

Scenic: The landscape ranges from lush tropical rain forest, eucalyptus forests, and fragile heath vegetation to unspoilt sandy beaches, bays, rocky headlands, and mangrove swamps. Mountains rise majestically from the coast to over 3,300 feet (1,000m). Many mountain streams are encountered, some with pools, which are good places to enjoy a cool swim.

Wildlife & Flora: The island is famous for the richness and variety of its habitats – extensive mangrove forests, saltpans, rain forest, freshwater swamps and sloping mountain rock pavements.

→

Mangrove forests fringe much of this tropical island's coastline. Boardwalks provide easy access for trekkers.

temperature and precipitation

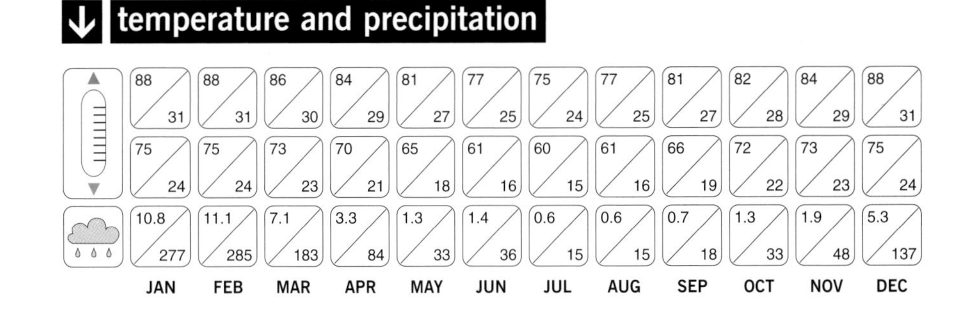

	JAN	FEB	MAR	APR	MAY	JUN	JUL	AUG	SEP	OCT	NOV	DEC
High (°F / °C)	88 / 31	88 / 31	86 / 30	84 / 29	81 / 27	77 / 25	75 / 24	77 / 25	81 / 27	82 / 28	84 / 29	88 / 31
Low (°F / °C)	75 / 24	75 / 24	73 / 23	70 / 21	65 / 18	61 / 16	60 / 15	61 / 16	66 / 19	72 / 22	73 / 23	75 / 24
Precipitation (in / mm)	10.8 / 277	11.1 / 285	7.1 / 183	3.3 / 84	1.3 / 33	1.4 / 36	0.6 / 15	0.6 / 15	0.7 / 18	1.3 / 33	1.9 / 48	5.3 / 137

THE AUTHORS

GENERAL EDITOR

BILL BIRKETT A leading climber and hill walker, and one of Britain's foremost mountain writers and photographers. Son of the rock climbing pioneer Jim Birkett, and a former civil engineer, Bill has spent a lifetime exploring wild places. His breathtaking photography illustrates numerous publications, and his own books include the best-selling and definitive work, *Complete Lakeland Fells*. He is a regular contributor to *The Great Outdoors* and *Climber* magazines, and is a member of the British Fell and Rock Club, the Climbers' Club, and the Outdoor Writers' Guild.

CONTRIBUTORS

JUDY ARMSTRONG Originally from New Zealand, she has now settled in England. She is an award-winning feature writer, specialising in adventure travel and outdoor pursuits. She returns regularly to New Zealand and has travelled widely – on foot, and by horse, boat and bus – through North and South America, Africa, Europe and Asia.

STEVE CALLEN Has been trekking and climbing the world's greatest mountain ranges for more than 20 years. He has been part of nine major expeditions to the Himalaya, and visits Pakistan, India and Nepal regularly. He thrives on the challenge of outdoor pursuits and has recently mountain biked across the Himalaya.

ALAN CASTLE Has trekked in over 20 countries, and written about his experiences in a dozen walking books. For the past 10 years he has led walking holidays in Europe, as well as making solo walks across the Alps, the French Pyrenees, and bush-walking in the mountains and rain forests of Australia and Tasmania.

JOHN CLEARE A professional photographer, writer, and lecturer, beside being a mountaineer, adventurer and wilderness traveller of wide experience. His work regularly takes him to the most remote corners of the world.

RICHARD GILBERT Has been hill walking and mountaineering for 45 years, and has written about his experiences in numerous books and magazines.

TOM HUTTON A freelance writer and photographer, who is equally at home walking, climbing, skiing or cycling. His passion for the outdoors, in particular remote or dramatic landscapes, is matched only by his love of nature.

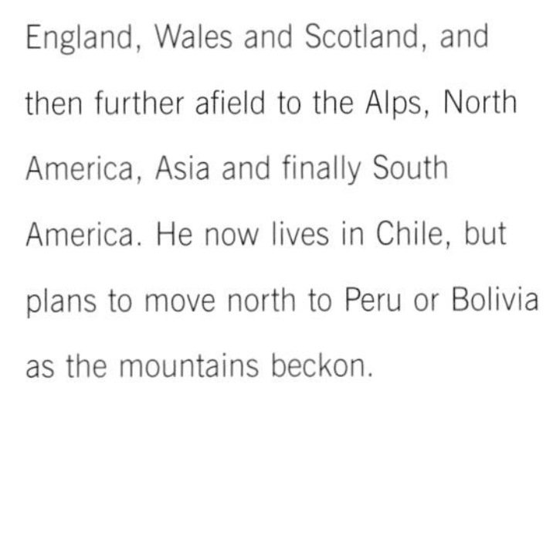

ANDREW SHEEHAN His love of the outdoors took him all over England, Wales and Scotland, and then further afield to the Alps, North America, Asia and finally South America. He now lives in Chile, but plans to move north to Peru or Bolivia as the mountains beckon.

JOHN MANNING The Deputy Editor of *The Great Outdoors*, Britain's premier magazine for hill walkers, trekkers and backpackers. He has walked in the Himalaya, the Andes, the Alps, the Arctic Circle and throughout Britain and Ireland, and is the winner of five awards for outdoor writing.

PHIL STONE A life-long love of mountains and wild places steered him to British Columbia. For six years he worked as a mountaineering guide on Vancouver Island before entering the world of adventure photography and outdoor writing. He now resides on Quadra Island, British Columbia.

BILL O'CONNOR A professional mountain guide, author and photographer with extensive trekking experience in East Africa. He spends much of the year ranging the mountains of the world on foot and ski looking for new ascents, untracked snow and remote places.

CHRIS TOWNSEND One of the world's most experienced long-distance hikers. He was the first person to hike the length of the Canadian Rockies, and has also hiked the Pacific Crest and Continental Divide Trails in the USA, south to north through Canada's Yukon Territory, and end to end through Norway and Sweden.

KEV REYNOLDS An extremely experienced trekker and climber, as well as a full-time writer, photo-journalist and lecturer. He has authored 35 books, including trekking guides to the Pyrenees, the Alps and the Himalaya; and is an active member of the Alpine Club, the Austrian Alpine Club, and the Outdoor Writers' Guild.

WALT UNSWORTH Has over 30 books to his credit and is President of the Outdoor Writers' Guild. An ardent enthusiast of adventure travel and mountain history; his book *Everest* is regarded as the definitive study of that great mountain.

TRAVELLER INFORMATION

AUSTRALIA
Australian Tourism Industry Association
P.O. Box E328
Canberra ACT 2600
Tel.: 00 61 2 6273 1000
www.aussie.net.au
www.australia.com

AUSTRIA
Austrian National Tourist Office
Margaretenstrabe 1
A-1040 Wien
Tel.: 00 43 1 58 72 000
www.austria-tourism.at

CANADA
Calgary Convention and Visitors Bureau
237-8 Avenue S.E.
Calgary AB
T2G 0K8
Tel. 00 1 403 263 8510
www.tourismcalgary.com

Tourism British Columbia
Parliament Buildings
Victoria BC
V8V 1X4
Tel.: 00 1 250 387 1642
www.tourism-vancouver.org/

CHILE
National Tourism Board of Chile
SERNATUR
P.O. Box 14082
Santiago
Tel.: 00 56 2 696 7141
www.seranatur.cl/

ENGLAND
English Tourist Board
Thames Tower
Black's Road
Hammersmith
London W6 9EL
Tel. 00 44 20 8846 9000
www.travelengland.org.uk/

FRANCE
Maison de la France
8 Avenue de l'Opera
Paris
Tel.: 00 33 1 42 96 10 23
www.maison-de-la-france.fr

GREECE
2 Amerikis Street
Athens 10564
Tel.: 00 30 1 327 1300
www.gnto.gr

INDIA
Government of India Tourist Office
88 Janpath
New Delhi 110 001
Tel.: 00 91 11 332 0005
www.tourisminindia.com/

ITALY
Ente Nazionale Italiano per il Turismo
Via Marghera no. 2
00185 ROMA
Tel.: 0039 6 49711
www.enit.it

KENYA
Kenya Tourist Board
P.O. Box 30630
Nairobi
Kenya
Tel.: 00 254 2 604 245
www.kenyatourism.org/

MALAYSIA
Tourism Malaysia
26th Floor Menara Dato' Onn
Putra World Trade Center
45 Jalan Tun Ismail
50480 Kuala Lumpur
Tel.: 00 3 293 5188
www.tourism.gov.my

MOROCCO
Moroccan National Tourist Office
31 Angle Avenue Al Abtal/Rue Oued Fes
Agdal
Rabat-Morocco
00 212 7 681 531
www.tourism-in-morocco.com

NEPAL
Nepal Tourism Board
Tourist Service Center
P.O. Box 11018
Bhrikiti Mandap
Tel.: 00 997 1 256 909
www.welcomenepal.com/

NEW ZEALAND
New Zealand Tourism Board
Fletcher Challenge House
89 The Terrace
P.O. Box 95
Wellington
Tel.: 00 64 4 472 8860
www.nztb.govt.nz

PAKISTAN
Pakistan Tourism Development Corporation
House No. 170-171
Street 36, F-10/1
Islamabad 44000
Tel.: 00 92 51 294 790
www.tourism.gov.pk/

PERU
PROMPERU
Edificio MITINCI 14th Floor
Calle 1 Oeste 050
Urb. Corpac
San Isidro
Tel.: 00 51 1 224 3113
www.peruonline.net

SCOTLAND
Scotland Information Centre
4 Rothesay Terrace
Edinburgh EH3 7RY
Tel.: 00 44 131 473 3600
www.holiday.scotland.net/

SOUTH AFRICA
SATOUR
442 Rigel Avenue
South Erasmusrand 0181
Private Bag X164
Pretoria 0001
Tel.: 00 27 124 826 200
www.satour.co.za
www.satour.org

SPAIN
TourSpain
Jose Lazaro Galdiano, no. 6
28036 Madrid
Tel. 00 34 901 300 600
www.spaintour.com/
www.tourspain.es/

SWEDEN
Swedish Travel & Tourism Council
Box 3030
Kungsgarten 36
10361 Stockholm
Tel.:00 46 87 25 55 00
www.gosweden.org
www.visit-sweden.com

SWITZERLAND
Switzerland Tourism
Toedistrasse 7
P.O. Box 695
Zurich 8027
Tel.: 0041 1 288 11 11
www.switzerlandtourism.ch/
www.switzerlandvacation.ch

USA
Arizona Office of Tourism
2702 North 3rd Street
Suite 4015
Phoenix
AZ 85009
Tel.: 00 1 888 520 3433
www.arizonaguide.com/

California Division of Tourism
P.O. Box 1499
Sacramento
CA 95812
Tel.: 00 1 916 322 2881
www.gocalif.ca.gov/

Colorado Travel and Tourism Authority
1672 Pennsylvania Street
Denver
CO 80203
Tel.: 00 1 303 832 6171
www.colorado.com
www.wyomingtourism.org

Vermont Department of Tourism & Marketing
6 Baldwin Street
Drawer 33
Montpelier
VT 05633-1301
Tel.: 00 1 802 828 3237
www.travel-vermont.com
www.1-800-vermont.com

INDEX

Page numbers in *italics* refer to illustrations

CREDITS

Key: *b* below; *t* top; *tl* top left

We would like to acknowledge the use of photography by the following persons: p1 Bill O'Connor; p2 Bill O'Connor; p4 Chris Townsend; p6 Chris Townsend; p7*t* Walt Unsworth, *b* Richard Gilbert; p8 Richard Gilbert; p9 Bill O'Connor; p10*t* Walt Unsworth, *b* John Manning; p11*t* Alan Castle, *b* John Cleare; p12*tl* Richard Gilbert; p17 Chris Townsend; p59 Andrew Sheehan; p79 Bill Birkett; p117 Steve Callen; p143 Tom Hutton; p165 Walt Unsworth. All photographs in the treks (pages 18–57, 60–77, 80–115, 118–141, 144–163, 166–185) were reproduced with the kind permission of each trek's author.

Bill Birkett, General Editor, would like to offer his heartfelt thanks to: All the contributors, many from the Outdoor Writers' Guild, who have helped me with this book, often as a personal favour amid crowded schedules. I am indebted to their huge talent, enthusiasm, expertise, spellbinding adventures and sheer professionalism. It has been my great privilege to work with each of them. Many thanks also to those who offered their expert advice, and guidance, and pointed me to others.

Disclaimer